This book provides an introduction to the general principles of nuclear magnetic resonance and relaxation, concentrating on simple models and their application.

The author has produced an introduction to the ideas and applications of nuclear magnetic resonance. The concepts of relaxation and the time domain are particularly emphasised. Some relatively advanced topics are treated, but the approach is graduated and all points of potential difficulty are carefully explained. An introductory classical discussion of relaxation is followed by a quantum mechanical treatment, but without making explicit use of the density operator. Only when the principles of relaxation are firmly established is the density operator approach introduced; and then its power becomes apparent. A selection of case studies is considered in depth, providing applications of the ideas developed in the text. There are a number of appendices, including one on random functions, to help the reader understand some of the more specialised aspects of the subject.

This treatment of one of the most important experimental techniques in modern science will be of great value to final-year undergraduates, graduate students and researchers using nuclear magnetic resonance, particularly physicists, and especially those involved in the study of condensed matter physics.

NUCLEAR MAGNETIC
RESONANCE AND RELAXATION

NUCLEAR MAGNETIC
RESONANCE AND RELAXATION

BRIAN COWAN
Royal Holloway University of London

CAMBRIDGE UNIVERSITY PRESS
Cambridge, New York, Melbourne, Madrid, Cape Town, Singapore, São Paulo

Cambridge University Press
The Edinburgh Building, Cambridge CB2 2RU, UK

Published in the United States of America by Cambridge University Press, New York

www.cambridge.org
Information on this title: www.cambridge.org/9780521303934

First published 1997
This digitally printed first paperback version 2005

A catalogue record for this publication is available from the British Library

Library of Congress Cataloguing in Publication data
Cowan, B. P., 1951–
Nuclear magnetic resonance and relaxation / Brian Cowan.
p. cm.
Includes bibliographical references and index.
ISBN 0 521 30393 1
1. Nuclear magnetic resonance. 2. Nuclear magnetic resonance –
Industrial applications. 3. Relaxation (Nuclear physics)
I. Title.
QC762.C69 1997
538′.362 dc21 96-46614 CIP

ISBN-13 978-0-521-30393-4 hardback
ISBN-10 0-521-30393-1 hardback

ISBN-13 978-0-521-01811-1 paperback
ISBN-10 0-521-01811-0 paperback

Contents

Contents   xv

Preface

This book started as a joint project between Michael Richards and myself. We discussed and planned the work in great detail but somewhere along the way Michael moved away from physics, drawing on his experience at Sussex University, to become a psychotherapist. Notwithstanding the plans we made, his withdrawal from the project has resulted in a book hardly recognisable from that originally envisaged. Since the publication of Anatole Abragam's encyclopaedic treatise on nuclear magnetism in 1961 the field has grown at such a pace that no single book could hope to cover its many aspects. However the coverage of this book is without doubt narrower and its content impoverished as a consequence of Michael's career change.

Mea culpa. I am guilty of the very sin of which I have accused others. The motivation for writing a book on NMR was that the various books available, since Abragam's, did not cover the material which I, my colleagues and research students required. It seemed that the new books comprised assorted collections of topics in NMR. Michael and I wanted to redress this, but in surveying the finished product I see no more than yet another assortment of topics. This time, however, it is *my* collection of topics. While the broader aim might not have been achieved my hope is that this assortment will have its appeal.

In the preface to Abragam's book in 1960 he wrote that he hoped his book would still be *the* book on the subject thirty years after its publication. That was prophetic; his wish has been amply fulfilled. So what sort of book is this intended to be? This book is very much a *monograph*; it is a personal (and coherent, I hope) view of topics which have been chewed over and digested, often in the company of persons mentioned below. Granted, the view may be skew; probably it is idiosyncratic. However, I particularly hope the book will be found to have pedagogical value. For this reason the concentration is on simple systems: spin half nuclei, monatomic species etc.

In this context the knowing reader might well be surprised to find a chapter entitled 'Dipolar lineshape in solids' in a book on NMR published in the 1990s. It is my view, however, that this well-defined topic provides a microcosm of much in many-body physics. The formalisms of quantum statistical mechanics are there applied to a system which is both simple enough for meaningful calculations to be performed, while being sufficiently complex for irreversible behaviour to be manifest. An added attraction is the exactly-solvable model introduced by Irving Lowe and subsequently simplified by Steve Heims and further by Bill Mullin. The culmination of the dipolar lineshape exposition occurs in a later chapter, in the case study of the fluorine resonance of calcium fluoride. There the painstaking measurements of Lowe over two decades are compared with a surprisingly wide range of 'theories'. Since some of the finest brains in the business have worked on this problem the survey of their achievements is highly instructive.

The knowing reader might also be surprised at the quantum treatment of relaxation in Chapter 5. There is no explicit mention, there, of the density operator; it is hiding in the background. This is made possible by working in the Heisenberg picture of quantum mechanics where the time independent density operator simply provides a Boltzmann distribution of states. I find this presentation much more acceptable to students for whom the ideas and procedures are new. Once the formalism is assimilated in familiar language it is possible to introduce the general methods of the density operator in a more agreeable and satisfactory manner, as in Chapter 9.

Solid ^{3}He is a particular research interest of mine. It is highly amenable to investigation by NMR and much has been learned from such studies. Notwithstanding this I have tried to prevent considerations of ^{3}He from permeating the book. Solid ^{3}He is a unique system because quantum exchange leads to appreciable atomic motion. From the NMR point of view this solid is more like a fluid! Furthermore it has the unusual feature that the motion has a (relatively) simple Hamiltonian description. I have restricted my explicit discussion of this exciting substance to the case study in Chapter 8. It is necessarily a potted version of the current position in that area, but I hope it gives the reader a flavour of that topic.

In keeping with the pedagogical aims and the concentration on simple systems, there is no mention of many of the newer topics in NMR such as multi-dimensional NMR and multiple quantum methods. In truth there were many things which I would have liked to have included, but the constraints of the finite amount of time available conspired against me. The book *could* have been more comprehensive; it just never would have been finished.

It is only fair to warn the reader of a number of eccentricities of presentation. Early on in my research career I was seduced by R. M. White's specification of the complex susceptibility as $\chi = \chi' + i\chi''$ instead of the common form involving $-i$. White's form makes particular sense when considering k-dependent susceptibilities, where the complex sinusoid is conventional written as $\exp[i(kx - \omega t)]$. All is fine until one considers experimental electrical observation. Then for consistency the impedance of an inductor becomes $-i\omega L$, in disagreement with the standard electric circuit convention which follows from using the complex sinusoid $\exp(i\omega t)$. I must also confess to a cavalier regard (or disregard) of time-ordered exponentials. My justification is that in the interaction representation, when expansions are used they are taken only to second order and it seemed expedient simply to write the pair products in their correct order when required. In these matters I beg the reader's indulgence.

The appendices have particular pedagogical significance. There are a number of original treatments which I have found to be useful. The physical interpretation of an autocorrelation function as an ensemble average weighted by initial values is a helpful aid to understanding. Also the treatment of canonical momentum in the presence of a magnetic field has been found acceptable by students who have not had the benefit of a course in classical Hamiltonian mechanics: a scandalous situation which exists in an increasing number of UK universities.

There is a small number of problems for the reader at the end of the book. These range in difficulty from the trivial to the impossible. Often they are included to take a specific point from the text further or to cause the reader to think about a particular matter in a different way. I urge the novice reader to look at the problems and to think about them, even if he or she does not work all of them through.

I have tried to use SI units throughout the book wherever it seemed sensible to do so. Electromagnetism, in particular, is beset by problems of units. The $B-H$ controversy continues as ever. Unfortunately many NMR practitioners have contributed to the confusion by writing H while quoting values in gauss! They should know better. In conformity with SI usage I have used B (measured in teslas) as the fundamental field, but to be clear I have included, in Box 1.1, a discussion of the relationship between the microscopic and the macroscopic magnetic fields.

NMR was discovered by physicists, but it soon became the province of the chemists. Of late biologists have displaced the chemists and soon the next revolution will be complete with the biologists relinquishing their position to the medics. Without doubt the potential for NMR in medical diagnostics is enormous. It is these changes which have dictated the content of many books in the field and it is partly as a reaction to this that I offer this book to physicists, and in particular to those using or contemplating using NMR as a research tool.

This book is the consequence of a learning process which has extended over many years and is the result of interactions I have had with numerous people over that time. In enunciating the roll of people to whom I owe a debt of gratitude I wish in no way to incriminate them or blame them for the shortcomings of this book. All errors of omission and of commission are my own. Many people have aided my understanding of NMR over the years; I pray I have done them justice, I fear I have not.

Michael Richards was my teacher at Sussex, and he became my good friend. He welcomed me into his research group and thus started my initiation into the mysteries of NMR. I am particularly grateful to him for the guidance and encouragement he has given me over the years. The subliminal psychotherapy was a bonus.

I was fortunate enough to spend a year in Abragam's laboratory in Paris. It was an exciting time during which I benefited from collaboration with Neil Sullivan and André Landesman as well as from many discussions with Maurice Goldman and the Great Man himself. Many ideas and approaches to NMR problems were crystallised during that year, even if they were not in accord with those of the 'Paris School'. In particular, my study of the works of Ryogo Kubo convinced me of the power and efficacy of his methods; if anything, it was that which 'changed my life'.

Roland Dobbs invited me to join his Low Temperature Group at Bedford College, later to become Royal Holloway and Bedford New College. He provided the facilities and the encouragement for me to pursue research in what was a genial college environment. In the succeeding years I have been privileged to have had a number of outstanding research students. In particular I would like to mention Mihail Fardis; some of the original material presented in the case study of solid ^{3}He in Chapter 8 was developed in collaboration with him.

If I can single out one visionary action of Roland Dobbs, it was his appointment of John Saunders. As a consequence the 'genial environment' has been transformed into the most stimulating and exciting of scientific atmospheres. John's intuition and drive are inspirational and collaboration with him has been *the* most rewarding aspect of life at Royal Holloway. He is both the best of friends and the sternest of critics and it is a pleasure to thank him for all aspects of help and support he has given me over the years.

I first met Bill Mullin when I was an undergraduate and he was on sabbatical leave at Sussex. We started talking about physics then and we have continued ever since. Subsequently I have enjoyed a transatlantic collaboration with him over many years. He read much of the manuscript with great care and pointed out numerous places where a clearer presentation was possible. He discovered many errors of fact and of grammar; I am grateful for the tutorial on dangling participles in particular.

I would like to thank my young friend Yoram Swirski for Figure 10.13 which came from his PhD thesis. I also thank my old friend Paul Tofts together with David MacManus, of the NMR Research Unit, Institute of Neurology, for Figure 10.14. Tom Crane has rescued me from the malign behaviour of all manner of computer infelicities for longer than I care to remember. His encyclopaedic knowledge of computer matters has proved invaluable; without his help I would have lost my sanity years ago. I thank Chris Lusher for discussions spanning a wide range of topics. I also thank Martin Dann who, among other things, writes ingenious software; the cover image came from one of his virtual instruments. The NMR work at Royal Holloway has benefited incomparably from the technical expertise of Alan Betts, our electronics engineer. It is a pleasure to acknowledge our fruitful collaboration over the past 18 years.

The writing of this book has been a long ordeal (far longer than I would ever have dreamed). As the years have passed the time spent on mindless administration and paper-pushing has expanded to fill what seems far more than the time available between research and teaching duties. Consequently it has become increasingly difficult to set aside the periods of uninterrupted time necessary for clear thinking and writing. In this respect I am forever grateful to Emil Polturak for extending to me the hospitality of his research group at the Technion: Israel Institute of Science and Technology, Haifa during the summer of 1994 as well as for other shorter periods. During those times I have been liberated from the concerns of everyday life and much of the latter part of the book was written in that peaceful and idyllic environment.

Louise, my wife and Abigail, my daughter, were kind enough not to complain of the time I spent away from them, mentally if not physically, while I was working on the book. To them I offer my heartfelt thanks. Their support was invaluable; their disbelief that the book would ever be finished was understandable.

I dedicate this book to the memory of my grandmother, Phoebe Davis, who died in December 1992 in her 96th year. She was a source of inspiration and encouragement to me for as long as I can remember. She never tired of asking me to explain my research to her and answering her questions was a continual challenge; she was very much my 'lady on the Clapham omnibus'.

Egham   BPC
1997

1

Introduction

1.1 Historical overview

1.1.1 The interaction of radiation with matter

Nuclear magnetic resonance (NMR) is one of a large range of phenomena associated with the interaction of electromagnetic radiation with matter. Other, more familiar, examples of this type of interaction include the attenuation of X-rays by lead, the visible emission and absorption spectra of atoms, microwave heating of food and radio frequency (RF) induction heating of metals. What distinguishes these examples is primarily the frequency of the electromagnetic radiation involved. X-rays have a typical frequency of around 10^{18} Hz, while the orange light of the sodium lamp corresponds to about 5×10^{14} Hz. Microwave cookers operate at frequencies of order 10^9 Hz and RF induction heaters use a frequency of order 10^6 Hz.

The experimental techniques of producing and detecting the radiation in each of these cases are clearly different. But there will be similarities in the theoretical description of the processes involved. The most comprehensive language will be that of quantum mechanics. Using quantum theory one explains these phenomena in terms of transitions between states of differing energy. The frequency f of the radiation is related to the energy difference ΔE of the states by the Einstein relation

$$\Delta E = hf,$$

where h is Planck's constant.

1.1.2 Nuclear magnetic resonance

In NMR the energy levels are those associated with the different orientations of the nuclear magnetic moment of an atom in an applied magnetic field. As we shall see, the spacings between these energy states correspond to radiation in the RF range. Thus transitions between the levels will be induced by an applied *RF* magnetic field.

It is instructive to point out some significant differences between NMR and other

1

forms of spectroscopy. Firstly, an external static magnetic field is nearly always applied. Secondly, it is the magnetic field in the incident radiation rather than the electric field which induces the transitions. Thirdly, as a consequence of this, the observable effects are smaller than electric phenomena by a factor of v/c where v is a typical velocity in the system and c is the speed of light.

Although the language of quantum theory lends itself directly to the description of the fundamental interaction, for many of the simpler phenomena associated with NMR a quantum description is not actually necessary. Much of the introductory material in this book is therefore discussed in terms of classical concepts, thus expediting understanding of the basic ideas without the complication of a full quantum treatment. For some of the more advanced topics, however, it will be seen that an understanding may be achieved only within the context of quantum mechanics.

1.1.3 The first observations of NMR

For NMR as defined in this introduction, credit for its first observation should go to Rabi and co-workers who used a beam of silver atoms. Detection of transitions was observed by noting the change in the fluxes of beams representing the different energy states of the nuclear magnetic moments. However, the term NMR has come to be used for experiments which differ in two important respects from those of Rabi. Firstly the detection of the transitions in NMR is usually through the energy absorbed from the RF field rather than through changes in the particle flux reaching a detector as in the beam experiments. Secondly, although perhaps more arbitrarily, the term NMR is commonly reserved for phenomena occurring in bulk matter rather than in a beam of essentially non-interacting atoms. As a consequence of these conventions the first observations of NMR are usually attributed to two groups: Purcell, Torrey and Pound, working on the east coast of America and Bloch, Hansen and Packard working independently on the west coast. Their almost simultaneous discoveries are reported in the same volume of the *Physical Review* published in 1946.

It is no coincidence that both this work and the observation of a similar effect involving the electron magnetic moment rather than the nuclear magnetic moment (called electron spin resonance, ESR) should have occurred at the end of a world war in which dramatic advances were made in RF and microwave technology.

1.1.4 Pulsed NMR

Regarding NMR as a branch of spectroscopy and the use of the word 'resonance' imply that the phenomenon is to be observed by irradiating a sample and observing the response of the system to the radiation as the frequency is varied. Furthermore, something dramatic is expected to occur in the neighbourhood of some resonant frequency. This indeed was the way that NMR was first observed. However, very soon after these early observations a number of people, notably Hahn in 1950, realised that similar information could be gathered by observing the response to short intense pulses of resonant RF radiation.

There is a duality here which is discussed in some detail in Section 1.3. We note here that an increasing majority of NMR experiments are now being carried out using pulse methods. Broadly speaking the reasons for this are that in many situations one has an improved signal-to-noise ratio by this method and, perhaps more significantly, information about interactions between nuclear spins, or between the molecules carrying the spins, can most readily be obtained by this method.

1.1.5 Scope of this book

The field of magnetic resonance has undergone a remarkable expansion since it was first discovered. From being the domain of the nuclear physicist it developed into a technique for the study of atomic motion in solids and fluids. It has become a powerful technique in chemical structural analysis and presently it is being developed into a technique for medical imaging.

The first monograph on NMR was written by E. R. Andrew and published in 1955. He refers to there being more than 400 published papers discussing work in the field of NMR. His book was able to summarise the majority of this work as well as to describe the basic ideas of NMR at a level suitable for a reader approaching the subject for the first time. Six years later A. Abragam's definitive treatise on NMR was published. Already the subject had diversified and expanded, there being then over 2000 published papers. Abragam's monograph is characterised by its breadth and depth of coverage; it is still the standard work. Thirty five years later such a comprehensive approach is impossible in a single volume. What we are attempting in this book is a description of the basic theory of NMR together with some applications which illustrate these basic ideas. The aim is to give readers a clear picture of simple NMR phenomena and in this way to equip them to tackle specialist papers or to pursue the topic further using an advanced text such as that by Abragam and other more specialist works.

Chapter 2 introduces some of the theoretical apparatus required for the understanding of NMR phenomena, while Chapter 3 is devoted to some of the experimental aspects. There we will encounter the basic structure of an NMR spectrometer and some of the instrumentational details will be discussed. Chapters 4 and 5 are concerned with the central topic of spin relaxation. In Chapter 4 we approach the problem from the classical point of view, when many of the important concepts will be introduced. Chapter 5 then takes up the discussion within the framework of quantum mechanics, allowing a fuller and more complete treatment of many aspects of relaxation.

Although numerous uses of NMR are discussed in these early chapters, serious applications start with Chapter 6 which considers NMR in solid systems. Chapter 7 then extends the discussion to fluids, considering both liquids and gases. Chapter 8 is devoted to a number of case studies, which consider various NMR phenomena in different materials and systems. Only in Chapter 9 do we introduce the sophistication of the density operator. This then allows a discussion of spin temperature in solids and a coherent treatment of relaxation and irreversibility is possible. Finally, in Chapter 10 we turn our attention to NMR imaging. At the end of the book are a number of appendices, covering specific matters in greater depth than is appropriate for inclusion in the body of the text.

1.2 Nuclear magnetism

1.2.1 Magnetic moments

Before beginning the basic theory in the next chapter a few preliminary remarks are appropriate to set the scene and to develop the language that will be used. In this section we shall start with a description of the phenomenon of nuclear magnetism.

From the time of Ampère, the magnetism of matter has been understood to arise from microscopic electric current loops. The most obvious mechanism for this is the orbits of the electrons. Using classical theory it is easy to see that an electron of charge e, orbiting with an angular momentum $\hbar L$, has associated with it a magnetic moment μ_L given by

$$\mu_L = e\hbar L/2m_e, \tag{1.1}$$

where m_e is the electron mass. (We follow the convention of measuring angular momentum in units of $\hbar$, Planck's constant divided by 2π; thus L is a dimensionless number.) We see that the magnetic moment is proportional to the angular momentum. When the vector nature of these quantities is considered it is also seen that the magnetic moment and the angular momentum are colinear. Problem 1.1 treats this. The specific property of electrons, $e/2m_e$ is known as the magnetogyric (or gyromagnetic) ratio. It is given the symbol γ and has a value of approximately $8 \times 10^{10} \, \text{s}^{-1} \, \text{T}^{-1}$.

However, in molecules the orbital angular momenta of the atoms tend to cancel so that there is rarely any resultant orbital magnetic moment. But electrons also have an intrinsic angular momentum $\hbar/2$ associated with their so-called spin, and electron spin is not always cancelled out; a molecule may have a total electron spin angular momentum. It was shown experimentally by Stern and Gerlach, and theoretically by Dirac (1930) that associated with spin momentum $\hbar S$ is a magnetic moment of magnitude μ_s which we write as

$$\mu_s = ge\hbar S/2m_e$$

to show similarity with Equation (1.1). The quantity g is referred to as the Landé splitting factor, or simply the g factor. Its value is close to 2 for free electrons – Dirac's theory tells us that for a free point particle of mass m and spin S

$$\mu = e\hbar S/m.$$

1.2.2 Nuclear magnetic moments

A similar situation exists for nuclei. Individual nucleons also have a spin of $\frac{1}{2}$ but their orbital moments are quenched by the interactions present in the nuclei. The spins of each nucleon are coupled together to lead to a total spin angular momentum of $\hbar I$ where I is an integral or half integral number. Stern and co-workers (Gerlach and Stern, 1924) were able to measure the small magnetic moment μ_I associated with $\hbar I$. We shall write this in the form

$$\mu_I = g_I e\hbar I/2m_p, \tag{1.2}$$

where m_p is the mass of the proton.

In early experiments g_I was a phenomenological dimensionless constant found to be between about 0.5 and 5. It has been one of the tests of theories of the nucleus to obtain values for g_I. Since I, like S, is of the order of unity while m_p is some thousand times m_e, we see that nuclear magnetic moments are typically of the order of one thousandth that of electrons.

We might mention in passing that the proton and the neutron, unlike the electron, are not regarded as fundamental particles. For this reason Dirac's theory of the magnetic moment is not applicable to them. The currently accepted constituents of the proton and the neutron, quarks, are fundamental particles and thus to them Dirac's theory does apply. Combining correctly the magnetic moments of the three quarks in the proton and the neutron yields the correct ratio of the proton and neutron magnetic moments. This assumes that the constituent quarks all have the same mass. The *ratio* of the magnetic moments is independent of the actual mass; the value of the mass may be inferred from the magnitude of the proton and neutron magnetic moments.

1.2.3 Energy splitting and resonance

An externally applied static magnetic field is usually an essential feature of magnetic resonance. In a static magnetic field $\mathbf{B}_0$ a nuclear spin angular momentum vector may exist in a number of different quantum states (Figure 1.1). In each distinct state the spin vector will possess a different orientation such that its component parallel to the field has values $m\hbar$ where m runs from $-I$ to $+I$ in steps of unity. Thus for $I = 3/2$, m takes the values $-3/2$, $-1/2$, $+1/2$, $+3/2$. The number m may be used as a label to distinguish the various quantum states and their energies, as we shall see.

The energy of a magnetic moment μ in a magnetic field $\mathbf{B}$ is given from classical electromagnetism as $-\mu \cdot \mathbf{B}$. So the energy of the mth state is given by

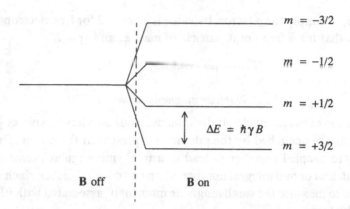

Figure 1.1 Energy levels for a nuclear spin of $I = 3/2$ in a magnetic field **B**.

$$E_m = -\gamma m\hbar B, \qquad (1.3)$$

where we have designated $g_I e/2m_p$ in Equation (1.2) as γ, the nuclear magnetogyric ratio. So if we now consider adjacent states, the energy difference is given by

$$\Delta E = \hbar\gamma B, \qquad (1.4)$$

and the Einstein relation tells us that transitions will be induced between the states – resonance will occur – when the applied radiation has an angular frequency ω ($= 2\pi f$) of

$$\omega = \gamma B. \qquad (1.5)$$

Equations (1.1) and (1.2) tell us that the magnetic moment varies inversely with the mass of the particle. Thus we see that while electron spin resonance is usually carried out in the microwave range (10^9–10^{11} Hz), NMR usually lies in the RF range (10^6–10^9 Hz).

The resonance condition, Equation (1.5), implies that the magnetogyric ratio γ is an important constant. It differs for different nuclei. In Table 1.1 we list values of γ for a few of the common nuclei.

1.2.4 Uses of NMR

We can already see from Equation (1.4) some of the reasons why NMR has come to be such a valuable tool in many branches of science. Mention has already been made of it allowing γ and therefore g_I to be measured accurately and compared with values predicted from theories of the nucleus. This, however, had largely been completed within ten years of NMR first being observed. Of much more current interest is the use of NMR through Equation (1.5) to measure and control magnetic fields. In a proton magnetometer NMR is performed on a sample such as water (using the two protons in

Table 1.1. *Magnetogyric ratios and abundances of some nuclei.*

Isotope	Spin I	Natural abundance %	Magnetogyric ratio $\gamma/2\pi \times 10^6$, i.e. MHz per tesla
^{1}H	1/2	100	42.58
^{2}H	1	0.02	6.54
^{3}He	1/2	0.0001	32.44
^{7}Li	3/2	92	16.55
^{13}C	1/2	1.1	10.71
^{19}F	1/2	100	40.06
^{23}Na	3/2	100	11.26
^{29}Si	1/2	4.7	8.46
^{35}Cl	3/2	75	4.17

the H_2O molecule) and the resonant frequency is measured with an accuracy up to 1 in 10^8. Since the proton γ is known to about this accuracy, magnetic fields can thus be accurately measured.

Many uses of NMR follow from the fact that ω in Equation (1.5) is determined by the field seen by the individual nuclei. This arises from the applied magnetic field together with a local field b_{loc} which is created by other particles in the sample. It may occur through diamagnetic screening effects from electrons in the sample. This will cause a shift in the observed resonance frequency and it is, in fact, the basis for the use of NMR as a technique for chemical structural analysis, discussed further in Chapter 5.

Often the local field is fictitious in the sense that the nuclei of interest are interacting with other parts of the system with an interaction energy of the form

$$E = - \mathbf{I} \cdot \mathbf{X}. \tag{1.6}$$

With no loss of generality we can replace the quantity $\mathbf{X}$ by a 'field'

$$\mathbf{b}_{loc} = \mathbf{X}/\hbar\gamma.$$

This fictitious local magnetic field may be a function of time within the sample. The main use of NMR today is to measure the temporal and spatial variation of $\mathbf{b}_{loc}$ and make deductions about the sample from this information. One of the main aims of this book is to show how this connection is made.

Concerning the concept of the local field we should point out that an important simplification has been made here. In condensed systems the magnetic moment and the spin angular momentum states should refer to the whole sample rather than an individual nucleus. The transitions considered then are induced between eigenstates of these macroscopic quantities. What we are doing is to set up the problem as though we were dealing with non-interacting, or at least weakly interacting, particles so that we can concentrate on one nucleus. Interactions with the other particles are then supposed to be equivalent to a local field which varies from nucleus to nucleus with time. This approach is the one taken in many of the topics discussed in this book.

1.3 Two views of resonance

1.3.1 The damped harmonic oscillator

All resonance phenomena including NMR share certain features in common. For this reason we shall examine some of the properties of resonant systems from a rather more general point of view. We hope this will facilitate the understanding of NMR while setting it in the context of other resonance phenomena.

We will consider a one-dimensional harmonic oscillator, the epitome of a resonant system. Let us denote the displacement of a unit mass from its equilibrium position by x. To be an oscillator it must be subject to a restoring force assumed proportional to the displacement; this we denote by $-\alpha^2 x$. Furthermore we consider a friction or damping force proportional to the velocity, which we write as $-2\beta\dot{x}$, using a dot to denote differentiation with respect to time. In the absence of an external driving force the equation of motion for the system takes the canonical form

$$\ddot{x} + 2\beta\dot{x} + \alpha^2 x = 0. \tag{1.7}$$

1.3.2 Time response

The equilibrium state of the harmonic oscillator occurs when the displacement x is zero and the system is stationary. If the oscillator suffers a disturbance it will return to the equilibrium state in a manner dictated by the equation of motion. Let us consider first the nature of this relaxation from the non-equilibrium to the equilibrium state.

In this case one is concerned with the free relaxation of the oscillator from some initial condition. We require the time response, the solution to the homogeneous Equation (1.7). Substituting a trial solution of the form $x = x_0 \exp(st)$ and solving for s, we find that provided $\alpha > \beta$ the behaviour of the relaxing harmonic oscillator has the form

$$x(t) = [A\cos(\omega_0 t) + B\sin(\omega_0 t)]\exp(-t/T) \tag{1.8}$$

where

$$\omega_0 = (\alpha^2 - \beta^2)^{\frac{1}{2}} \text{ and } 1/T = \beta.$$

There is an oscillation at the angular frequency ω_0 which is decaying away exponentially with a time constant T (Figure 1.2). The constants A and B depend on the nature

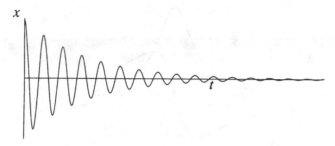

Figure 1.2 Relaxation of a harmonic oscillator.

of the initial conditions: the state of the system at time $t = 0$. In the case in which the oscillator is released from a displacement of x_0 at $t = 0$ the evolution takes the form

$$x(t) = x_0\cos(\omega_0 t)\exp(-t/T).$$

We see that the behaviour of the oscillator is characterised by two independent quantities, related to the two parameters: α, the restoring force, and β, the friction force. These are reflected in the measurable quantities, the angular frequency ω_0 and the damping time T.

1.3.3 Frequency response

There is another way of studying the behaviour of our oscillator. We may examine how the system responds to 'irradiation' by an oscillatory excitation. We consider an applied force per unit mass of $F_0\cos(\omega t)$. In a conventional spectroscopy experiment the response $x(t)$ is studied as the excitation frequency ω is varied. The complete equation of motion in this case is the inhomogeneous equation

$$\ddot{x} + 2\beta\dot{x} + \alpha^2 x = F_0\cos(\omega t)$$

which has the steady state solution

$$x(t) = F_0\left[\frac{(\alpha^2 - \omega^2)}{(\alpha^2 - \omega^2)^2 + (2\beta\omega)^2}\cos(\omega t) + \frac{2\beta\omega}{(\alpha^2 - \omega^2)^2 + (2\beta\omega)^2}\sin(\omega t)\right]. \quad (1.9)$$

Recalling that the excitation is given by $F_0\cos(\omega t)$ we see that the in-phase response per unit force, denoted by $\chi'(\omega)$, is given by

$$\chi'(\omega) = \frac{(\alpha^2 - \omega^2)}{(\alpha^2 - \omega^2)^2 + (2\beta\omega)^2} \quad (1.10a)$$

and the quadrature response per unit force, $\chi''(\omega)$, by

$$\chi''(\omega) = \frac{2\beta\omega}{(\alpha^2 - \omega^2)^2 + (2\beta\omega)^2}. \quad (1.10b)$$

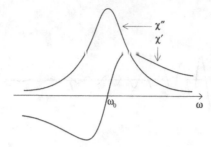

Figure 1.3 In-phase and quadrature parts of the response of an oscillator to a sinusoidal force.

The quantities $\chi'(\omega)$ and $\chi''(\omega)$ are known respectively as the in-phase and quadrature components of the generalised susceptibility. They are plotted in Figure 1.3 and one observes the characteristic shapes for a resonance, particularly the bell shape of $\chi''(\omega)$, with its peak at resonance. On resonance $\chi'(\omega)$ goes through zero. These generalised susceptibilities, particularly $\chi''(\omega)$, are the quantities that are investigated in the traditional methods of spectroscopy.

1.3.4 Linearity and superposition

The parameters which characterise the oscillator: the frequency and the relaxation time, may equally well be obtained from the steady state frequency response, Equation (1.9), or from the time response, Equation (1.8). In fact the connection between the frequency domain response and the time domain response is fundamental. Both the exponential decay of Equation (1.8) and the characteristic bell shaped response of Equation (1.9) are a consequence of the particular equation of motion for the system. A more complex equation of motion would give a different steady state response and a different time behaviour, but a simple interrelationship would persist.

Following Equation (1.10) and the introduction of the generalised susceptibility, we may write the steady state response of an arbitrary system to an oscillating force $f_0\cos(\omega t)$ as

$$x(t) = f_0[\chi'(\omega)\cos(\omega t) + \chi''(\omega)\sin(\omega t)]. \tag{1.11}$$

Implicit in this is the assumption that the system is linear or that the force is sufficiently small that deviations from linearity may be neglected; $x(t)$ is written as being linear in f. Our task now is to relate this response to the relaxation following a transient excitation, and it is specifically the linearity condition which comes to our aid since it guarantees that the response to the sum of a number of forces is the sum of the separate responses to the individual forces.

1.3.5 Fourier duality

Using elementary Fourier synthesis (Appendix A deals with this) we may construct an arbitrary force from a distribution of various oscillating waves. The response to this force is constructed using the same distribution: from the responses to the various oscillating waves. If the arbitrary force $F(t)$ is given by

$$F(t) = \int_0^\infty f(\omega)\cos(\omega t)d\omega, \tag{1.12}$$

where $f(\omega)$ gives the distribution of cosine waves in the force, then the corresponding time response is given by

$$\int_0^\infty f(\omega)[\chi'(\omega)\cos(\omega t) + \chi''(\omega)\sin(\omega t)]d\omega.$$

In particular for a unit shock excitation, which we may represent by a delta function force (Appendix A)

$$F(t) = \delta(t),$$

since $\delta(t)$ may be expressed as

$$\delta(t) = \frac{1}{\pi}\int_0^\infty \cos(\omega t)d\omega,$$

the delta function force then is simply the Fourier transform of a constant. Thus in this case we have a 'white' distribution

$$f(\omega) = 1/\pi$$

and the transient response is then given by

$$X(t) = \frac{1}{\pi}\int_0^\infty [\chi'(\omega)\cos(\omega t) + \chi''(\omega)\sin(\omega t)]d\omega.$$

This expression may be inverted (Appendix A) to give the in-phase and quadrature components of the generalised susceptibility as the cosine and sine transforms of the transient response, or relaxation function, $X(t)$:

$$\chi'(\omega) = \int_0^\infty X(t)\cos(\omega t)\mathrm{d}t,$$

$$\chi''(\omega) = \int_0^\infty X(t)\sin(\omega t)\mathrm{d}t.$$

The steady state response and the transient response are then simply related by Fourier transformation which is, of course, as it should be. The cosine is the Fourier transform of the delta function, so the response to a cosine excitation is the Fourier transform of the response to a delta function excitation.

We see that the steady state response and the transient response – the resonance and the relaxation – the frequency domain and the time domain, are complementary pairs. Exactly the same information is contained within each member of a pair (within the restriction of linearity of course). This then gives us two ways of looking at resonant systems in general and NMR in particular. In conventional spectroscopy it is the frequency response, $\chi'(\omega)$ and/or $\chi''(\omega)$, which is studied. We now see that exactly the same information may be obtained from *time domain spectroscopy* where $X(t)$, the relaxation following a transient excitation, is studied. Within the context of NMR this technique, to be discussed further in the following sections, is better referred to as nuclear magnetic relaxation (NMR!). Thus the title of this book.

1.4 Larmor precession

1.4.1 Spin equation of motion

Central to magnetic resonance is the existence of a collection of magnetic moments and the presence of a magnetic field. We shall start by considering the dynamics of such a system at the simplest level possible. This classical description will be a first step towards understanding magnetic resonance and relaxation as introduced in the previous sections. Also, it will connect with the quantum discussion of Section 1.2 where we spoke in terms of the energy difference of adjacent quantum states and the transitions between them.

A magnetic moment μ in a static magnetic field $\mathbf{B}$ will experience a torque $\mu \times \mathbf{B}$. By the rotational analogue of Newton's Second Law:

torque = rate of change of angular momentum,

we obtain the equation of motion for the spin angular momentum:

$$\mu \times \mathbf{B} = \hbar \dot{\mathbf{I}}.$$

But since the magnetic moment is proportional to and parallel to the angular momentum,

$$\mu = \gamma\hbar I$$

we have the equation for μ,

$$\dot{\mu} = \gamma\mu \times \mathbf{B}.$$

1.4.2 Precession

For an assembly of nuclear spins we introduce the magnetisation vector $\mathbf{M}$, the total magnetic moment per unit volume. Adding the equations of motion for each spin (of the same species) we obtain the equation of motion for $\mathbf{M}$,

$$\dot{\mathbf{M}} = \gamma\mathbf{M} \times \mathbf{B}. \tag{1.13}$$

This can be solved easily once the direction of $\mathbf{B}$ has been specified. By convention the magnetic field is assumed to be along the z axis. Thus we write

$$\mathbf{B} = B_0\mathbf{k},$$

where $\mathbf{k}$ is the unit vector in the z direction.

In component form the equations of motion are then

$$\left.\begin{aligned}
\dot{M}_x &= \gamma B_0 M_y \\
\dot{M}_y &= -\gamma B_0 M_x \\
\dot{M}_z &= 0,
\end{aligned}\right\} \tag{1.14}$$

which have the solution

$$\left.\begin{aligned}
M_x(t) &= M_x(0)\cos(\gamma B_0 t) \\
M_y(t) &= -M_x(0)\sin(\gamma B_0 t) \\
M_z(t) &= M_z(0), \text{ a constant,}
\end{aligned}\right\} \tag{1.15}$$

assuming the initial condition of $M_y(0) = 0$. These equations describe a rotation in the x–y plane, and a constant component in the z direction. The magnetisation component in the x–y plane has a constant magnitude $M_x(0)$ and it rotates in the clockwise (negative) direction with an angular velocity of $\omega_0 = \gamma B_0$. This is precisely the NMR resonant (angular) frequency as discussed in Section 1.2.3, Equation (1.5).

In Figure 1.4 we show the full behaviour of the magnetisation vector. The length of the $\mathbf{M}$ vector remains constant and the angle it makes with the applied magnetic field is constant. The motion of $\mathbf{M}$ about $\mathbf{B}$ is similar to gyroscope motion in the earth's gravitational field. Such motion is called *precession*, and the behaviour of the magnetisation vector in an applied magnetic field is referred to as *Larmor precession*.

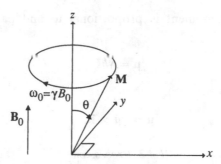

Figure 1.4 Larmor precession of the magnetisation vector **M** about **B**$_0$.

The frequency of the precession is called the Larmor frequency. Thus it follows that in Table 1.1 the right-hand column gives the precession frequency for the specified nuclei in a magnetic field of 1 T.

In terms of the magnitude of the magnetisation **M** and its inclination θ from the field **B** we may express the evolution of the components of **M** as

$$M_x(t) = M\sin(\theta)\cos(\omega_0 t),$$
$$M_y(t) = -M\sin(\theta)\sin(\omega_0 t),$$
$$M_z(t) = M\cos(\theta).$$

We see that the transverse magnetisation, made up of M_x and M_y, continues to ring indefinitely. This is similar to our oscillator of Section 1.3 when the damping constant β is zero, which means that the resonance is infinitesimally narrow. The similarity is not complete, however, since now we have two components M_x and M_y rather than the single x. The longitudinal magnetisation M_z remains constant, a reflection of the conservation of magnetic energy $-M_z B_0$.

1.4.3 Quantum versus classical treatment

In both the quantum and the classical views we have found that the Larmor frequency plays a fundamental role. Nevertheless, at first glance the two views seem somewhat different. In the quantum treatment it is the energy equivalent of the resonance frequency that is fundamental: RF quanta of the requisite energy are absorbed, causing energy level transitions. On the other hand, according to the classical approach, it is the actual frequency that is important: the magnetic moment vectors are precessing at this particular rate.

This duality of approaches mirrors precisely the twin independent discoveries of NMR. The Harvard team of Purcell, Torrey and Pound fundamentally were spectroscopists. They regarded the effect of a magnetic field on an individual magnetic moment in terms of splitting of its energy levels. The NMR experiment then involved the stimulation of transitions between these levels. On the other hand the Stanford University group, comprising Bloch, Hansen and Packard, thought about the gross effect of a magnetic field on a collection of magnetic moments: reorienting the total magnetic moment, thus their picture in terms of observable (and therefore classical) quantities.

The unification of these apparently different pictures relates to the dual role of energy in physics. The elementary interpretation of energy as work done is connected with the internal state of the system. You can do a bit more work on the system and its internal state will have changed correspondingly; this is the spectroscopist's point of view. However, the more advanced interpretation of energy follows from the Hamiltonian formulation of dynamics. There one learns that it is the energy function which actually determines the time evolution of the system: Hamilton's equations in classical mechanics and Heisenberg's equation (or the time dependent Schrödinger equation) in quantum mechanics. Thus the inextricable connection between work done and time evolution – the spectroscopic approach of Purcell at Harvard and the reorientation view of Bloch at Stanford.

1.4.4 Behaviour of real systems – relaxation

In practice things do not behave quite like this – precessing continuously. When a specimen is placed in a magnetic field **B** it will achieve a magnetisation **M** given by

$$M = (\chi_0/\mu_0)B. \tag{1.16}$$

Box 1.1 Magnetic B and H fields

According to the SI rules the magnetic susceptibility χ is defined as the ratio of the magnetisation **M** to the applied magnetic **H** field:

$$\chi = M/H.$$

Now the **B** field is related to the **H** field by

$$B = \mu_0(H + M).$$

Thus the ratio **M/B** is related to the magnetic susceptibility through

$$M = \frac{\chi}{1 + \chi} \frac{B}{\mu_0}.$$

In practice χ is very small for paramagnetic systems, a value of order 10^{-9} being typical, as shown in Chapter 2. Under such circumstances the denominator $(1 + \chi)$ may be replaced by unity without any significant error. This then gives the expression used in the text:

$$M = \chi B/\mu_0.$$

There is, however, a further point which must be considered, relating to the magnetogyric ratio and the Larmor frequency. As pointed out in Section 1.2.4, the frequency of precession is determined by the *local* field: the microscopic field experienced at the site of the nuclear spin. This is not the same as the macroscopic field **B**, which is in reality some spatial average within the material. One would expect the microscopic field to vary wildly within the material. Conventionally one estimates the microscopic field at a nuclear site by imagining a hole within the material, surrounding the nucleus. The local field is then the sum of the external field applied plus the effect of the polarised material. For a spherical hole one finds (Robinson 1973)

$$B_{loc} = B_{ext} - \tfrac{2}{3}\mu_0 M,$$

which is appropriate for a lattice of cubic symmetry or a random distribution of spins. For lattices of lower symmetry the coefficient of **M** will be different. But since, as we argued above, the susceptibility is very small, the magnetisation will be negligible compared with B/μ_0. Thus usually the term in **M** may be ignored, and the local magnetic field identified with the externally applied field.

Equation (1.16) assumes the absence of ferromagnetism and that the system responds linearly. It is also assumed that the system is isotropic. The constant χ_0 is the static magnetic susceptibility and μ_0 is the permeability of free space.

There are two things to be said about such a real system when placed in a magnetic field. Firstly the magnetisation parallel to the applied magnetic field grows to its equilibrium value; but in Section 1.4.2 we learned that M_z remains constant! Secondly, any magnetisation perpendicular to the applied field must go to zero; but Section 1.4.2 showed that the transverse magnetisation rotates continuously at the Larmor frequency!

Whatever the initial state of a real system, the magnetisation will relax to its equilibrium value $M_0 = (\chi_0/\mu_0)B_0$ along the z axis; M_z relaxes to M_0, and M_x and M_y relax to zero (with some Larmor precession as well). The relaxation, which was not accounted for in the microscopic calculations of the previous section, occurs as a consequence of *interactions* amongst the nuclear spins and between them and their environment. These questions will be discussed in more detail in the next chapter in Section 2.2. The nature of these relaxation processes is a central concern of NMR and of much of this book.

For the present we simply comment that as explained for the model oscillator of Section 1.3, studying relaxation gives the same information as the study of resonance.

1.5 Typical signal size

1.5.1 The induced voltage

We conclude this chapter with some practical considerations. We shall examine the question of how easy it is to observe NMR. In particular we will look at the size of a typical NMR signal. This is important both for giving a further feel for the experimental problems of NMR and because many modern applications of NMR involve unfavourable situations where signals may be rather poor. In these cases a good understanding of the factors affecting signal size is particularly important. Chapter 3 is devoted to a fuller treatment of such practical considerations.

We shall consider the detection of Larmor precession, i.e. observation of the transient response of a typical specimen. Let us take a millilitre of water placed in a magnetic field of 1 T. First we note from Table 1.1 that $\gamma/2\pi$ for the protons in the water molecules is about $42.6 \times 10^6 \, \mathrm{s}^{-1} \, \mathrm{T}^{-1}$, so that the Larmor precession frequency is then 42.6 MHz in this field.

Now let us surround our sample with a close-fitting coil as in Figure 1.5. The precessing magnetisation vector **M** will, by induction, generate a voltage v in the coil:

$$v = \text{rate of change of flux enclosed by coil wire}$$
$$= NA\dot{B},$$

where N is the number of turns of the coil and A its area of cross section. B is the magnitude of the magnetic field produced by the precessing magnetisation. It is given approximately by

$$\mathbf{B} \approx \mu_0 \mathbf{M}$$

where **M** is the magnetisation along the axis of the coil. (The precise factor relating **B** and **M** involves the demagnetisation or shape factor of the specimen. In practice it will be somewhere between $\frac{1}{2}$ and $\frac{1}{3}$.) We have then the approximate relation for the induced voltage,

$$v \approx \mu_0 NA\dot{M}. \tag{1.17}$$

1.5.2 Initial conditions

But what is the magnitude of the precessing magnetisation? In the language of Section 1.3 it depends on the strength of the transient excitation. In practice, as we shall see in Section 2.5.4, we can arrange for an initial condition to be produced whereby the entire

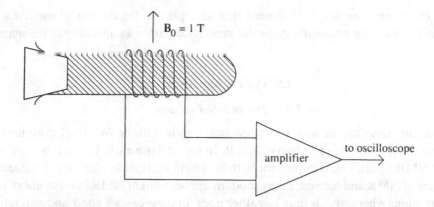

$B_0 = 1\,\text{T}$

amplifier

to oscilloscope

Figure 1.5 Basic elements of an NMR system.

equilibrium magnetisation $M_0 = (\chi_0/\mu_0)B_0$ is tipped over into the transverse plane, where it will precess at the Larmor frequency. Thus we take the precessing magnetisation to be

$$M(t) = M_0\cos(\omega_0 t)$$
$$= (\chi_0/\mu_0)B_0\cos(\omega_0 t).$$

Upon differentiation and substitution into Equation (1.17) we obtain the voltage induced by the precession,

$$v \approx -NAB_0\chi_0\omega_0\sin(\omega_0 t).$$

However, since

$$\omega_0 = \gamma B_0,$$

we obtain for the magnitude of the induced voltage

$$|v| \approx NA\chi_0\omega_0^2/\gamma \tag{1.18}$$

or

$$|v| \approx NA\chi_0 B_0^2\gamma,$$

which is quite simple to estimate.

1.5.3 A typical system

Let us put some numbers into this expression. The cross-sectional area A will be approximately $1\,\text{cm}^2$, or $10^{-4}\,\text{m}^2$. A typical coil of these dimensions at $42\,\text{MHz}$ would have something like ten turns. The nuclear magnetic susceptibility of water (see next chapter) at room temperature due to its protons is approximately 3.8×10^{-9} (no units as χ is dimensionless). In summary

$$N = 10,$$
$$A = 10^{-4}\,\mathrm{m}^2,$$
$$\chi_0 = 3.8 \times 10^{-9},$$
$$\gamma = 2.7 \times 10^8\,\mathrm{s}^{-1}\,\mathrm{T}^{-1},$$
$$B = 1\,\mathrm{T},$$
$$\omega_0/2\pi = 42.6\,\mathrm{MHz}.$$

Substituting these quantities into our formula we obtain an induced voltage of approximately 1 mV.

1.5.4 Other considerations

A millivolt is certainly a measurable quantity although in practice other considerations must be taken into account. We may, for instance, resonate the coil with a capacitor to magnify the observed voltage. Also, thus far we do not know anything about the duration of the precession signal, nor have we considered the electrical noise that will always be superimposed on the signal.

Electrical noise is a randomness in current or voltage due to such factors as the thermal agitation of the current-carrying electrons. Ultimately it is the ratio of the signal voltage to the noise voltage which determines whether a signal may be observed or not. These questions will be discussed further in Chapter 3.

2

Theoretical background

2.1 Paramagnetism and Curie's Law

2.1.1 Paramagnetism

The static magnetic susceptibility was introduced in Section 1.4.4. It was explained there that when a substance is placed in a static magnetic field B_0 it acquires a magnetisation M_0. If the substance is isotropic then M_0 will be parallel to B_0. Furthermore, for reasonably small B_0 the magnetisation will be proportional to the applied magnetic field. Precisely how small will be considered in Section 2.1.5. We are, of course, excluding ferromagnetism. Under these conditions we write, as Equation (1.16),

$$M_0 = (\chi_0/\mu_0)B_0, \tag{2.1}$$

where χ_0 is the static magnetic susceptibility of the system and μ_0 is the permeability of free space, defined in SI units to be $4\pi \times 10^{-7}\,\mathrm{H\,m^{-1}}$. The factor μ_0 appears because fundamentally χ is defined as the ratio of the magnetisation to the *magnetic H field* not the B field, as we saw in Box 1.1. This definition results in χ being dimensionless.

We saw in the previous chapter that the size of the NMR free precession signal is proportional to the nuclear spin magnetic susceptibility, and in calculating the expected signal size from a water sample the susceptibility χ was simply quoted. In this section we shall consider the mechanism for the occurrence of such magnetisation, and an expression for χ will be obtained. Knowing the equilibrium magnetisation of a specimen in a magnetic field, we know the state towards which it will relax following a disturbance.

In a magnetic field it is energetically favourable for a magnetic dipole to orient itself along the field – the state of minimum energy. However there is a competing factor. The thermal agitation of the atoms tends to randomise the orientations – the law of entropy increase. The equilibrium state which thus results is a balance between the ordering

20

tendency of energy minimisation and the disordering tendency of entropy maximisation. The magnetisation which results from this mechanism is known as *paramagnetism*. It follows that the magnetisation of a paramagnetic substance increases as its temperature decreases.

2.1.2 Calculation of magnetisation

For simplicity let us consider a collection of particles of spin $I = \frac{1}{2}$. In this case each particle has two spin states, $m = +\frac{1}{2}$ and $m = -\frac{1}{2}$ which we refer to as 'spin up' and 'spin down'. Let us denote the number of particles per unit volume with spin up by $N_\uparrow$ and the number per unit volume with spin down by $N_\downarrow$. Furthermore, we denote the total number of particles per unit volume by N_v. Then

$$N_v = N_\uparrow + N_\downarrow. \tag{2.2}$$

In a magnetic field $\mathbf{B}_0$ the up states and the down states are separated by an energy $\Delta E = \hbar \gamma B_0$. So by Boltzmann's principle the ratio $N_\uparrow / N_\downarrow$ is given by

$$\begin{aligned} N_\uparrow / N_\downarrow &= \exp(\Delta E/kT) \\ &= \exp(\hbar \gamma B_0/kT), \end{aligned} \tag{2.3}$$

where k is Boltzmann's constant and T is the absolute temperature.

The net magnetisation $\mathbf{M}_0$ is proportional to the excess of up spins over down spins,

$$M_0 = \mu(N_\uparrow - N_\downarrow), \tag{2.4}$$

where μ is the magnetic moment of a single particle. So evaluating $N_\uparrow$ and $N_\downarrow$ we find

$$N_\uparrow = \frac{N_v}{\exp(-\hbar \gamma B_0/kT) + 1}$$

and

$$N_\downarrow = \frac{N_v}{\exp(+\hbar \gamma B_0/kT) + 1}.$$

The magnetisation is then found to be

$$M_0 = N_v \mu \left[\frac{1 - \exp(-\hbar \gamma B_0/kT)}{1 + \exp(-\hbar \gamma B_0/kT)} \right]$$

or

$$M_0 = N_v \mu \tanh(\hbar \gamma B_0/2kT). \tag{2.5}$$

We see that in general the magnetisation is not proportional to the applied magnetic

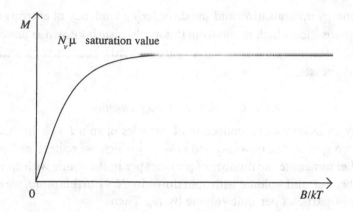

Figure 2.1 Magnetisation of a paramagnet.

field. Although for low fields the magnetisation response will be linear, at high fields the magnetisation will saturate, corresponding to all dipoles being oriented along the magnetic field. Clearly no further magnetisation can be achieved. Figure 2.1 shows the relationship between magnetic field and magnetisation.

2.1.3 Curie's Law

When the magnetic field is sufficiently small and/or the temperature is sufficiently high then the magnetisation is proportional to the applied field. In such cases this linear response may be obtained by expanding the hyperbolic tangent in Equation (2.5). To first order we obtain

$$M_0 = \frac{N_v \mu \hbar \gamma}{2kT} B_0. \tag{2.6}$$

Now since the magnetic moment μ is related to the spin I which in our example has been chosen specifically to be $I = \frac{1}{2}$, we have

$$\mu = \gamma \hbar I = \gamma \hbar / 2,$$

so that

$$M_0 = \frac{N_v \hbar^2 \gamma^2}{4kT} B_0. \tag{2.7}$$

This equation is the embodiment of *Curie's Law*: that for a paramagnet the magnetisation is proportional to the applied field and inversely proportional to the absolute temperature. This is often written as

$$M_0 = C\frac{B_0}{T}.$$

and the constant C is referred to as the *Curie constant*.

Box 2.1 Dimensional considerations relating to paramagnetism

In the above model for paramagnetism we see that the magnetic field B and the absolute temperature T occur only in the dimensionless combination $\hbar\gamma B/kT$ – the magnetisation is essentially a homogeneous function of B/T. This tells us that in the linear region, where M is proportional to B, M *must* also be proportional to $1/T$. Away from linearity, if the dependence on B is known, the variation with T may be inferred and *vice versa*.

2.1.4 Magnetic susceptibility

In the linear region of paramagnetism it is most convenient to use the concept of magnetic susceptibility, the constant of proportionality between M_0 and B_0/μ_0. The susceptibility is given, from Equations (2.1) and (2.7), by

$$\chi_0 = \frac{\mu_0 N_v \hbar^2 \gamma^2}{4kT}.$$

For spin I greater than $\frac{1}{2}$ the calculation is a little more complicated. It is considered in Problem 5.2. The result then takes the form

$$\chi_0 = \frac{\mu_0 N_v \hbar^2 \gamma^2 I(I+1)}{3kT},$$

which, of course, reduces to our calculated expression when $I = \frac{1}{2}$.

2.1.5 Conditions for linearity

We are, of course, now in a position to say how small B_0 must be to ensure linearity. It is the condition that the hyperbolic tangent in Equation (2.5) may be approximated by its expansion to first order. Its argument must thus be much smaller than 1, or

$$\hbar\gamma B_0 \ll 2kT. \tag{2.8}$$

In practice this turns out not to be too restrictive. For the case of protons in water at room temperature we find that B_0 should be less than 7000 T!! It is therefore mainly at low temperatures that deviations from linearity are found.

But by low temperatures we really do mean *very* low. In realistic fields the *nuclear* paramagnetic susceptibility of copper, for instance, obeys Curie's Law well at temperatures as low as 1 mK. Thus measurement of magnetisation is a common technique of thermometry at such temperatures

2.2 Relaxation

2.2.1 Equilibrium states

The existence of states of equilibrium is a fact of life, and it is to this fact that the subject of thermodynamics owes its utility; the laws of thermodynamics describe the equilibrium behaviour of complex systems with large numbers of particles in terms of a small number of variables.

We have seen already that the calculation of equilibrium properties of a system is, in principle, a relatively simple procedure involving the Boltzmann distribution function, or something similar. But how is this equilibrium state approached? In general this is a complex question and it is not often possible to obtain an explicit expression describing the return of a system to equilibrium following a disturbance. Nuclear spin relaxation is a nice exception and much of this book is an exploration of aspects of this problem.

2.2.2 The relaxation process

Let us now look further at the question of magnetic relaxation. We will clarify what are the important features of the process and what is the appropriate language with which to describe them.

In a static magnetic field the equilibrium state of a spin system displays two features of particular relevance to us. Firstly there is a magnetisation parallel to the applied magnetic field and secondly there is zero magnetisation in the plane perpendicular to the field. These features must be compared with the equations of motion for the magnetisation vector and their solution, Equation (1.25). Experimentally the magnetisation achieves its equilibrium value along the B_0 direction, but Equation (1.25) says that the magnetisation in this direction is a constant! Experimentally any magnetisation in the transverse plane will tend to zero, but Equation (1.25) predicts that the transverse magnetisation will continue to precess about B_0 indefinitely!

How do we explain these seeming contradictions? Clearly the mathematical analysis used in the lead up to Equation (1.25) is inadequate in some respect; it is missing some important feature, and particularly from the point of view of studying the transient response, the absent feature is the very relaxation process itself.

The flaw in the model was the implicit assumption that all nuclei see exactly the same magnetic field, namely B_0. It was this point which enabled the field to be factorised out of the equations of motion for the magnetisation, leading to the subsequent simple solution. In practice, as was mentioned in Section 1.2, there are other contributions to the local field. Each spin sees a slightly different magnetic field and, moreover, these fields may change with time.

2.2.3 Spin interactions

Of the various sources of these fields there are two of particular relevance. Firstly the magnet producing the $\mathbf{B_0}$ field will not be perfect. There will always be a spread of values of $\mathbf{B_0}$ throughout the specimen. Secondly, and of more physical importance, the nuclear dipoles *produce* magnetic fields as well as responding to them. Other smaller apparent fields may arise from the hyperfine interaction with orbital electrons and possible electric quadrupole interactions with crystal fields. These fields will be experienced by the nuclear spins. If the spins are moving then the fields seen will fluctuate with time.

These different and varying fields cause the spins to be precessing at different and varying frequencies. It is the consequent incoherence in the motion of the spins which results in dephasing and thus in relaxation. In summary, then, the relaxation processes observed in the dynamics of a spin system are due to the interactions of the spin dipoles with their environment and with themselves.

2.2.4 Spin–lattice and spin–spin relaxation

We can take this a little further and introduce some terminology. A distinction must be made between the relaxation of magnetisation parallel to the $\mathbf{B_0}$ field: the approach towards $\mathbf{M_0} = (\chi_0/\mu_0)\mathbf{B_0}$, and the decay of magnetisation in the transverse plane to zero. The former process involves a flow of energy, a change of the spin magnetic energy density $-\mathbf{M \cdot B}$. Since total energy must be conserved this process will involve exchange of energy with other degrees of freedom in our system. In NMR jargon these other degrees of freedom are referred to as the 'lattice' and thus we call this process *spin–lattice relaxation*. Also, since it is the longitudinal component of the magnetisation which is varying, this is also referred to as *longitudinal relaxation*.

On the other hand the relaxation of transverse magnetisation involves no energy exchange. This should then be an 'easier' process. The mechanism here may be understood rather simply as the destructive interference in the precessing magnetisation as each spin precesses at a slightly different rate. This will be due at least in part to the extra fields seen from other spins and we thus speak, in the transverse case, of *spin–spin relaxation* or *transverse relaxation*.

2.2.5 The Bloch equations

We may quantify these relaxation processes through the introduction of *relaxation times*. The spin–lattice relaxation time, conventionally denoted by T_1, characterises the relaxation of magnetisation parallel to $\mathbf{B_0}$, while the spin–spin relaxation time, T_2, characterises the relaxation in the plane perpendicular to the $\mathbf{B}_0$ field.

In a purely phenomenological approach, relaxation terms may be added to the precession equations for the magnetisation, Equations (1.13), giving

$$\left.\begin{aligned}
\dot{M}_x &= \gamma |\mathbf{M} \times \mathbf{B}|_x - M_x/T_2 \\
\dot{M}_y &= \gamma |\mathbf{M} \times \mathbf{B}|_y - M_y/T_2 \\
\dot{M}_z &= \gamma |\mathbf{M} \times \mathbf{B}|_z + (M_0 - M_z)/T_1.
\end{aligned}\right\} \tag{2.9}$$

These are known as the Bloch equations after Felix Bloch who first proposed them in 1946. The quantity M_0 is the equilibrium value of M_z; remember that M_x and M_y relax to zero while M_z relaxes to M_0. Clearly, relaxation described by these equations, where the rate of return is proportional to the deviation, will follow an exponential law. In the absence of relaxation the solution to these equations was given, for a constant field $\mathbf{B}_0$, in the z direction by Equations (1.15). Including the effects of T_1 and T_2, the magnetisation components now evolve as:

$$\left.\begin{aligned}
M_x(t) &= M_x(0)\cos(\omega_0 t)\exp(-t/T_2) \\
M_z(t) &= M_0 - [M_0 - M(0)]\exp(-t/T_1).
\end{aligned}\right\} \tag{2.10}$$

Such behaviour for the magnetisation is shown in Figures 2.2 and 2.3.

Whenever the observed relaxation does not proceed exponentially, the above description is, of course, invalid, certainly quantitatively, and the relaxation times no longer have the precise definitions as above. In such cases T_1 and/or T_2 are usually used simply as a rough measure of the time scale for the relaxation process. (Problem 2.1 looks at a more precise approach.)

Fortunately it is often observed that the magnetisation does indeed relax in an exponential manner (particularly in fluids). T_1 and T_2 are then good parameters for characterising such systems. One of our main tasks in the following chapters will be to relate these relaxation times to the behaviour and the environment of the constituent spins. As a preliminary, however, we must consider further the elementary dynamics of spins.

In the next section we proceed with a comparison of the quantum and the classical approach to these problems.

2.3 Quantum and classical descriptions of motion

2.3.1 The Heisenberg equation

In this section we shall resume our discussion of the dynamics of nuclear spins which was started in Section 1.4. We have mentioned already that for many applications treated in this book a classical description is adequate, the refinements of quantum theory being unnecessary. In justification, we start this section with a discussion of the

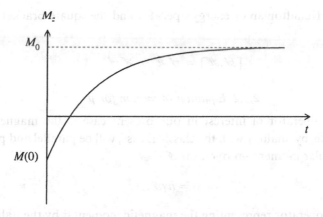

Figure 2.2 Longitudinal relaxation.

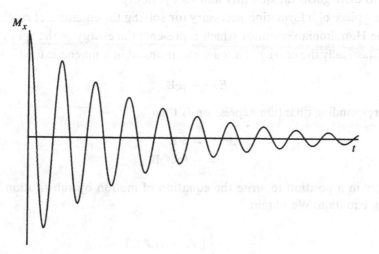

Figure 2.3 Transverse relaxation.

quantum equations for spin angular momentum and their relation to the classical 'gyroscope' equations, Equations (1.14).

In quantum mechanics observable quantities are represented by *operators*. For an arbitrary operator $\mathscr{A}$ representing the observable A the equation of motion, known as the Heisenberg equation, is given by

$$\frac{\mathrm{d}\mathscr{A}}{\mathrm{d}t} = -\frac{\mathrm{i}}{\hbar}[\mathscr{A},\mathscr{H}], \qquad (2.11)$$

where $\mathcal{H}$ is the Hamiltonian or energy operator and the square bracket denotes the *commutator*,

$$[\mathcal{A}, \mathcal{H}] = \mathcal{A}\mathcal{H} - \mathcal{H}\mathcal{A}.$$

2.3.2 Equation of motion for μ

The observable operator of interest in our present case is the magnetic moment operator **μ**, which, by analogy with the classical case, will be parallel and proportional to the spin angular momentum operator $\mathcal{I}$,

$$\boldsymbol{\mu} = \hbar\gamma\mathcal{I}.$$

(We denote the operator representing the magnetic moment **μ** by the italic symbol *μ*, and the operator representing the angular momentum **I** by the script symbol $\mathcal{I}$, when necessary to distinguish the quantity and its operator).

The other piece of information necessary for solving the equation of motion is the form of the Hamiltonian operator which represents the energy of the spin magnetic moment. Classically the energy of a magnetic moment in a magnetic field is given by

$$E = -\boldsymbol{\mu} \cdot \mathbf{B},$$

so the corresponding quantum expression is then

$$\begin{aligned}\mathcal{H} &= -\boldsymbol{\mu} \cdot \mathbf{B} \\ &= -\hbar\gamma\mathcal{I} \cdot \mathbf{B}.\end{aligned}$$

We are now in a position to write the equation of motion by substitution into the Heisenberg equation. We obtain

$$\frac{d\mathcal{I}}{dt} = -\frac{i}{\hbar}[\mathcal{I}, -\hbar\gamma\mathcal{I} \cdot \mathbf{B}]$$

$$= i\gamma[\mathcal{I}, \mathcal{I} \cdot \mathbf{B}].$$

At this stage we are working with the (dimensionless) spin angular momentum operator $\mathcal{I}$ rather than the magnetic moment **μ** because we shall want to make use of the commutation relations for angular momenta. Let us look at the equation of motion for one component of $\mathcal{I}$, say the x component. This is

$$\frac{d\mathcal{I}_x}{dt} = i\gamma[\mathcal{I}_x, \mathcal{I}_x B_x + \mathcal{I}_y B_y + \mathcal{I}_z B_z]$$

or

$$\frac{\mathrm{d}\mathscr{I}_x}{\mathrm{d}t} = i\gamma B_x[\mathscr{I}_x, \mathscr{I}_x] + i\gamma B_y[\mathscr{I}_x, \mathscr{I}_y] + i\gamma B_z[\mathscr{I}_x, \mathscr{I}_z].$$

2.3.3 Evaluation of commutators

These angular momentum commutators are well known in quantum mechanics. As any text book on the subject, for example Cohen-Tannoudji, Diu and Laloë (1977), will show

$$[\mathscr{I}_x, \mathscr{I}_y] = i\mathscr{I}_z,$$ and cyclic permutations thereof.

Since the commutator of an operator with itself is zero the first term of the equation of motion vanishes. Substituting in the results for the second and third commutator we find that the equation of motion for $\mathscr{I}_x$ reduces to

$$\frac{\mathrm{d}\mathscr{I}_x}{\mathrm{d}t} = \gamma(\mathscr{I}_y B_z - \mathscr{I}_z B_y). \tag{2.12}$$

At this point we can make a number of observations. Firstly Planck's constant has cancelled out, telling us that the resultant equation is not inherently quantum mechanical. Secondly, since the equation is linear in $\mathscr{I}$, the same equation holds for the magnetic moment μ on multiplying through by γ. Thirdly we see that the right-hand side of Equation (2.12) is the x component of the cross product of $\mathscr{I}$ with $\mathbf{B}$,

$$(\mathscr{I}_y B_z - \mathscr{I}_z B_y) = |\mathscr{I} \times \mathbf{B}|_x.$$

We finally obtain the equation of motion for the magnetic moment operator μ:

$$\frac{\mathrm{d}\mu}{\mathrm{d}t} = \gamma\mu \times \mathbf{B}.$$

This equation for the quantum mechanical magnetic moment operator is identical to the classical equation derived for the magnetic moment, Equation (1.13).

2.3.4 Expectation values

There is one further step to go in establishing the equivalence of the classical and quantum descriptions of the behaviour of a magnetic moment in a magnetic field. If a quantum system is in a state s, described by a wave function $\psi_s(x)$, then the mean or expectation value of an operator, say $\mathscr{A}$, is given by the integral

$$A = \langle\mathscr{A}\rangle = \int \psi_s^*(x)\mathscr{A}\psi_s(x)\mathrm{d}x, \tag{2.13}$$

where x is the set of whatever coordinates are being used to describe the state. It should be noted that, as is often the case in quantum mechanics, by 'integration' we include summation over discrete variables as well as integration over continuous ones. This is neatly accounted for using the Dirac notation for the above expression:

$$\langle \mathscr{A} \rangle = \langle s | \mathscr{A} | s \rangle.$$

So performing such an operation on the equation of motion for the quantum operator μ we find

$$\frac{d}{dt} \langle \mu \rangle = \gamma \langle \mu \rangle \times \mathbf{B}.$$

Thus for a magnetic moment in a magnetic field we have the key result that:

The quantum equation for the expectation value of the operator μ is identical to the classical equation for the quantity μ.

It should be noted that the final step in this demonstration relied on the quantum equation of motion being linear in μ and the fact that there were no other operators multiplying μ. In general, quantum and classical equations do not have equivalent forms – otherwise quantum theory would tell us nothing new!

Now that the result of this section is established we shall, except where clarity requires, henceforth no longer make a typographical distinction between an operator and its expectation value. Thus $\mathbf{I}$ will usually be used for both the spin angular momentum operator and the expectation value of spin angular momentum. It should be clear from the context which meaning is intended.

2.4 The rotating frame

2.4.1 The equivalence principle

We have already mentioned that one feature common to (almost) all NMR phenomena is the application of a static magnetic field $\mathbf{B}_0$, and resulting from this field is the Larmor precession at the angular frequency $\omega_0 = \gamma B_0$. There is, however, another feature common to these systems: the use of oscillating magnetic fields. At some stage or another in most NMR experiments a sinusoidal magnetic field is applied in the plane perpendicular to $\mathbf{B}_0$.

In a steady state or continuous wave experiment (frequency domain) a very small oscillating field is applied and the magnetisation response is observed as the frequency is varied. On the other hand in a pulsed NMR experiment (time domain) a large resonant – i.e. at ω_0 – oscillating field is applied in a short burst to prepare the system in the required non-equilibrium state. These procedures will be discussed fully in the next section.

The aim of this section is to provide the mathematical basis for a simple physical model describing the motion of spins in a large static field together with transverse oscillating fields. To this end we introduce the concept of the *rotating frame*; we propose to look at the system through the eyes of someone who is rotating rather than stationary with respect to the system. Intuitively this is a natural thing to do, since we know that Larmor precession is always occurring; what we want to know is what *else* is happening. A very clear account of this procedure is given in the paper by Rabi, Ramsey and Schwinger (1954).

A static magnetic field $\mathbf{B}_0$ causes a magnetic moment to precess at a frequency $\omega_0 = \gamma B_0$. So an observer rotating at the same rate ω_0 would observe the magnetic moment to be stationary; he would not see the effect of the field. This is analogous to the equivalence principle of the theory of relativity. If the cables of an elevator are cut the occupants do not know whether they are accelerating to the ground or if gravity has simply been turned off!

In a rotating reference frame a static magnetic field may thus effectively disappear. We now turn to the mathematical demonstration of this.

2.4.2 Transformation to the rotating frame

In a stationary reference frame the time derivative of a vector μ is given in terms of its components by

$$\frac{d\mu}{dt} = \frac{\partial \mu_x}{\partial t}\mathbf{i} + \frac{\partial \mu_y}{\partial t}\mathbf{j} + \frac{\partial \mu_z}{\partial t}\mathbf{k}.$$

But if the reference is moving then the basis vectors will not be constant; the time derivative of μ has an extra contribution.

$$\frac{d\mu}{dt} = \left(\frac{\partial \mu_x}{\partial t}\mathbf{i} + \frac{\partial \mu_y}{\partial t}\mathbf{j} + \frac{\partial \mu_z}{\partial t}\mathbf{k}\right) + \mu_x\frac{\partial \mathbf{i}}{\partial t} + \mu_y\frac{\partial \mathbf{j}}{\partial t} + \mu_z\frac{\partial \mathbf{k}}{\partial t}. \tag{2.14}$$

The part in brackets is the rate of change of μ as seen in the moving frame – the frame from which the unit vectors $\mathbf{i}$, $\mathbf{j}$ and $\mathbf{k}$ appear stationary. We denote this rate of change by the partial derivative $\partial\mu/\partial t$

$$\frac{\partial \mu}{\partial t} = \frac{\partial \mu_x}{\partial t}\mathbf{i} + \frac{\partial \mu_y}{\partial t}\mathbf{j} + \frac{\partial \mu_z}{\partial t}\mathbf{k}.$$

Turning to the last three terms of Equation (2.14) we shall consider the particular case in which the reference frame is rotating about the z axis with angular velocity ω. Viewed from the fixed frame the unit vectors are varying as

$$\mathbf{i}(t) = \mathbf{i}(0)\cos(\omega t) + \mathbf{j}(0)\sin(\omega t),$$
$$\mathbf{j}(t) = -\mathbf{i}(0)\sin(\omega t) + \mathbf{j}(0)\cos(\omega t),$$
$$\mathbf{k}(t) = \mathbf{k}(0).$$

So the time derivatives of the basis vectors are then

$$\frac{\partial \mathbf{i}}{\partial t} = \omega\mathbf{j}, \quad \frac{\partial \mathbf{j}}{\partial t} = -\omega\mathbf{i}, \quad \frac{\partial \mathbf{k}}{\partial t} = 0.$$

This gives the total rate of change of the vector μ as

$$\frac{d\mu}{dt} = \frac{\partial \mu}{\partial t} + \omega(\mu_x\mathbf{j} - \mu_y\mathbf{i}),$$

which may be expressed in terms of the vector cross product:

$$\frac{d\mu}{dt} = \frac{\partial \mu}{\partial t} + \omega\mathbf{k} \times \mu.$$

Now since the quantity ω is the *magnitude* of the angular velocity, and the z axis is the *direction* of the angular velocity, the product $\omega\mathbf{k}$ is the angular velocity vector. Denoting this by ω

$$\omega = \omega\mathbf{k}$$

we have for the case of rotation about an arbitrary axis

$$\frac{d\mu}{dt} = \frac{\partial \mu}{\partial t} + \omega \times \mu. \tag{2.15}$$

2.4.3 The fictitious field

Let us now apply this result to the equation of motion for a magnetic moment in a magnetic field, Equation (1.13),

$$\frac{d\mu}{dt} = \gamma\mu \times \mathbf{B},$$

in a frame of reference rotating at angular velocity ω. The rate of change of μ as seen from the rotating frame $\partial\mu/\partial t$ is given by

$$\frac{\partial \mu}{\partial t} = \frac{d\mu}{dt} - \omega \times \mu$$

$$= \gamma\mu \times \mathbf{B} - \omega \times \mu$$

or

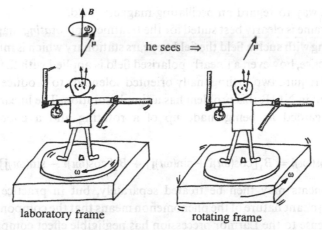

he sees ⟹

laboratory frame   rotating frame

Figure 2.4 Rotating and fixed frames of reference. If the observer is rotating at the same angular velocity as the precessing spin then he sees a stationary magnetic moment. Since the precession of the spin is caused by the applied magnetic field, to this rotating observer there appears to be no field.

$$\frac{\partial \boldsymbol{\mu}}{\partial t} = \gamma \boldsymbol{\mu} \times (\mathbf{B} + \boldsymbol{\omega}/\gamma).$$

In other words, in a frame rotating with angular velocity $\boldsymbol{\omega}$ a magnetic moment *sees* an effective magnetic field $\mathbf{B} + \boldsymbol{\omega}/\gamma$.

$$\mathbf{B}_{\text{eff}} = \mathbf{B} + \boldsymbol{\omega}/\gamma. \tag{2.16}$$

We can regard Larmor precession in this way: in a frame rotating at angular velocity $\omega_0 = -\gamma/B_0$ a magnetic moment appears stationary, i.e. *it* is rotating at angular velocity $-\gamma B_0$. (Recall that in Section 1.4 we saw that Larmor precession was in the negative rotational sense.)

In general we see that in a rotating frame there appears an extra 'fictitious' magnetic field. It is usual to choose the frame rotating at the same rate as the applied transverse field. In this frame the effective field is then stationary. This is indicated in Figure 2.4.

2.5 Rotating and oscillating fields

2.5.1 Rotating and counter-rotating components

We are now in a position to turn our attention to the effect of oscillating magnetic fields, to see their effect in NMR experiments. We shall make use of the idea of the rotating frame, introduced in the previous section, and to this end we start by considering the most convenient way to regard an oscillating magnetic field.

The rotating frame is clearly best suited for the treatment of *rotating* magnetic fields. In a frame rotating with such a field the field appears stationary which is much easier to deal with. In practice, however, a linearly polarised field is applied with the aid of a coil system. It would require two orthogonally oriented solenoids to produce a circularly polarised field, but in NMR this problem has a simple solution. The linearly polarised field may be regarded as being made up of a rotating and a counter-rotating component,

$$2B_1\cos(\omega t)\mathbf{i} = B_1[\cos(\omega t)\mathbf{i} + \sin(\omega t)\mathbf{j}] + B_1[\cos(\omega t)\mathbf{i} - \sin(\omega t)\mathbf{j}].$$

The two components may then be treated separately, but in practice this is not necessary. The resonant nature of the phenomenon means that the component rotating in the opposite sense to the Larmor precession has negligible effect compared to that moving with the precession; only fields near resonance need be considered. Thus the normal procedure is to use a linearly polarised field and simply to ignore the counter-rotating component.

2.5.2 *The effective field*

Let us now specify the problem in hand. We are considering an assembly of similar magnetic moments in a static magnetic field $\mathbf{B}_0$ and a second field $\mathbf{B}_1$ rotating at angular speed ω in the plane transverse to $\mathbf{B}_0$. The total magnetic field is then given by

$$\mathbf{B} = B_1[\cos(\omega t)\mathbf{i} + \sin(\omega t)\mathbf{j}] + B_0\mathbf{k}.$$

In an arbitrary frame of reference rotating about the z axis at an angular speed of ω', our transformation rule, Equation (2.16), tells us that the apparent or effective field will be

$$\mathbf{B}_{\mathrm{eff}} = B_1\{\cos[(\omega - \omega')t]\mathbf{i} + \sin[(\omega - \omega')t]\mathbf{j}\} + (B_0 + \omega/\gamma)\mathbf{k},$$

and in particular in the reference frame moving at the same rate as the rotating $\mathbf{B}_1$ field the apparent field is stationary, as shown in Figure 2.5.

$$\mathbf{B}_{\mathrm{eff}} = B_1\mathbf{i} + (B_0 + \omega/\gamma)\mathbf{k}. \tag{2.17}$$

To describe the motion of a magnetic moment in this combination of magnetic fields is now an easy matter. Having transformed to a frame from which the total magnetic field appears stationary, in this frame the magnetic moment will simply precess around the effective field at a rate $\omega = \gamma B_{\mathrm{eff}}$.

The magnetic moment vector $\boldsymbol{\mu}$ lies along the surface of a cone. The tip of the vector moves around the base of the cone at the 'effective' Larmor frequency, the effective field lying along the axis of the cone, as in Figure 2.6. Meanwhile in the laboratory frame the cone is rotating about the z axis synchronously with the rotating B_1 field.

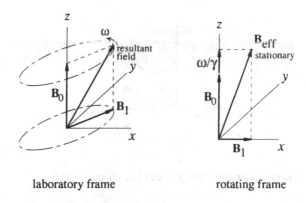

laboratory frame   rotating frame

Figure 2.5 Local field in the laboratory and rotating frames.

2.5.3 Resonant transverse field

As we have mentioned, a particularly important example of applied rotating magnetic fields is where the field is on resonance; then it rotates with the same angular velocity as the precessing magnetic moments. In this case the effective field

$$\mathbf{B}_{eff} = B_1\mathbf{i} + (\mathbf{B}_0 + \omega_0/\gamma)$$

becomes simply

$$\mathbf{B}_{eff} = B_1\mathbf{i}, \tag{2.18}$$

since

$$\omega_0 = -\gamma\mathbf{B}_0.$$

The effective field is now in the transverse plane. The $\mathbf{B}_0$ field has been completely eliminated. This has important uses.

2.5.4 90° and 180° pulses

Let us consider an assembly of magnetic moments placed in a steady magnetic field $\mathbf{B}_0$. When thermal equilibrium has been established there will be a magnetisation $\mathbf{M}_0$,

$$\mathbf{M}_0 = \frac{1}{\mu_0}\chi_0\mathbf{B}_0.$$

What happens if we now apply a resonant rotating field in the transverse plane? In the rotating frame all that the magnetisation sees is the effective field perpendicular to itself. So $\mathbf{M}$ will then simply precess around $\mathbf{B}_{eff}$, moving in a circle in the vertical plane; the

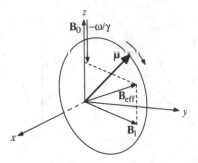

Figure 2.6 Magnetisation vector precessing about the effective field.

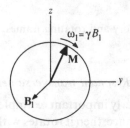

Figure 2.7 Magnetisation precessing in the x–y plane.

conical hat has become flat as in Figure 2.7. The angular velocity of the precession of the magnetisation vector in the vertical plane is $\omega_1 = \gamma B_1$. Thus if the rotating field is applied for a time t such that

$$\gamma B_1 t = \pi/2,$$

then the magnetisation vector will have been rotated through an angle of $\pi/2$ radians or 90°, starting from the z axis, and it will be left lying in the transverse plane.

This is just what is needed to initiate the study of transverse relaxation, since the **M** vector, now in the transverse plane, will precess at the Larmor frequency while its magnitude decays with a characteristic time T_2. Such a burst of resonant rotating field of duration

$$t_{90} = \pi/(2\gamma B_1) \qquad (2.19)$$

is referred to as a $\pi/2$ or a 90° pulse.

A pulse of twice this duration would rotate the magnetisation through 180°. It would invert **M** to $-$**M**. Such a pulse, of length

$$t_{180} = \pi/(\gamma B_1), \qquad (2.20)$$

is called a π or 180° pulse.

Both a 90° pulse and a 180° pulse disturb the equilibrium value of $\mathbf{M}_z$ and so they can be used to initiate longitudinal relaxation. There are other uses for these sorts of pulses, as we shall see.

2.6 The dynamic magnetic susceptibility

2.6.1 Complex susceptibility

The simple harmonic oscillator was discussed in Chapter 1 as a paradigm of a resonant system. We were able to draw a number of very general conclusions from the analysis of that case, in particular the Fourier transform relation between the transient behaviour and the frequency response. In this section we review and extend that discussion.

We saw that the response $x(t)$ to an oscillatory force $f_0\cos(\omega t)$ could be written as

$$x(t) = f_0[\chi'(\omega)\cos(\omega t) + \chi''(\omega)\sin(\omega t)], \tag{2.21}$$

where $\chi'(\omega)$ and $\chi''(\omega)$ are the in-phase and the quadrature response functions or generalised susceptibilities. Furthermore we saw that the response $X(t)$ to a transient or delta function force was given by

$$X(t) = \frac{1}{\pi} \int_0^\infty [\chi'(\omega)\cos(\omega t) + \chi''(\omega)\sin(\omega t)]d\omega. \tag{2.22}$$

Inverting this expression, the frequency functions $\chi'(\omega)$ and $\chi''(\omega)$ were thus seen to be the Fourier cosine and sine transforms of the transient function $X(t)$:

$$\chi'(\omega) = \int_0^\infty X(t)\cos(\omega t)dt, \tag{2.23}$$

$$\chi''(\omega) = \int_0^\infty X(t)\sin(\omega t)dt. \tag{2.24}$$

This Fourier transform relation may be displayed in a more compact and symmetrical manner through the introduction of a complex notation. Defining a complex dynamic susceptibility $\chi(\omega)$ as

$$\chi(\omega) = \chi'(\omega) + i\chi''(\omega) \tag{2.25}$$

we see that using Equations (2.23) and (2.24) this function may be written in terms of the transient response function $X(t)$

$$\chi(\omega) = \int_0^\infty X(t)\exp(i\omega t)dt. \tag{2.26}$$

Now the principle of causality implies that $X(t)$ is zero for negative times: there can be no response before the delta function force is applied. Thus the lower limit of the integral may be extended to minus infinity, giving an expression which is then the usual complex Fourier integral transform (Appendix A). We thus have the Fourier transform pair

$$\left.\begin{aligned} \chi(\omega) &= \int_{-\infty}^{\infty} X(t)\exp(+i\omega t)dt, \\[2mm] X(t) &= \frac{1}{2\pi} \int_{-\infty}^{\infty} \chi(\omega)\exp(-i\omega t)d\omega. \end{aligned}\right\} \tag{2.27}$$

A number of important consequences follow from the fact that the observed response $X(t)$ is fundamentally a real quantity. Taking the real part of Equation (2.27) we recover the original Equation (2.22) provided we generalise $\chi(\omega)$ to negative frequencies through

$$\chi(-\omega) = \chi^*(\omega), \tag{2.28}$$

i.e.

$$\chi'(-\omega) = \chi'(\omega)$$

and

$$\chi''(-\omega) = -\chi''(\omega).$$

In the spirit of the complex representation we may regard the oscillatory force $f_0\cos(\omega t)$ as the real part of a complex exponential $f_0\exp(-i\omega t)$. Since the real part is being taken it is immaterial whether i or $-$i is used in the exponential. However, the form adopted here conforms with the particular conventions chosen for Fourier transforms (Appendix A). From the linearity of our equations it follows that the response $x(t)$ may be expressed as the real part of the corresponding complex response:

$$x(t) = \mathrm{Re}\, f_0\chi(\omega)\exp(-i\omega t).$$

This is analogous to the use of complex exponentials in AC circuit theory. Frequently it is left understood that the real part is being taken, whereupon one says simply:

Response to $f_0\exp(-i\omega t)$ is $x(t) = f_0\chi(\omega)\exp(-\omega t)$.

Or, in terms of the harmonic components $x(\omega)$ and $f(\omega)$,

$$x(\omega) = f(\omega)\chi(\omega).$$

2.6.2 *Dynamic magnetic susceptibility*

There is an important way in which the harmonic oscillator of Chapter 1 and the previous section differs from the phenomena associated with NMR. The harmonic oscillator model was one-dimensional whereas the precession of transverse magnetisation occurs in two dimensions. For this reason we shall examine some of our previous results with a view to generalising them in a way suitable for the treatment of NMR.

Let us consider a magnetic field of magnitude b rotating clockwise in the x–y plane at an angular velocity ω. In terms of its components B_x and B_y we may write

$$\left. \begin{array}{l} B_x(t) = b\cos(\omega t), \\ B_y(t) = -b\sin(\omega t). \end{array} \right\} \tag{2.29}$$

The response to this, in the linear case, will be a rotating magnetisation, essentially a generalisation of Equation (2.21):

$$M_x(t) = \frac{b}{\mu_0}[\chi'(\omega)\cos(\omega t) + \chi''(\omega)\sin(\omega t)], \tag{2.30}$$

$$M_y(t) = \frac{b}{\mu_0}[-\chi'(\omega)\sin(\omega t) + \chi''(\omega)\cos(\omega t)], \tag{2.31}$$

although here μ_0 is introduced so that the χ have the dimensions of magnetic susceptibility. The form of Equation (2.31) then follows from the rotational invariance of the system – that M_y rotates into M_x in a quarter of the cycle period, i.e. in a time $\pi/2\omega$. In other words:

$$M_x(t) = M_y(t + \pi/2\omega).$$

Using this expression Equation (2.31) may be obtained directly from Equation (2.30).

In this two-dimensional/rotational treatment note that the hermiticity condition on $\chi(\omega)$, Equation (2.28), certainly does not apply. Here the sign of ω has a physical meaning, as may be seen from Equation (2.29): it gives the direction of rotation. Since, in the presence of a magnetic field, the Larmor precession has a well-defined direction it follows that both χ' and χ'' are different from zero only in the vicinity of $\omega = \omega_0$, with no comparable behaviour around $\omega = -\omega_0$.

Let us turn now to the complex representation of these results in Equations (2.30) and (2.31). A vector moving in a plane may be described by a complex function: the real part denoting the x component and the imaginary part the y component, as in an Argand diagram. This use of complex variables is slightly different from that of the previous section treating the one-dimensional case. There the imaginary part was introduced as an auxilliary quantity solely to simplify the mathematics. Here both real and imaginary components represent physical quantities.

We shall be using a complex magnetic field $B(t)$ and a corresponding magnetisation $M(t)$:

$$B(t) = B_x(t) + iB_y(t),$$
$$M(t) = M_x(t) + iM_y(t).$$

The magnetic field, as described by Equation (2.29), is rotating at a frequency ω

$$B(t) = b\exp(-i\omega t)$$

and we obtain the resultant magnetisation from Equations (2.30) and (2.31)

$$M(t) = \frac{b}{\mu_0}\{\chi'(\omega)[\cos(\omega t) - i\sin(\omega t)] + \chi''(\omega)[\sin(\omega t) + i\cos(\omega t)]\},$$

which may be simplified to

$$M(t) = \frac{b}{\mu_0}\{\chi'(\omega) + i\chi''(\omega)\}\exp(-i\omega t).$$

However, we have already introduced the complex susceptibility function, Equation (2.25):

$$\chi(\omega) = \chi'(\omega) + i\chi''(\omega),$$

so that we have

$$M(t) = \frac{b}{\mu_0}\chi(\omega)\exp(-i\omega t).$$

Thus we may summarise the results of this section by:

The response to field

$$B(t) = b\exp(-i\omega t),$$

is magnetisation

$$M(t) = \frac{b}{\mu_0}\chi(\omega)\exp(-i\omega t).$$

This is similar to the panel summarising the results of the previous section. There the imaginary part of χ represented the response component which was 90° out of phase with the excitation, here it gives the magnetisation component at right angles to the rotating magnetic field.

The quantity χ used here is the dynamic (complex) magnetic susceptibility. Within the limitations of linear response this function will describe the transverse magnetisation of an NMR system. Thus, for example, the Bloch equations introduced in Section 2.2.5 may be solved for χ' and χ'', which will be taken up in Problem 2.2. The following section considers the behaviour of the transverse magnetisation in the time domain.

2.6.3 The transverse relaxation function

We now consider the dynamic magnetic susceptibility from the point of view of the time domain. In particular we shall examine the response of our system to a transient or delta function burst of magnetic field. Since, as discussed in Appendix A, a delta function is made up from a superposition of complex exponentials of all frequencies with equal weights

$$\delta(t) = \frac{1}{2\pi} \int_{-\infty}^{\infty} \exp(-i\omega t) d\omega,$$

it follows that the response to a delta function excitation is a similar superposition of the responses to the individual complex exponentials:

The response to field

$$B(t) = \frac{\theta}{2\pi\gamma} \int_{-\infty}^{\infty} \exp(-i\omega t) d\omega$$

is magnetisation

$$M(t) = \frac{\theta}{2\pi\gamma\mu_0} \int_{-\infty}^{\infty} \chi(\omega)\exp(-i\omega t) d\omega.$$

The explanation for the constant θ/γ multiplying the delta function for $B(t)$ appears in Box 2.2. The integral of χ is recognised as the transient response function, although here, in the two-dimensional/rotational case, both the real and the imaginary parts of $B(t)$ and $M(t)$ have a physical meaning so that $X(t)$ is no longer required to be real.

Box 2.2 Effect of a magnetic field impulse

Let us take the impulse of a magnetic field to be generated by the solenoid wound around the specimen, with its axis in the x direction. A burst of current in the coil then produces an impulsive magnetic field along the x direction

$$B_x(t) = \theta\delta(t)/\gamma$$

according to our equations above.

The effect of this magnetic field will be to cause the equilibrium magnetisation (in the z direction) to be rotated around the x axis, producing some magnetisation in the y direction. The angle through which the magnetisation is tipped is given by

$$\gamma \int B_x(t)\mathrm{d}t,$$

since γB_x is the instantaneous frequency of precession about the x axis. Since the integral of the delta function is unity, the angle of tip is simply θ, thus the expression for the magnetic field impulse. Immediately following application of the magnetic field impulse there will be transverse magnetisation given by

$$M_y = M_0\sin\theta.$$

The evolution of the (complex) transverse magnetisation following the transient excitation is given by

$$M(t) = \frac{\theta}{\gamma\mu_0} X(t). \tag{2.32}$$

There are a number of restrictions on the form of this transverse relaxation which follow from the physics of our system. The initial transverse magnetisation will be perpendicular to the magnetic field impulse which caused it. Thus at zero time $X(t)$ must be pure imaginary:

$$\int_{-\infty}^{\infty} \chi'(\omega)\mathrm{d}\omega = 0;$$

the area under χ' is zero: equal areas above and below the axis. Since, from Problem 1.2, χ'' is proportional to energy dissipation, the imaginary part of the dynamic susceptibility must always be greater than zero.

The causality condition, that no response occurs before the magnetic field is applied, causes $X(t)$ to be zero for negative times. Thus the function is discontinuous at $t = 0$. This will cause problems if we should investigate the behaviour of the transverse magnetisation by expanding in powers of time: differentiating $M(t)$ at the origin, as indeed we shall do in Chapter 6. This mathematical inconvenience can be circumvented by symmetrising $M(t)$ about the origin. That is we shall consider a function that is equal to $M(t)$ for positive times, and with a suitable continuation to negative times.

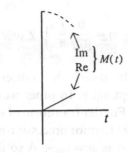

Figure 2.8 Real and imaginary parts of complex magnetisation in vicinity of $t = 0$.

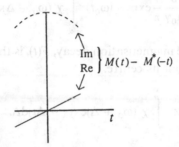

Figure 2.9 Symmetrised form for the complex magnetisation in the vicinity of $t = 0$.

Essentially we shall add the mirror image. In fact this is the shape of a spin echo, as we shall discover in Chapter 4. A spin echo grows from nothing to its peak value and then returns to zero. The shape is that of two back-to-back free induction decays.

As $t \to 0$ we saw above that $M(t)$ becomes pure imaginary; its real part goes to zero while its imaginary part tends to its maximum value, shown in Figure 2.8.

Thus to symmetrise these functions by continuation through the origin we take the combination $M(t) - M^*(-t)$ whose behaviour is shown in Figure 2.9.

Since $M(t)$ and $M^*(-t)$ are given by

$$M(t) = \frac{\theta}{\mu_0 \gamma} \frac{1}{2\pi} \int_{-\infty}^{\infty} \chi(\omega) \exp(-i\omega t) d\omega,$$

$$M^*(-t) = \frac{\theta}{\mu_0 \gamma} \frac{1}{2\pi} \int_{-\infty}^{\infty} \chi^*(\omega) \exp(-i\omega t) d\omega,$$

we then have

$$M(t) - M^*(-t) = \frac{\theta}{\mu_0 \gamma} \frac{i}{\pi} \int_{-\infty}^{\infty} \chi''(\omega)\exp(-i\omega t)d\omega, \qquad (2.33)$$

indicating that the transverse relaxation is the Fourier transform of the *imaginary* part of the dynamic magnetic susceptibility. In other words the transverse relaxation is proportional, essentially, to the Fourier transform of the energy absorption lineshape.

One immediately notes that the Larmor precession may be factorised from Equation (2.33). If the frequency is written as $\omega = \omega_0 + \Delta$ so that Δ is the deviation from the Larmor frequency then the transverse response, Equation (2.33), becomes

$$M(t) - M^*(-t) = \frac{\theta}{\mu_0 \gamma} \frac{i}{\pi} \exp(-i\omega_0 t) \int_{-\infty}^{\infty} \chi''(\omega_0 + \Delta)\exp(-i\Delta t)d\Delta. \qquad (2.34)$$

The envelope of the transverse magnetisation decay, $F(t)$, is then given by the Fourier transform of χ'' evaluated about its centre.

$$F(t) \propto \int_{-\infty}^{\infty} \chi''(\omega_0 + \Delta)\exp(-i\Delta t)d\Delta. \qquad (2.35)$$

An important inference which we can make from this is that if χ'' is symmetric about ω_0 it follows that the imaginary sine integral will vanish. Then $F(t)$ will be real, meaning that throughout the decay the magnetisation remains parallel to its initial direction (in the rotating frame).

2.6.4 Response to linearly polarised field

The complex magnetic susceptibility discussed in the previous section relates to the response of our system to a rotating magnetic field. However, as previously discussed, in most experimental arrangements it is a linearly polarised field which is actually applied. We must therefore consider what the response to this sort of excitation will be.

We may state immediately the form of the response to a linearly oscillating magnetic field. Since a linearly polarised field may be regarded as the sum of equal rotating and counter-rotating components, the response will be the sum of the corresponding rotating and counter-rotating magnetisations. In general these will not be equal. The response to

$$\mathbf{B}(t) = b\cos(\omega t)\mathbf{i}$$

$$= \frac{b}{2}[\cos(\omega t)\mathbf{i} + \sin(\omega t)\mathbf{j}] + \frac{b}{2}[\cos(\omega t)\mathbf{i} - \sin(\omega t)\mathbf{j}]$$

is thus

$$M_x(t) = \frac{b}{2\mu_0}\{[\chi'(\omega) + \chi'(-\omega)]\cos(\omega t) + [\chi''(\omega) - \chi''(-\omega)]\sin(\omega t)\},$$

$$M_y(t) = \frac{b}{2\mu_0}\{-[\chi'(\omega) + \chi'(-\omega)]\sin(\omega t) + [\chi''(\omega) - \chi''(-\omega)]\cos(\omega t)\}.$$

Comparing these results with the response of the one-dimensional oscillator, Equation (2.21), we observe that in the general case each component of the system is equivalent to a one-dimensional oscillator of real susceptibility component

$$\chi'(\omega) + \chi'(-\omega)$$

and imaginary component

$$\chi''(\omega) - \chi''(-\omega).$$

Note that such a form has now recovered the 'hermiticity' condition, Equation (2.28).

We now use arguments similar to those presented in Section 2.5.1, namely that because of the resonant nature of the phenomenon under investigation, the effects of the counter-rotating magnetisation may be ignored (in the vicinity of the rotating component resonance). In that case we conclude that in the presence of an oscillating magnetic field

$$B_x(t) = b\cos(\omega t),$$

parallel to this field there is an oscillating magnetisation

$$M_x(t) = \frac{b}{2\mu_0}[\chi'(\omega)\cos(\omega t) + \chi''(\omega)\sin(\omega t)]. \tag{2.36}$$

Apart from the factor of $\frac{1}{2}$ we observe that, expressed in this form, the excitation and response are identical to those of the one-dimensional oscillator treated in Section 2.6.1. In particular the magnetisation response here corresponds to Equation (2.21), the displacement response of that section.

3

Detection methods

3.1 The CW method

3.1.1 Frequency and time domains

We have seen in Chapter 1 that the essential features of a resonance phenomenon may be investigated by looking at the system's response to a small sinusoidal disturbance of varying frequency. This is called the *continuous wave* (CW) method and is the technique traditionally used in most branches of spectroscopy. Alternatively, as we also saw in Chapter 1, we may look at the time response to a transient excitation, this being called the *pulse* method in NMR.

Results of the two methods have been shown to be equivalent, one response being the Fourier transform of the other. However, if we look at the history of the subject we find that studying the frequency response (CW) was the method used almost exclusively for the first ten years of NMR, i.e. 1946–56. In Andrew's book (Andrew, 1955) for instance 27 pages are devoted to CW experimental methods while pulse methods are treated in a page and a half. Erwin Hahn is usually thought of as the father of pulsed NMR due to his important paper in 1950. Other important contributors include Carr and Purcell (1954) and Torrey (1952).

Since Andrew's book was written, pulsed NMR has to a large extent eclipsed CW NMR. There are two main reasons for this. Firstly, it is much easier to measure the relaxation times T_1 and T_2 using pulse methods. In fact the long T_2 values found in some liquids can only be measured by the pulse technique because magnet inhomogeneity masks the weaker effects of spin–spin coupling. This point will be taken up again in the following chapter. Secondly, as will be shown below, certain features lead to a more favourable signal-to-noise ratio in pulse methods.

One important development in the 1960s which further favoured the use of pulse methods was the rediscovery of the so-called Cooley–Tukey algorithm for fast Fourier transformation (Cooley and Tukey, 1965). Before then the computer calculation of a Fourier transform was a very time-intensive process. Perhaps the single most common use of NMR is as an analytical tool in chemistry. Much detailed information about molecular structure is contained in the resonance line. For instance, the proton resonance in a complex organic molecule may include a great deal of fine structure relating to the interaction of the nuclear spins with the electrons of the molecule. This fine structure can be recognised and analysed in the frequency domain in terms of splittings, shifts, strengths and widths of the peaks. Hence it was not until the development of the fast Fourier transform that high resolution spectroscopy could usefully be carried out from the time domain, with all the consequent instrumentational advantages.

The main emphasis of this book is on the pulse method of NMR; accordingly this chapter concentrates on experimental aspects of the pulsed NMR spectrometer. However, there are some cases, particularly in physics, where the CW method is advantageous. For this reason and also for comparison and for completeness we shall treat both in this chapter. We start by considering the detection of CW NMR signals.

3.1.2 Detection of CW NMR

In both the pulse and the CW techniques an NMR signal is produced by a precessing magnetisation in the specimen under investigation. In the pulse case this magnetisation results following the application of intense RF 'pulses' which create a non-equilibrium state. The precessing magnetisation of CW NMR occurs in the presence of, and in response to, a weak oscillating transverse magnetic field.

The precessing magnetisation will induce a voltage in a coil placed around the specimen and the function of the reception side of any spectrometer system is fundamentally to detect this voltage signal. Recall that the magnitude of a typical such voltage was estimated for pulsed NMR in the final section of Chapter 1, and found to be as much as a millivolt in favourable circumstances. Microvolt signal levels are more typical.

The magnitude of a CW signal could be calculated in a similar way. A voltage is induced in the coil by the precessing magnetisation, the relation between the two being determined by various geometrical factors concerning the size of the specimen and the disposition of the coil. The precessing magnetisation results from the oscillating (or rotating) transverse magnetic field, the ratio of the two being the dynamic susceptibility. Thus the signal voltage can be found.

There is, however, a simpler and more straightforward way of treating the detection of CW NMR signals. Let us turn directly to the dynamic magnetic susceptibility. Although this is defined as the ratio of magnetisation to the applied magnetic field, it is also the factor by which the inductance of a coil is enhanced. An inductor of inductance L_0 when embedded in a medium of susceptibility χ has its inductance changed to

$$L = L_0(1 + \chi).$$

The factor $1 + \chi$ is what is usually referred to as the relative permeability of the medium.

In a practical arrangement the coil is not usually embedded in the specimen; it is wound around it. Thus not all the magnetic field of the coil permeates the medium; some is in the surrounding space. The enhancement of inductance is then not complete and one writes

$$L = L_0(1 + \eta\chi),$$

where the dimensionless quantity η, the 'filling factor' quantifies this effect. Similarly, a small specimen in a larger coil would be accounted for by a reduced filling factor.

We have seen in Chapter 2 how the dynamic susceptibility embodies a full description of the linear magnetic response of a system. We now see, in the CW case, that susceptibility is reflected in the inductance of a coil around the specimen.

The frequency dependent dynamic susceptibility $\chi(\omega)$ has been seen to be a complex quantity. This has important consequences for the observed inductance L of the coil, which are best considered in terms of the complex impedance of the inductor: (the $-i$ factor here is referred to in the Preface)

$$Z = -i\omega L + r,$$

where r is the resistance associated with the inductor. With a specimen inside the coil this becomes

$$Z = -i\omega L_0(1 + \eta\chi) + r$$

and writing χ in terms of its real and imaginary parts

$$\chi = \chi' + i\chi''$$

we obtain

$$Z = -i\omega L_0(1 + \eta\chi') + \omega L_0\eta\chi'' + r.$$

The imaginary part of the susceptibility has introduced a real term in the impedance. In other words, inserting a specimen into a coil causes it to appear as a changed inductance in series with an increased resistance. This is indicated in Figure 3.1.

From the above consideration we see that there is a direct electrical way in which CW NMR may be detected. We shall explore this further.

3.1.3 Q meter detection

There are a number of contributions to r, the series resistance of the inductor. Apart from the resistance of the windings there are other sources of loss such as the

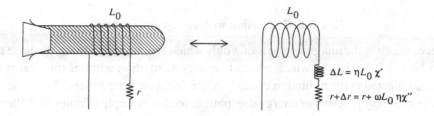

Figure 3.1 Effect of NMR specimen on electrical properties of sample coil.

electromagnetic radiation from the coil at high frequencies and possible losses in the wire insulation and the former. There is also the skin effect whereby at higher frequencies the current flows only in a layer at the surface of the wire. In all, the equivalent series resistance r has various contributions which may be somewhat ill-defined; r is essentially a phenomenological quantity.

It proves convenient to quantify the loss in the coil, instead of using the resistance r, in terms of the dimensionless parameter, the Q factor. For this case the Q factor is defined as

$$Q = \omega L/r,$$

i.e. it is the ratio of the reactive impedance to the resistive impedance – or how much it is like an inductor to how much like a resistor.

We see that NMR phenomena have the effect of changing the Q factor of the coil wound around the specimen. If Q_0 is the natural Q of the coil, corresponding to an equivalent series resistance r of

$$r = \omega L_0/Q_0,$$

then, since the effect of the specimen is to add a further resistance Δr

$$\Delta r = \omega L_0 \eta \chi'',$$

we see that the inverse Q is increased by

$$\frac{1}{Q} = \frac{1}{Q_0} + \eta \chi''$$

or

$$\Delta \frac{1}{Q} = \eta \chi''.$$

Box 3.1 Connection with energy dissipation

Observe that it is the imaginary part of χ only which affects the Q factor. We have seen in Problem 1.2 that it is χ'' which relates to energy dissipation and it is this energy loss which is reflected in the reduction of the Q factor. In fact a more general definition of Q is made precisely in terms of energy dissipation; see for example Bleaney and Bleaney (1976).

Since the change in $1/Q$ is very small we may approximate the above equation as

$$\Delta Q = - Q^2 \eta \chi''.$$

Thus we see how measurement of the coil's Q factor may be used to detect the NMR absorption as reflected in the imaginary part of the susceptibility. Let us now turn our attention to a practical implementation of Q meter detection.

If the lossy inductor is resonated with a parallel capacitor then this circuit combination appears as a resistor of resistance R given by Q times the coil's inductive impedance (Q^2 times the coil's effective series resistance, see Problem 3.1):

$$R = Q\omega L.$$

At this point we shall ignore the difference between L and L_0; it is very small and of no consequence for future discussions since it enters only as a higher order correction. To obtain the Q factor it is this on-resonance resistance which must be measured. This may be done in the following way.

Figure 3.2 shows a schematic arrangement for the detection of NMR absorption. Because the magnitude of the large impedance Z_s is much greater than R (the parallel impedance of the tuned circuit at its resonant frequency $\omega_0 = (LC)^{-\frac{1}{2}}$), the signal generator may be regarded as supplying a constant current I to the tuned circuit.

In the absence of magnetic resonance, there will be a voltage $V = IR$ on the coil. If B_0 is swept through the NMR frequency of the specimen, R changes from $Q\omega L$ to $(Q + \Delta Q)\omega L$, i.e. the coil voltage changes fractionally by

$$\frac{\Delta V}{V} = \frac{\Delta Q}{Q} = - Q\eta\chi'' \tag{3.1}$$

and the output voltage from the detector in Figure 3.2 will be the inverted $\chi''(\omega)$ curve.

Two important approximations are made here. We have ignored the effects of χ' on the inductance which will also change the circuit impedance near magnetic resonance, and we have assumed linearity of response which implies that the observed signal in Figure 3.2 will increase indefinitely in proportion to increasing the signal generator

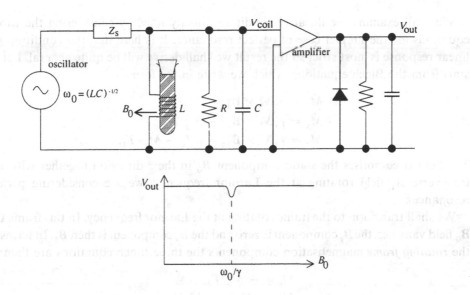

Figure 3.2 Q meter detection of NMR.

output. The effects of χ' are shown in Problem 3.2 to be only very small; the problem of linearity is now taken up.

3.1.4 Saturation

The size of the absorption signal voltage is seen to be proportional to the applied oscillator voltage amplitude V_0. Essentially, this follows from the linearity assumption that the precessing magnetisation (which induces the voltage) is proportional to the oscillating transverse magnetic field B_1 (produced by the current from the oscillator).

It would follow then that the NMR absorption signal could be made arbitrarily large through increasing the output level of the oscillator. Clearly this cannot be the case. At some stage non-linearities must become important. We must investigate this further since it is the appearance of such non-linearities which places an upper limit on the size of the excitation permissible, and thus on the size of the absorption signal observable.

We shall treat the non-linearity of the response in CW NMR within the framework of the Bloch equations, introduced in Section 2.2.5. Although the Bloch equations do not provide an adequate description of all systems in all circumstances, they do provide at the very least a representation of NMR systems in terms of the phenomenological quantities, the spin–lattice and the spin–spin relaxation times T_1 and T_2. Later parts of this book will demonstrate the validity of the Bloch equations in many cases, but even where they are not entirely correct we shall in this section adopt them as a relatively simple model of the basic phenomena.

We shall examine the linearity condition directly now, by considering the Bloch equations for the simpler case of perfect resonance. It is here that the condition for linear response is most strict so the result we shall derive will be quite general. Let us start from the Bloch equations, which we write in the form

$$\dot{M}_x = \gamma |\mathbf{M} \times \mathbf{B}|_x - M_x/T_2,$$
$$\dot{M}_y = \gamma |\mathbf{M} \times \mathbf{B}|_y - M_y/T_2,$$
$$\dot{M}_z = \gamma |\mathbf{M} \times \mathbf{B}|_z + (M_0 - M_z)/T_1.$$

The field $\mathbf{B}$ comprises the static component B_0 in the z direction together with the transverse B_1 field rotating at the Larmor frequency (we are considering perfect resonance).

We shall transform to the frame rotating at the Larmor frequency. In this frame the $\mathbf{B}_0$ field vanishes, the B_y component is zero and the B_x component is then B_1. In terms of the *rotating frame* magnetisation components the three Bloch equations are then

$$M_x' = - M_x/T_2,$$
$$M_y' = \gamma M_z B_1 - M_y/T_2,$$
$$M_z' = - \gamma M_y B_1 + (M_0 - M_z)/T_1.$$

Since we are looking for the equilibrium state of the system the three time derivatives must be set to zero. From these equations we then obtain

$$\left. \begin{array}{l} M_x = 0, \\ M_y = T_2 \gamma M_z B_1, \\ M_z = \dfrac{M_0}{1 + \gamma^2 B_1^2 T_1 T_2}. \end{array} \right\} \tag{3.2}$$

The first equation confirms that there is no in-phase response on resonance. The transverse response is contained in the second equation, for M_y. The seeming linearity of M_y in the excitation B_1 neglects the dependence of M_z on B_1: substituting for M_z from the third equation we obtain for M_y

$$M_y = \frac{T_2 \gamma M_0 B_1}{1 + \gamma^2 B_1^2 T_1 T_2}.$$

The linearity of response and excitation then follows when B_1 is absent from the denominator of this expression. Thus we obtain the linearity condition:

$$\gamma^2 B_1^2 T_1 T_2 \ll 1. \tag{3.3}$$

Also, we see that this is equivalent to

$$(M_0 - M_z)/M_z \ll 1. \tag{3.4}$$

i.e. the z component of magnetisation should not be reduced too much from its equilibrium value M_0.

Problem 2.2 treated the general solution to the Bloch equations. That solution may be expressed as the real and the imaginary parts of the dynamic susceptibility in terms of the relaxation times T_1 and T_2; we quote the expressions:

$$\chi'(\omega) = \frac{\chi_0 \omega_0 T_2^2 (\omega - \omega_0)}{1 + (\omega - \omega_0)^2 T_2^2 + \gamma^2 B_1^2 T_1 T_2}$$

and

$$\chi''(\omega) = \frac{\chi_0 \omega_0 T_2}{1 + (\omega - \omega_0)^2 T_2^2 + \gamma^2 B_1^2 T_1 T_2}.$$

Observe that these equations indicate non-linear behaviour since the magnitude of the susceptibility depends on the strength of the applied field B_1. It is only in the limit

$$\gamma^2 B_1^2 T_1 T_2 \ll 1,$$

when the term involving B_1 may be ignored in the denominators, that the observed susceptibility is independent of the strength of the excitation. It is only in this limiting case that the frequency response takes on the characteristic Lorentzian form.

For a physical understanding of the linearity condition let us turn to a consideration of the energies involved. The quantity $B_1 \dot{M}_x$ represents the rate of flow of energy from the rotating magnetic field into the spin system. The quantity $B_0 \dot{M}_z$ represents the rate of flow of energy out of the spin system: the spin–lattice relaxation. In the steady state the flow in is balanced by the flow out. What is important from the linearity point of view is that in this steady state the spin system should not be disturbed significantly from its equilibrium state – in particular, that the Boltzmann distribution population of the spin states (recall the derivation of Curie's Law, Section 2.1) should not be appreciably disturbed. It is precisely this requirement which is contained in Equation (3.4).

We see that the effect of a large B_1 is to reduce the z component of magnetisation. It does this by tending to equalise the populations of the up and the down spin states. This phenomenon is known as *saturation*. Thus to maintain linearity of response it is necessary to avoid saturation.

This then allows us to estimate the typical maximum signal voltage observable in a CW NMR experiment. For a coil of N turns and area A, Faraday's Law states that

$$V = NA\omega B,$$

where V is the voltage on the coil at frequency ω and B is the field in the coil (assumed here to be uniform).

Equation (3.1) now gives (ignoring the minus sign)

$$\Delta V = NA\omega BQ\eta\chi'' = 2NA\omega B_1 Q\eta\chi'',$$

bearing in mind that B_1 is the amplitude of the correctly rotating component. The signal is seen to increase with the strength of the oscillating field B_1. However, there is a maximum realistic value for B_1 set by the saturation condition, which we take as:

$$\gamma^2 B_1^2 T_1 T_2 = 1,$$

i.e.

$$B_1 = \frac{1}{\gamma(T_1 T_2)^{\frac{1}{2}}}.$$

In Problem 3.3 it is shown that with this degree of saturation

$$\chi''_{max} = \chi''(\omega_0) = T_2 \omega_0 \chi_0 / 2.$$

Hence

$$(\Delta V)_{max} = NA\omega_0^2 Q\eta\chi_0 (T_2/T_1)^{\frac{1}{2}}/\gamma. \qquad (3.5)$$

This expression bears a striking similarity to the estimated size of a pulsed NMR signal treated in Section 1.5.2, Equation (1.18). The filling factor η was not considered there, but it would be included in a more realistic calculation. The factor Q appears here because of the voltage magnification effected by resonating the coil; it would also appear in the pulsed case under similar conditions. The outstanding difference is the factor $(T_2/T_1)^{\frac{1}{2}}$. Now for non-viscous fluids one generally has $T_1 = T_2$ so that then the pulsed and CW NMR signal sizes are identical. In general

$$T_2 \lesssim T_1,$$

so that we conclude that the optimal signal size from a CW experiment is expected to be less than or equal to that from a pulsed experiment under identical circumstances.

However, as hinted in Section 1.5.4, signal size is generally not the prime consideration. There are various sources of electrical noise that can be important and ultimately one must examine the signal-to-noise ratio for the various experimental considerations. This will be pursued in Section 3.4.

3.2 The pulsed NMR spectrometer

3.2.1 Outline of a pulsed NMR system

A pulsed NMR spectrometer is usually a more complicated instrument than a CW spectrometer because it has to carry out at least two quite separate functions. The specimen has to be subjected to intense short bursts of pulses of RF radiation

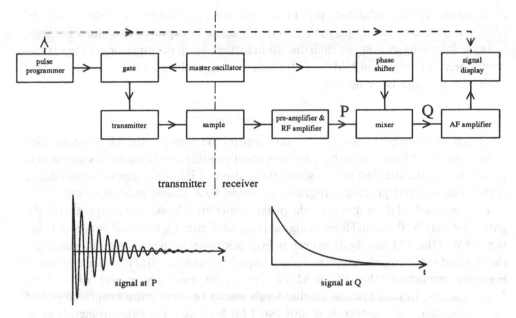

Figure 3.3 Block diagram of a basic coherent pulsed NMR spectrometer.

(transmitter function) and small signals have to be amplified, processed and displayed (receiver function). It is easy to observe pulsed NMR in say a water sample on simple home-built equipment comparable to a cheap short wave radio transmitter and receiver. However, many other applications are much more sophisticated and commercial spectrometers offering many refinements are now available. This is particularly true in chemical applications.

Figure 3.3 shows a block diagram of a basic pulsed NMR spectrometer, such as would be used for looking at large signals in liquid samples such as protons in water at fields of around 1 T. The master oscillator services both transmitter and receiver functions. Ordinarily it is set at the Larmor frequency of the spins of interest corresponding to the field of the magnet.

In transmit mode the RF gate, under control of the pulse programmer, switches on and off the signals from the master oscillator. These bursts of RF are amplified by the transmitter unit to a level sufficient to produce the RF pulses to manipulate the sample magnetisation in the required way: the 90° and 180° pulses, for example.

In receiver mode the weak signal induced in the NMR coil by the precessing magnetisation is amplified by the pre-amplifier to a level suitable for demodulation by the demodulator/mixer. The mixer provides *synchronous* demodulation, which we explain later. The result of demodulation is that the high frequency Larmor oscillation is removed leaving only the amplitude of the envelope of the precession signal as shown in the figure. The demodulated output is filtered and amplified by the 'audio frequency' (AF) amplifier and then displayed on an oscilloscope or captured in some other way.

In the following sections we shall discuss in further detail the functions of the various components of the pulsed NMR spectrometer. Some of the possible enhancements of each part will also be indicated.

3.2.2 Pulse programmer

The function of the pulse programmer is to control the timing of the RF pulses applied to the sample. At this stage we have not discussed possible uses of complex sequences of pulses, but at the simplest level a single 90° pulse will initiate a free induction decay (FID). The simplest practical programmer would be a double pulse generator.

The amplitude of the pulses from the programmer must be such as to operate the RF gate. This may be the signal level of digital integrated circuits, the most common being the $+5$ V of the TTL family. However, in high frequency systems where the pulses are shorter and matching of output and input impedances to the 50 Ω resistance of cables becomes important, the 5 V level can result in excessive power dissipation: $V^2/R = 0.5$ W. In such systems smaller levels would be used, pulse programmer and gate being designed together. It is thus usual for high frequency spectrometers to be built as integrated systems while at lower frequencies the versatility of the modular approach is possible.

Typically, the widths of pulses from a simple pulse generator would be variable in the range 1 µs to 1 ms with spacings in the range 10 µs to 10 s. The sequence would have a repetition rate of 1 mHz to 10 kHz. It is relatively straightforward to construct such a pulse sequencer from integrated circuit TTL monostables such as the 74121 together with a few gates and other components.

We turn now to consider some extra features which might be used in more sophisticated experiments. Complex pulse sequences of up to six or so pulses of variable width and spacing may be used. Such sequences may also continuously recycle. While it is possible to build a hard-wired circuit to generate such sequences the complexity and versatility required is best realised by a software implementation: a microprocessor or computer based sequencer where the details of the required pulse trains are stored as a program. In such a scheme pulse timings are digitally derived from a clock rather than defined by the RC time constants of the monostables described above. The clock may be phase locked to the master oscillator or it may simply be the master oscillator itself.

The simple spectrometer described in Figure 3.3 is *coherent*. By this we mean that RF pulses are produced by gating an oscillator which is continually running at the Larmor frequency ω_0. This implies that successive pulses will be applied along the same axis in the frame rotating with ω_0. These pulses will produce transverse magnetisation precessing at ω_0 with a phase well-defined in relation to oscillator voltage.

In some experiments it is important to apply successive RF pulses along *different*

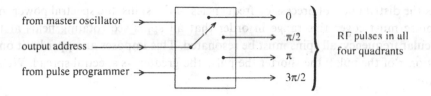

Figure 3.4 Four-channel gate.

axes in the rotating frame. This can be achieved by gating a phase-shifted carrier, with the rotation of axis corresponding to the phase shift used. Full versatility is obtained by having a four-channel facility capable of generating all four quadrants. A schematic outline of such a facility is shown in Figure 3.4.

If the NMR signals are being captured with the aid of a computer then it requires only a simple addition to use the same computer to control the pulse sequences. It is then possible for the computer also to generate the logic pulses to drive the RF gate directly.

3.2.3 The transmitter

The purpose of the transmitter is to amplify the RF pulses to the desired power level. The transmitter amplifier is fed from the RF gate unit mentioned in the previous section. It is important that the on/off ratio of the gate is at least $\sim 10^6$ so that there is not even weak irradiation of the sample between pulses since this might cause saturation or otherwise influence the magnetisation vector.

The transmitter is thus essentially a variable gain power amplifier. The amplitude and duration of the output pulses will vary greatly in different applications. The product of B_1 and its duration τ is related to the angle of rotation ϕ of the RF pulse as we saw in Section 2.5.4:

$$\phi = \gamma B_1 \tau. \tag{3.6}$$

What factors determine the required magnitude of the B_1 field and hence the output power of the transmitter? In order to rotate the magnetic moments, the applied B_1 field must dominate the other transverse fields the spins see. Thus a sufficiently large B_1 field effectively *decouples* the spins and they evolve as independent entities. This will happen when

$$B_1 > \Delta\omega/\gamma,$$

where $\Delta\omega/\gamma$ is the spread of local B_z fields present in the sample.

This condition on B_1 can be looked at from an alternative point of view. Regarding

$\Delta\omega$ as the distribution of precession frequencies of the spins, the spectral power in the RF pulse must cover this range in order that all spins see rotating fields at their particular frequency; all spins must be resonated. This imposes an upper limit on the duration τ of the pulse; the shorter the pulse the greater its spectral spread. We then require

$$\tau < 1/\Delta\omega.$$

But since τ is related to B_1 and the angle of rotation through Equation (3.6), we have

$$\tau = \phi/\gamma B_1.$$

Thus the condition is

$$B_1 > \phi\Delta\omega/\gamma$$

for all spins to be resonated and thus rotated. Then since the angle ϕ is usually of the order of 1 rad this reduces to the earlier result.

The range of local fields $\Delta\omega/\gamma$ varies considerably between solids and non-viscous fluids, as will become apparent in the next chapter. For the simple water experiment the spread will be determined by the magnet inhomogeneity and we might well have

$$\Delta\omega/\gamma \sim 10^{-5}\,\text{T}$$

for an applied B_0 field of 1 T. Thus a B_1 field of only 10^{-4} T would be sufficient and this would only require a few tens of volts on a coil of volume about 1 ml (Problem 3.4). Typically this could be produced by an amplifier delivering a few watts of power, and common solid-state devices such as V-MOS field effect transistors can be used.

In solids $\Delta\omega/\gamma$ is determined by interspin interactions, and it may be over 10^{-3} T, requiring B_1 fields of some 10^{-2} T and hence a few thousands of RF volts. This is at present best achieved with vacuum tube devices. Peak power requirements of such an amplifier would be many kilowatts.

3.2.4 The receiver

As we have seen, the signals from the proton resonance in water at 1 T are several millivolts in magnitude. To observe them on an oscilloscope we need only an amplifier with a bandwidth of a few megahertz centred on the Larmor frequency of 42.6 MHz. It is, however, important for the amplifier to recover within a few microseconds from the effects of the transmitter pulse. Demodulation, i.e. extraction of the signal amplitude envelope from the Larmor carrier oscillation, may be effected by simple diode detection.

When weaker signals are involved it is necessary to use an amplifier with higher gain. But this has the effect of amplifying the noise in the circuit. This may be overcome by reducing the bandwidth of the receiver down to a few kilohertz. This is as problem similar to the reception of broadcast amplitude modulated radio signals. The information is in the modulation and narrow band amplification followed by demodulation is required. This can, however, present practical problems since the centre frequency of the amplifier must be varied according to the frequency of the chosen signal. The solution, employed in almost all radio receivers, is to use a heterodyne system.

In a heterodyne receiver the incoming amplitude modulated signal is multiplied by a pure sinusoidal signal from a local oscillator. Let us represent the incoming signal by

$$S(t)\cos(\omega_s t) \tag{3.7}$$

where $S(t)$ is the envelope modulation signal and ω_s is the carrier (angular) frequency. We denote the signal from the local oscillator by $\cos(\omega_1 t)$. Upon multiplication, using the trigonometrical identity, we then have

$$S(t)\cos(\omega_s t)\cos(\omega_1 t) = S(t)[\cos(\omega_s + \omega_1)t + \cos(\omega_s - \omega_1)t]/2. \tag{3.8}$$

In other words the effect of multiplication of the AM signal by the local oscillator voltage is to produce two AM signals, one at the *sum* frequency $\omega_s + \omega_1$ and one at the *difference* frequency $\omega_s - \omega_1$.

The local oscillator is so arranged that its frequency is changed with that of the tuned signal so as to maintain the difference frequency constant. The multiplied signal goes to a narrow-band amplifier whose pass band is centred on this difference frequency. The result is that the sum frequency is rejected since it falls outside the bandwidth; only the difference frequency is amplified. Thus the major part of the signal amplification takes place at this *intermediate frequency* (IF) and it is a much easier task to make a high gain, narrow band amplifier with a fixed centre frequency. After IF amplification the signal may be demodulated with simple diode rectification.

It is possible to implement a pulsed NMR reception system in this way and indeed this is how some commercial spectrometers operate. However, the question of coherence must be considered in relation to reception. The phase of the received signal indicates the angular position of the transverse magnetisation vector in the rotating frame and the position following an RF pulse is related to the phase of the pulse, derived from the master oscillator. Thus to keep phase information it is necessary for the local oscillator to be coherent with the master oscillator. In other words they must both be derived from the same source. This is possible using an arrangement of frequency dividers and mixers (multipliers), but it is rather cumbersome.

There is an alternative to the conventional heterodyne method which overcomes some of the above difficulties. It is simply to have a zero frequency IF. In such a scheme the master oscillator and the local oscillator are at the same frequency; they can be one

and the same. In this case the effect of multiplication in the mixer, as described by
Equation (3.8), becomes

$$S(t)\cos(\omega_0 t)\cos(\omega_0 t) = S(t)[\cos(2\omega_0 t) + 1]/2.$$

There is the sum frequency, now at twice the Larmor frequency. But the difference
oscillation is a constant. The high frequency term is removed with a low pass filter and
what is left is simply the demodulated signal $S(t)$. Thus in this scheme the demodulation
is provided as a bonus. The narrow-band amplification is now provided by the audio
frequency amplifier, i.e. the pass band is centred at zero frequency. The receiver
bandwidth is just that of the audio amplifier which can be reduced as low as is required
without distorting the signal. The simple spectrometer of Figure 3.3 uses this approach.

The AF amplifier has two purposes: to raise the signal to a suitable display level and
to limit the bandwidth to the minimum acceptable level. If the signal is to be digitised it
is important to provide a level close to the maximum range of the digitisers otherwise
unnecessary digital noise will be introduced. The bandwidth needs to be variable so
that the smallest value which removes noise but does not distort the signal is chosen.
This value will be approximately $1/T_2$.

3.2.5 Display of signals

In CW NMR the sweep (in frequency or field) through resonance is often sufficiently
slow to display the spectrum directly on a chart recorder. In pulsed NMR the signals
require bandwidths up to many kilohertz and faster recorders are required. The water
signal lasts for about a millisecond and has a favourable signal-to-noise ratio so it can
be displayed on a low quality oscilloscope triggered by the pulse which initiated the
signal. For a permanent record it was customary in the past to photograph the
oscilloscope trace. Alternatively a transient recorder or digitiser may be used. This
works, on receiving a trigger, by dividing the signal into a large number of time
channels in each of which the signal is digitised, i.e. it is an analogue-to-digital
converter. The digitised voltages in the channels form a semipermanent record of the
signal which can be fed out slowly onto a chart recorder, saved to disc, or passed to a
computer for further processing. A digital storage oscilloscope is essentially a transient
recorder with automatic display and is very useful as part of an NMR spectrometer.

It is now a relatively simple matter to average a number of repeated signals with a
view to improving the signal-to-noise ratio. As will be discussed in Section 3.4.2
averaging improves the signal-to-noise ratio by the square root of the number of
repetitions so an S/N of 100 can be achieved with 10^4 repetitions starting with a
marginally detectable signal ($S/N \sim 1$). However, if T_1 is about 0.1 s the averaging time
will be about an hour and this places severe constraints on the stability requirements of
the spectrometer. This point is taken up below.

3.3 Magnets

3.3.1 Basic requirements

In order to observe the signals from protons in water, the magnet must produce a field of about 1 T which is constant in space and time: in space – because any variation in field over the sample volume will show up by a shortening of the FID (destructive interference, see Chapter 4); in time – since the Larmor frequency must not fluctuate. To see the water signal clearly we need a homogeneity of about 1 in 10^5 over 1 ml. In this case the signal will last for about $(\Delta\omega)^{-1} = 10^5\omega_0^{-1} \approx 0.4$ ms. While searching for the signal (sweeping the field) it is desirable that the field stays constant at least to within the linewidth $\Delta\omega$. This implies a time stability of 1 in 10^5 over many seconds. These requirements can be met easily with a permanent magnet or an iron core electromagnet energised by a stabilised constant current power supply.

As soon as we contemplate more sophisticated experiments, the requirements of the magnet become very much more stringent in a number of directions.

3.3.2 Homogeneity

If we want to see the natural linewidth of the proton resonance in water $(T_2 \approx 3\,\text{s})$ we need to reduce the magnet linewidth below 0.1 Hz, i.e. fractional homogeneity better than 3 parts in 10^9 at 1 T. This can only be achieved by special methods. A figure of about 1 in 10^6 over 1 ml can be achieved in electromagnets and superconducting magnets without adjustable correction coils.

Approximately two orders of magnitude of improvement can be obtained with a set of up to 18 shim coils which correct for the various spatial derivatives of B_z; see Problem 3.7. For a superconducting magnet some of these can be superconducting current loops which after energising can be left with a persistent stable current flowing.

A final order of magnitude improvement can be obtained by spinning the specimen in the magnetic field at an angular frequency greater than the residual line width.

3.3.3 Time stability

To obtain stable NMR signals, B_0 must fluctuate less than the linewidth during the time of measurement. When averaging very weak signals, this can be as long as an hour. So the most stringent requirement is a drift rate of less than 10^{-9} per hour. This cannot be achieved in an electromagnet by stabilising the magnet current. Instead, NMR itself is used to lock the field to the centre of the resonance of one nucleus (B), I_A being the spin of the nucleus being studied. The stability of the A resonance is thus governed by the oscillators used to supply ω_A and ω_B, and with crystal control this can be better than 10^{-9} over short and long periods. The ^{2}H resonance in a deuterated solvent is often used as the B species.

In superconducting magnets the problem is quite different. The power needed to energise a 5 T solenoid is less than 1 kW (compared with up to 30 kW for a conventional electromagnet) and this can easily be stabilised sufficiently to locate a resonance. High stability can then be relatively easily achieved by opening a superconducting bridge connecting magnet output and input terminals and reducing the power supply current to zero. This leaves the current circulating continuously and exclusively in the superconducting material.

In fact because of the difficulty of making perfect joints between the filaments of superconductor, the joints are slightly ohmic leading to the current and field decaying with a time constant of L/R. To achieve decay rates of, say, 10^{-8} per hour, L/R has to be about 3×10^{12} s. This is now achievable. Nevertheless superconducting magnets do need attention, perhaps every month, to reset the field and replenish the cryostat with liquid helium.

3.3.4 Maximum field

There are two attractions to working at higher fields: signal sensitivity increases as ω increases (see Section 3.4) and in certain high resolution work, so does the resolution. The reason for this is that in complex molecular spectra certain line *shifts* are proportional to B_0 while certain *splittings* are independent of B_0. To spread out the lines sufficiently to avoid overlap sometimes needs fields of over 1 T.

The largest NMR magnets known currently to the author can produce about 10 T corresponding to some 426 MHz in protons and 107 MHz in ^{13}C. The largest fields obtainable with electromagnets are about 3 T and with superconducting solenoids 100 T. Among other obstacles to the use of such fields for NMR is restricted access to the magnetic field. Technical difficulties associated with building low noise amplifiers in the gigahertz region, the skin effect (Bleaney and Bleaney, 1976) and other features of samples conspire to reduce the benefits of higher fields.

3.3.5 Large volume magnets

Until about 1970, few NMR measurements were performed on samples much larger than 100 ml. The advantages in signal-to-noise ratio of a larger specimen are not great (see Section 3.4) and the cost of the magnet rises dramatically with increasing size if the same fractional homogeneity across the sample is to be achieved. The situation changed remarkably during the 1970s because of the development of NMR imaging. In these experiments, to be discussed further in Chapter 10, objects are placed in the field and a three-dimensional image with one-to-one correspondence with points in the object is obtained. The image will map the spin density of the object or the relaxation times. The main application of this technique is to living tissue since it provides a non-invasive microscopic probe which has diagnostic value without the harmful effects of X-rays.

In the current generation of such instruments whole human bodies are passed through a magnet and a thin transverse 'slice' of the body is imaged onto an array of perhaps 64 × 64 pixels. A magnet with a bore of about 30 cm is needed. Since field gradients are employed in these procedures to resolve spatially separate points it might seem that good homogeneity is not important. However, the larger the gradient the greater the receiver bandwidth required; this in turn increases the noise power. The smallest usable gradient is determined by the linewidth of the spins in one slice and this is determined by a combination of intrinsic linewidth ($\sim$ 10 Hz for protons in living tissue) and magnet inhomogeneity. Hence it is desirable to reduce the magnet linewidth to below 10 Hz in a 30 cm slice. This is not so difficult to achieve because this work has to be carried out at low fields. The RF penetration depth of living tissue, caused by its electrical conductivity, is such that whole body imaging is usually carried out below 10 MHz. Magnets of about 0.1 T for this work are currently made from water-cooled air core coils. If imaging experiments come to be carried out on other nuclei, for example ^{31}P, with smaller magnetogyric ratios, a higher field could be used and superconducting magnets will then probably be found advantageous for imaging as they are for most other applications of NMR.

3.4 The sample probe and noise considerations

3.4.1 Introduction

The sample probe and the way in which it is coupled to the pre-amplifier are in many ways the most critical parts of an NMR apparatus. To display the previously discussed water signal is a relatively easy matter since we have millivolts of signal available, but in more sophisticated experiments optimising sample size and shape and the coil's coupling to the preamplifier can be essential. Where signal averaging is used (Section 3.4.2) an improvement in signal-to-noise ratio by other means by a factor of 1.4 will cut acquisition time by half, say from 40 to 20 s. This may be crucial if the specimen is a human body required not to breathe during the averaging process!

The sample probe receives pulses from the transmitter which must generate B_1 fields for short times (10^{-6}–10^{-3} s). The response to this, in the form of a time varying transverse magnetic moment, has to be converted into a signal voltage and fed to the pre-amplifier. One coil can do both these jobs provided its axis is perpendicular to the static B_0 field.

In the following sections we shall discuss how to optimise the sometimes conflicting requirements for the probe assembly. We have mentioned previously that electrical noise is an important consideration when deciding what signals can be observed, and we now explain the main features of this phenomenon.

3.4.2 *Electrical noise*

The term *electrical noise* has a special meaning which is slightly different from the common perception. It does *not* refer to such things as mains hum and interference from switching circuits and passing vehicles. These are all avoidable in principle, although often with difficulty. There is other noise, originating in the electrical components themselves, which is a consequence of fundamental laws (Robinson, 1974). Such noise is unavoidable: all that can be done is to reduce its harmful effect so far as is possible.

The current-carrying electrons in a circuit have a random element to their motion due to their thermal energy. The Equipartition Theorem of statistical thermodynamics tells us that in thermal equilibrium at temperature T each degree of freedom has $kT/2$ of energy in its random motion, where k is Boltzmann's constant. Thus in a resistor, for example, at any one moment there can be slightly more electrons closer to one end than the other. This would be observed as a potential difference and, in practice, a small fluctuating voltage is indeed seen. This is referred to as *Johnson noise*, the mean square magnitude of the voltage in a frequency band Δf being given by

$$\langle v^2 \rangle = 4kTR\Delta f.$$

In this expression R is the magnitude of the resistance.

We see that the Johnson noise is proportional to absolute temperature, as might well be expected. It is also proportional to the resistance; it is related to energy dissipation, a special case of the Fluctuation Dissipation Theorem (Reif, 1965).

Another cause of noise is the finite charge carried by each electron. Rather than a steady stream, current arrives as the 'drips on a tin roof'. This is known as *shot noise*. Shot noise is fundamentally a fluctuation in the current flowing and its magnitude in a frequency band Δf is given by

$$\langle i^2 \rangle = 2eI\Delta f,$$

where I is the current flowing and e is the electron charge.

Both shot noise and Johnson noise have a power density which is independent of frequency; they are both 'white' noise. This is in contrast to the so-called $1/f$ noise whose power density increases with decreasing frequency. There seem to be many unrelated causes of low frequency noise, but they all seem to exhibit, at least approximately, the characteristic $1/f$ dependence of noise power.

Since noise is an essentially random process, the average noise voltage is zero; it is equally likely to be positive or negative. As we saw above, it is the mean square voltage which has a finite value. Thus when talking of a noise voltage of a certain magnitude we are really referring to a *root mean square* (RMS) value. Obviously one cannot add such noise voltages from unrelated sources. It is the powers, or the mean square voltages, that must be added, thus the rule that noise voltages add by squares.

This has an important consequence for signal averaging. A common way of reducing the noise in a measurement is to repeat the reading a number to times and then take the average value. If one adds N readings the signal increases N-fold and the *square* of the noise increases N-fold. The noise thus increases by the square root of N, so on dividing by N to evaluate the average we see that the noise has been reduced by the factor $1/N^{\frac{1}{2}}$.

Signal averaging can now be incorporated into signal processing using microprocessors or microcomputers. An NMR signal produced in response to an RF pulse may last for about 1 ms. The signal could be divided into 1024 (this is 2^{10}, a convenient computer number) time channels of about 1 μs each, with the addressing starting a controllable delay time after the initiating pulse. The signal voltage in each channel is sampled (often for only a small fraction of the channel width time) and digitised with say eight-bit precision. During the delay before the second sequence (typically for a time of $5T_1$ to allow for M_z to recover to within 1% of M_0 – but see Section 3.4.6), the channel memories are emptied and stored ready for the next signal. Since no useful information can be gathered at frequencies higher than the Nyquist frequency $1/2\tau$ (where τ is the channel width) it is important to incorporate a low pass filter of this upper frequency to eliminate high frequency noise.

3.4.3 The NMR probe

We have seen that at its simplest level the NMR probe consists simply of a coil wound around the specimen. As soon as more sophisticated experiments are considered, more complex probe arrangements may be necessary. Often the receiver and transmitter requirements may differ. The inherent conflict in the two roles that the coil plays can be resolved by having separate receiver and transmitter coils. For instance a large transmitter coil with few turns may be desirable for producing a homogeneous B_1 field. On the other hand the receiver coil needs to be close-fitting and its number of turns may be determined by signal-to-noise considerations. By placing the coils orthogonal to each other (and to B_0) they can be usefully decoupled from each other so that the high transmitter voltages are kept away from the low signal reception voltages. It is difficult, however, to reduce the cross coupling below 1% so that other protection methods are still required (Section 3.4.5).

It is very common to resonate the sample coil L_s with a parallel capacitor $C_s = 1/\omega^2 L_s$. This enhances the signal by a factor Q, the quality factor of the coil, but it does not in itself increase the signal-to-noise ratio since the coil noise voltage is also increased by the same factor. Series resonance is also sometimes used. The three circuit configurations are shown in Figure 3.5.

The main difference in the three circuits is the impedance presented to the pre-amplifier. This will affect the final signal-to-noise ratio of the displayed signal

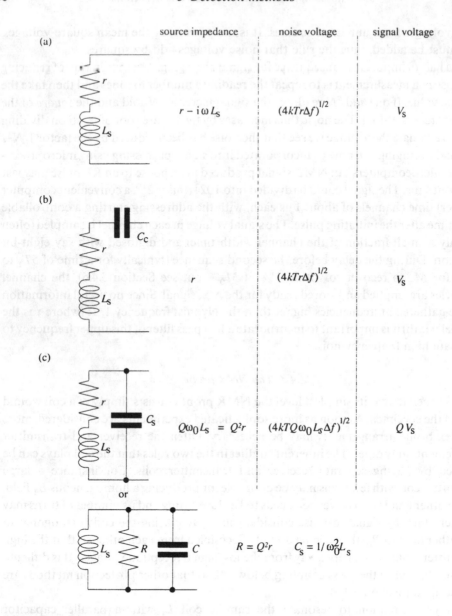

Figure 3.5 NMR sample probe circuits: (a) non-resonant; (b) series resonant; (c) parallel resonant.

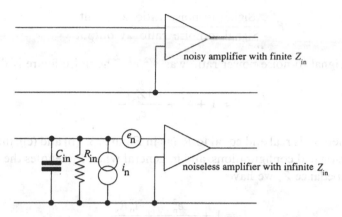

Figure 3.6 Representation of noise and input characteristics of amplifiers.

because the noise introduced by the amplifier will be a function of the probe impedance. The amplifier noise is conventionally expressed in terms of an equivalent noise source at its input. Most generally there will be both a voltage noise source e_n and a current noise source i_n. This is shown in Figure 3.6. Note that e_n and i_n are conventionally measured in $V\,Hz^{-\frac{1}{2}}$ and $A\,Hz^{-\frac{1}{2}}$ respectively since for white noise sources noise *powers* are proportional to bandwidth.

The amplifier current noise produces a noise voltage across the probe impedance Z_s of $i_n Z_s$. There are then three contributions to the noise and we add the squares of the voltages

$$v_n^2 + e_n^2 + i_n^2 |Z_s|^2.$$

Since Z_s may be complex we must take only its magnitude and by convention we omit the averaging brackets around the currents and voltages. The first term represents noise from the probe and the other two terms are the amplifier contributions. Assuming that the amplifier input impedance is very much greater than the probe impedance Z_s then all of this voltage appears at the amplifier input. If the amplifier has a voltage gain of G then the signal-to-noise power ratio at its output is given by

$$\frac{G^2 V_s^2}{G^2 (v_n^2 + e_n^2 + i_n^2 |Z_s|^2)}.$$

Note that the signal-to-noise ratio is independent of the amplifier gain.

It is useful to introduce the concept of the *noise factor F* for an amplifier. This is defined as the signal-to-noise power ratio in the absence of amplification divided by the signal-to-noise power ratio at the amplifier output:

$$F = \frac{\text{Signal-to-noise ratio at input}}{\text{Signal-to-noise ratio at output}}.$$

The original signal-to-noise power ratio was V_s^2/v_n^2. The noise figure is then

$$F = 1 + \frac{e_n^2}{v_n^2} + \frac{i_n^2|Z_s|^2}{v_n^2}.$$

In the case where Z_s is real and equal to R_s (as in Figures 3.5(b) and (c)), the series and the parallel resonated configurations, and in general when R_s denotes the real part of the source impedance Z_s we have:

$$F = 1 + \frac{e_n^2}{4kT_sR_s} + \frac{i_n^2R_s}{4kT_s},$$

where T_s is the temperature of the coil around the specimen.

We see that a perfect, noiseless amplifier would have a noise factor of unity. Real amplifiers have an F greater than this, and we see that the value of the noise factor depends on the impedance of the device feeding the amplifier. Since there is one term proportional to R_s: the current contribution, and one term inversely proportional to R_s: the voltage contribution, there must be an optimal source resistance which minimises F.

The minimum noise factor is easily found to be

$$F_{\min} = 1 + e_n i_n/2kT_s$$

which occurs when:

$$R_s = R_{\text{opt}} = e_n/i_n.$$

For modern low noise FET amplifiers in the 1–100 MHz range, $e_n \sim 3\,\text{nV}\,\text{Hz}^{-\frac{1}{2}}$ and i_n increases from $10^{-14}\,\text{A}\,\text{Hz}^{-\frac{1}{2}}$ to $10^{-12}\,\text{A}^{-\frac{1}{2}}\,\text{Hz}$ as the frequency increases from 1 to 100 MHz. Hence R_{opt} is in the range 3–300 kΩ which is the typical range of values for the impedance of a parallel tuned circuit, $Q\omega L$. This is why parallel resonance is so often used in NMR. $F_{\min}$ varies with the above parameters from 1.3 to 1.003. Even for the worst case this corresponds to the signal-to-noise voltage ratio becoming only 15% worse than it would be with a noiseless amplifier. It is worth pointing out, however, that in low temperature experiments where T_s may be 4 K or even lower, the demands on the amplifier are much stronger and special techniques such as cooling the pre-amplifier or the use of SQUIDS become worth considering.

One disadvantage of superconducting magnet systems is that the solenoid axis is usually vertical. To use a vertical cylindrical specimen, which is useful for sample changing, saddle coils as in Figure 3.7(a) are needed. With conventional electromagnets

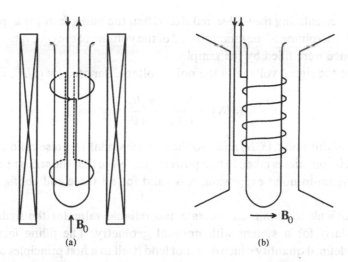

Figure 3.7 Sample coil configuration for use in superconducting solenoids and iron core electromagnets: (a) superconducting solenoid; (b) iron core electromagnet.

a horizontal B_0 field is usual and a solenoid receiver coil wound on the sample cylindrical axis is possible as in Figure 3.7(b).

3.4.4 Signal-to-noise ratio

We conclude from the previous section that for frequencies below about 100 MHz and using room temperature probes, the noise introduced by the amplifier electronics may be ignored. The dominant source of noise is the Johnson noise of the tuned NMR coil. The resistance of the coil assembly is R_s so that the resultant noise voltage is

$$v_n = (4kT_sR_s\Delta f)^{\frac{1}{2}},$$

where Δf is the bandwidth of the spectrometer, determined by the audio frequency amplifier, given that this is much smaller than the width of the coil resonance. In determining whether a signal is observable or not its magnitude must be compared with this noise voltage; we need a signal-to-noise ratio of at least unity.

At the end of Chapter 1 we obtained expressions for the size of the precession signal following a 90° pulse. The magnitude of the voltage was shown to be

$$v = NA\omega\chi_0B_0,$$

where A is the cross-sectional area of the coil which has N turns. Now we know that this expression is not entirely correct since it does not allow for the failure of some of the lines of B originating from the nuclear dipoles to intersect all the receiver coil turns. This failure is usually accounted for by the introduction of a filling factor η as we did in

Section 3.12 in calculating the CW signal size. Often the filling factor is approximated by the ratio of the volume of the sample to $\sqrt{2}$ of the volume of the coil; η would only be unity if all space were filled by the sample.

The ratio of the signal voltage to the noise voltage, denoted by (S/N), is given by:

$$(S/N) = \frac{NA\eta\omega\chi_0 B_0}{(4kT_s r_s \Delta f)^{\frac{1}{2}}}.$$ (3.9)

As in the rest of this book, (S/N) refers to the *voltage* signal-to-noise ratio although in other works it is sometimes taken to be a power ratio. Note that the series resistance r_s is used in this signal-to-noise expression; it is valid for all three coil configurations in Figure 3.5.

It can be difficult to obtain an accurate and reliable value for the signal-to-noise ratio, particularly for a system with unusual geometry. The filling factor η is a somewhat ill-defined quantity which does not lend itself to a first principles calculation. To obtain a more accurate expression for the signal-to-noise ratio it is better to proceed rather differently (Hoult and Richards, 1976).

The Reciprocity Theorem of electromagnetism states that a time varying magnetic moment $\mathbf{m}(t)$ will induce a voltage V in a conductor given by

$$V = -\frac{\partial}{\partial t}[\mathbf{B}^1 \cdot \mathbf{m}(t)],$$ (3.10)

where $\mathbf{B}^1$ is the field that would be produced at the location of $\mathbf{m}$ by unit current flowing in the conductor. Using this equation, the problem of evaluating the NMR signal size reduces to the relatively straightforward task of finding the field profile of the coil. Thus an accurate value for the signal-to-noise ratio may be found for a given experimental arrangement with the aid of a modest microcomputer.

A direct analytic evaluation is possible when B_1 and the magnetisation may be assumed homogeneous over the specimen. This is a nice example which serves to demonstrate the utility of the method. Following a 90° pulse there is a magnetic moment in the transverse plane

$$M_0 v_s \cos(\omega t) = v_s (B_0 \chi_0/\mu_0)\cos(\omega t),$$

where v_s is the volume of the specimen. The voltage induced in the coil is found from Equation (3.10) as

$$V = -\frac{\partial}{\partial t}\left[\frac{B^1 B_0 \chi_0 v_s}{\mu_0}\cos(\omega t)\right]$$

and upon differentiation, the induced signal voltage amplitude is found to be

$$V = \omega\chi_0 B_0 B^1 v_s/\mu_0,$$

leading to the following expression for the signal-to-noise ratio:

$$(S/N) = \frac{\omega \chi_0 B_0 B^1 v_s}{\mu_0 (4kT_s Fr_s \Delta f)^{\frac{1}{2}}}. \tag{3.11}$$

The possible amplifier noise contribution has also been incorporated via the noise factor F. This expression is an improvement over Equation (3.9) since all the geometric factors are contained in the B^1 factor.

Some of the parameters of Equation (3.11) are interrelated and it is not possible to exhibit the dependence of (S/N) on the variables of interest because in different experiments different constraints apply. However, a small number of points are of importance in a wide variety of applications:

(a) If a signal averaging is employed it is necessary to wait a time of the order of T_1 between readings to allow for the recovery of the system from the previous pulse. Then in a time t it is possible to take $\sim t/T_1$ traces, resulting in a signal-to-noise improvement of $\sim (t/T_1)^{\frac{1}{2}}$. The receiver bandwidth is usually determined by the linewidth $1/T_2$, hence (S/N) is proportional to $(tT_2/T_1)^{\frac{1}{2}}$. (See, however, Section 3.4.6.)

(b) Writing $r_s = \omega L_s/Q$ we find that $(S/N) \propto (Q\omega^3)^{\frac{1}{2}}$ since $B_0 = \omega_0/\gamma$. This has led to statements that $(S/N) \propto \omega^{3/2}$, or $\omega^{7/4}$ if we note that for frequencies well below the coil self-resonance frequency $Q \propto \omega^{1/2}$. These dependencies follow only if Δf is determined by the sample T_2 and not by magnet inhomogeneity.

To obtain the largest signal voltage rather than signal-to-noise ratio we need to obtain the largest B^1. However, we must limit the coil inductance to a maximum value $1/(\omega^2 C_{min})$ where C_{min} is determined by cable and amplifier capacitances. For a short close-fitting solenoid coil

$$B^1 \sim \frac{\mu_0 N}{l} \sim \frac{\mu_0 NA}{v_s}, \quad L_s \sim \frac{\mu_0 N^2 A}{l} \sim \frac{\mu_0 N^2 A^2}{v_s},$$

where the coil is of N turns, cross-sectional area A and length l. This applies only at low frequencies when one can work with high impedance circuits (at high frequencies everything is usually matched to $50\,\Omega$). Hence:

$$v_{max} = \omega \chi_0 B_0 NA = \omega_0 \chi_0 B_0 (L_s v_s/\mu_0)^{\frac{1}{2}} = \chi_0 B_0 \left(\frac{v_s}{C_{min}\mu_0}\right)^{\frac{1}{2}}.$$

This shows that as far as signal magnitude is concerned, if one wants to reduce frequency, it is possible to keep the signal up by increasing the volume of the specimen.

If the dominant source of noise is current noise from the amplifier or elsewhere (e.g. interference picked up by the receiver coil), then (S/N) is independent of Q since both noise and signal voltages are now proportional to Q, the circuit impedance at parallel resonance being $Q\omega L$.

(c) To obtain an estimate of the relative strength of different nuclei we note the factor $\omega_0 \chi_0 B_0$. We saw in Equation (2.7) that Curie's Law gives a static susceptibility $\chi_0 \propto N_v \gamma^2 I(I + 1)$, N_v being the number of spins per unit volume and I the spin of the nuclei under observation. So the signal voltage, s, is proportional to

$$\omega_0{}^2 \gamma N_v I(I + 1),$$

showing that signal sensitivity for different nuclei is proportional to $\gamma N_v I(I + 1)$ at constant frequency or $\gamma^3 N_v I(I + 1)$ at constant field. If, however, the isotope of interest is present at fractional concentration ζ in the naturally occurring element, signal sensitivity is usually given as $\zeta \gamma I(I + 1)$ or $\zeta \gamma^3 I(I + 1)$. Thus from Table 1.1 we note that the ^{13}C signal will be smaller than the proton signal by about 7000 at constant field or 140 at constant frequency, assuming equal densities of carbon and hydrogen atoms.

3.4.5 Recovery time

We have referred several times to the conflicting requirements of the sample coil during transmission and reception, and we now discuss two possible solutions to this problem.

During transmission, the coil quality factor should be low to ensure it recovers quickly at the end of the RF pulses. A parallel resonant coil has a characteristic decay time τ of $2Q/\omega$ and in decaying from say $100\,V$ to a signal level of $10\,\mu V$, a time of 16τ is needed. For a Q of 100 at 43 MHz this implies a recovery time of $12\,\mu s$. Our water signal is expected to decay in 1 ms so little information would be lost. However, at 4 HMz the recovery time would be $120\,\mu s$ and for solid samples, T_2 is often in the microsecond range and essential information would be lost unless the recovery time is shortened.

A second factor is the need to protect the pre-amplifier from the high voltages present on the coil during transmission. Both these requirements can be met by using diodes which can be regarded as switches that are *on* for voltages (of the appropriate polarity) of greater than 0.5 V, and *off* for other voltages. Figure 3.8 shows a pair of antiparallel diodes D_1 which conduct during the pulse and thereby allow the high voltage onto the coil. The diodes D_2 also limit the voltage entering the pre-amplifier to less than 0.5 V. When the pulse is finished the circuit rings down to about half a volt very quickly because the $50\,\Omega$ resistor greatly reduces the coil quality factor. Subsequent ringing down is slower and can only be speeded up with *active* damping, using, for instance, an appropriately biassed field effect transistor (as a voltage variable resistor) gradually to change the coil damping in a controlled manner.

When the pulses have decayed to below 0.5 V, the free induction signal develops

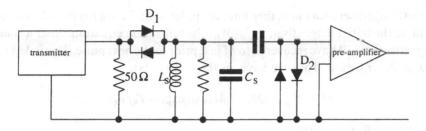

Figure 3.8 Use of diodes as passive switches for transmission/reception.

across the coil as it is now in the high Q parallel resonance mode, both pairs of diodes being effectively switched off.

If space and transmitter power allow, it may be more convenient to use separate coils for transmitting and receiving. In that case the receiver coil will be of high Q and close fitting around the specimen. The transmitter coil can be larger and in the form of a Helmholtz pair (to provide a homogeneous B_1 field) with its axis mutually perpendicular to both $\mathbf{B_0}$ and the receiver coil axis. This reduces the voltage on the receiver coil during transmitter pulses but as orthogonality cannot easily be achieved to better than about a degree of arc, the protection diodes D_2 are still required.

3.4.6 The Ernst angle

In Section 3.1.4 we saw that pulsed NMR gave a signal enhancement of the order of $(T_1/T_2)^{\frac{1}{2}}$ but we pointed out that the important consideration was the signal-to-noise ratio. Subsequently the analysis of noise sources was equally applicable to both pulse and CW NMR. Then we saw in Section 3.4.2 how the signal-to-noise ratio may be enhanced by signal averaging. Now following a 90° pulse it takes a time of approximately $5T_1$ for M_z to return to within 1% of its equilibrium value M_0. If one does not wait this long and/or if the pulse angle is smaller then the signal will be smaller. But on the other hand the signal averaging can be done more frequently. It was realised by Ernst and Anderson in 1966 that although the signal might be smaller, the signal-to-noise ratio can be enhanced by purposely using a smaller tipping angle and waiting a shorter time in order to accumulate data more quickly. In this way pulsed NMR can achieve a further enhancement of signal-to-noise over that which can be obtained by the CW method.

Let us consider the accumulation of data for signal averaging, where the tip angle is θ and the repetition time is T_R. Recall that following a disturbance the z component of magnetisation recovers towards its equilibrium value M_0 according to Equation (2.10):

$$M_z(t) = M_0 - [M_0 - M(0)]\exp(-t/T_1),$$

where $M(0)$ is the initial value of the magnetisation. If M is the z component of magnetisation just before a pulse, then after the pulse it will have a magnitude of $M\cos\theta$. This will be the initial magnetisation $M(0)$ in the relaxation equation. After a time T_R the magnetisation will have recovered to M just prior to the next pulse. This is $M_z(T_R)$ in the relaxation equation. Thus we have, in the steady state,

$$M = M_0 - (M_0 - M\cos\theta)\exp(-T_R/T_1),$$

which may be solved to give

$$M = M_0 \frac{1 - \exp(-T_R/T_1)}{1 - \cos\theta\exp(-T_R/T_1)}.$$

Following the pulse of angle θ the transverse magnetisation will be $M\sin\theta$, which is

$$M_{trans} = M_0 \frac{[1 - \exp(-T_R/T_1)]\sin\theta}{1 - \cos\theta\exp(-T_R/T_1)}.$$

Now the signal-to-noise enhancement is proportional to the square root of the number of averages. So in a given time it is proportional to $T_R^{-\frac{1}{2}}$ so that the signal-to-noise ratio is then

$$(S/N) \propto \frac{[1 - \exp(-T_R/T_1)]\sin\theta}{[1 - \cos\theta\exp(-T_R/T_1]T_R^{\frac{1}{2}}}. \tag{3.12}$$

If the repetition time is fixed then there is an optimum tip angle which will maximise (S/N). By differentiating Equation (3.12) with respect to θ and setting the result to zero we find that

$$\cos\theta = \exp(-T_R/T_1) \tag{3.13}$$

gives the tip angle which maximises (S/N) for a given T_R. This is known as the Ernst angle.

Full optimisation with respect to both T_R and θ of the signal-to-noise ratio actually gives the non-physical condition that the repetition time should be as short as possible, as Figure 3.9 shows. What we have plotted in this figure is the normalised (S/N) expression from Equation (3.12) as a function of T_R/T_1 after having substituted in the optimum tip angle (the Ernst angle) for each value of T_R/T_1.

There are various practical reasons, however, why the repetition time should not be too short. The delay intervals must always be long compared with the pulse lengths. Furthermore, if there is any remnant transverse magnetisation at the instant of the initiation pulse then spurious signals can occur in addition to the main signal. On the other hand there is an added benefit of shorter pulses since then the spectral width is that much greater.

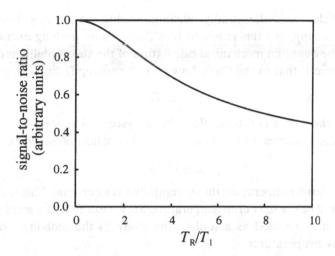

Figure 3.9 Optimal signal-to-noise ratio.

3.5 Refinements to the CW method

3.5.1 Uses of CW NMR

Without doubt pulsed NMR has many advantages over the older CW technique. Probably the considerations of Ernst and Anderson (1966), discussed in the previous section, provided the final 'nail in the coffin' of CW NMR. There are, however, a number of applications, particularly in physics, where the CW method remains the technique of choice. It turns out, particularly at the lowest temperatures (mK and below), that the intense pulses of RF radiation have undesirable effects. The RF pulses themselves can cause unacceptable heating of the specimen. Furthermore the magnetic forces on the transmitter coil during a pulse and even those within the specimen can cause vibrations of components in the B_0 field which induce currents, causing further heating. Another application is in delicate, small signal situations where the RF pulse causes movement of the receiver coil. Since the coil will move in the presence of the B_0 field a spurious EMF will be induced which might mask the true NMR signal. For these and for a variety of other reasons it is sometimes desirable to dispense with intense RF fields and return to the CW method. We therefore discuss a number of aspects of CW NMR in this section.

We have already mentioned Curie's Law, which can be exploited as a technique for thermometry. Since the susceptibility increases as the temperature decreases, this is a method particularly useful at low temperatures. But by the same token *deviations* from Curie's Law are of interest. As an example, Curie's Law does not apply to mobile fermions such as liquid ^{3}He at low temperatures. The derivation of the Curie Law assumed *distinguishable* particles. Now considering a solid, this is no problem since even if the particles are fundamentally indistinguishable, they may be distinguished by the *sites* they occupy, but this is not so in a fluid. There nothing distinguishes the particles and the quantum mechanical calculation of the susceptibility must take this into account. Recall that in the Curie Law case the susceptibility was given by

$$\chi = C/T,$$

where C was the Curie constant. Now for a system of mobile fermions at low temperatures this becomes, to within a numerical constant (Reif, 1965),

$$\chi = C/T_F$$

where T_F is the Fermi temperature; the susceptibility is a constant. This result holds for temperatures well below the Fermi temperature. So, for example, nuclear susceptibility measurements may be used as a tool in the study of the mobility/localisation of fermions at low temperatures.

We know that in a pulsed NMR experiment the initial height of the free precession signal is proportional to the equilibrium magnetisation, and thus it is proportional to the magnetic susceptibility. This provides a simple way of measuring magnetic susceptibility and it is often used. However, nuclear magnetic susceptibility may also be found by CW methods.

In Appendix A we encounter the Kramers–Kronig relations. These provide integral relationships between the real and the imaginary parts of the magnetic susceptibility. A special case of one of these relations is

$$\chi'(0) = \frac{1}{\pi} \int_{-\infty}^{\infty} \frac{\chi''(\omega)}{\omega} d\omega. \tag{3.14}$$

Since the imaginary part of χ vanishes at zero frequency it follows that $\chi'(0)$ is equal to the static susceptibility, which we have denoted by χ_0. Furthermore since the resonance peak of χ'' is located entirely in the vicinity of $\omega = \omega_0$ we can take the $1/\omega$ factor out of the integral as $1/\omega_0$, giving

$$\chi_0 = \frac{1}{\pi\omega_0} \int_{-\infty}^{\infty} \chi''(\omega) d\omega. \tag{3.15}$$

Recall that the absorption line shape is essentially the imaginary part of the susceptibility. From this we see that by sweeping through the resonance absorption line and integrating it, the static magnetic susceptibility may be obtained.

3.5.2 The Robinson oscillator

The Q meter detection circuit described in Section 3.1.3 suffers from a number of disadvantages. The oscillator must be tuned to precisely the same frequency as the tank

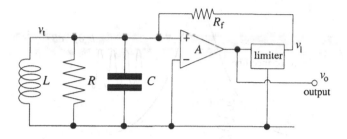

Figure 3.10 Robinson oscillator.

circuit, and there must be no drift between the two. One way of circumventing this difficulty is to make the tank circuit the frequency-determining element of the oscillator. Problem 3.2 considered an example of such an oscillator, but that circuit was not practical since it relied on having a precisely determined gain. The circuit shown in Figure 3.10 overcomes this problem. It was proposed for NMR applications by Robinson in 1959, but it has subsequently become widely used in many diverse applications. The unusual part of the circuit is the limiter. This provides stability for the amplitude, as we shall see.

The positive feedback will ensure that the circuit oscillates so long as the gain A of the amplifier is adequate. Let us assume it is so. The circuit will oscillate when there is zero phase shift around the loop. If there is no phase shift in the amplifier or in the limiter then there must be no phase shift at the parallel LRC network. In other words the impedance of the network must be real and this happens when $\omega^2 = LC$, the resonant frequency of the tank circuit. Let the voltage amplitude of this oscillation on the tank circuit be v_t. Then since the gain of the amplifier is A, the output voltage amplitude v_o will be Av_t. This voltage is fed into the limiter. The action of the limiter is to clip the amplitude of the signal to the level v_1. Thus the output of the limiter is a square wave at the resonant frequency.

The feedback resistor R_f is large. It is much larger than the resonant resistance of the tank circuit R, so that it may be assumed that R_f delivers a constant current square wave of amplitude $i = v_1/R_f$ to the tank circuit. This square wave is made up of its fundamental component, at the resonant frequency, plus the various harmonics. Now the impedance of the tank at the harmonics is very small since their frequencies are away from resonance. But the impedance at the fundamental is large. Thus only the fundamental develops a voltage across the tank. So the tank voltage is sinusoidal, at the resonant frequency, and its amplitude v_t will be $iR = v_1R/R_f$. Then the output voltage of the circuit will be Av_1R/R_f.

The conclusion is that the output from the circuit is a sinusoidal signal whose amplitude is proportional to the on-resonance resistance of the tank circuit. Since the

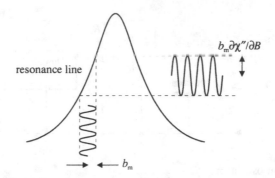

Figure 3.11 Modulated sweep through the resonance.

resistance of the circuit is proportional to the Q factor of the tank, we see that this is a convenient way of measuring Q.

The Robinson oscillator has a further advantage when sweeping through a resonance line. When we discussed the Q meter circuit of Section 3.1.3 we saw that the line was swept through by varying the magnetic field B_0. This can be inconvenient since it might involve varying the large currents in the main electromagnet. But with the self-oscillating system one has the option of varying the resonant frequency of the tank circuit. Sometimes one uses a motor-driven variable capacitor to effect this. Alternatively it is possible to use a varicap diode, whose capacitance is determined by the bias voltage applied to it.

3.5.3 Lock-in detection of CW NMR

In both Q meter schemes described the signal appears as the amplitude of a 'carrier wave' at the Larmor frequency. This must be demodulated to give the amplitude which can then be measured or recorded. In Figure 3.2 a diode demodulator is shown. The output signal is then a very small voltage variation superimposed on the larger level resulting from the RF amplitude before the detector. A diode detector circuit could be used at the output from a Robinson oscillator; the same considerations would apply. In practical set-ups one would subtract a constant amount at the signal frequency or 'back off' the signal level before detection, but the fundamental requirement is that a very small DC voltage change must be amplified to a suitable level to be measured.

Amplification of DC voltages is problematic because of $1/f$ noise. This always dominates at low frequencies, and the problem is compounded by $1/f$ noise from the diode detector. A solution is to use lock-in detection of a *modulated* signal. The effect of modulation is shown in Figure 3.11. A small oscillatory field of frequency ω_m and magnitude b_m is added to the field sweeping through the resonance. In the limit that the magnitude b_m is much less than the width of the line it can be regarded as traversing an essentially straight portion of the curve whose slope is the derivative of the curve at that point. (A fuller mathematical treatment of this will follow in the next section.) As can be seen from the figure, the result is that a small signal of frequency ω_m and magnitude $b_m \partial \chi'' / \partial B$ will be superimposed on the output. This could be fed through a narrow band amplifier having a centre-frequency of ω_m. The output signal in this case is the *derivative* of the line profile rather than the profile itself. However, the advantage is that the amplification of the signal is being performed at a frequency (hopefully) way above the $1/f$ knee of the amplifier. When the signal is at an adequate level it could then be demodulated using a diode detector.

There is an alternative method of demodulation which, in effect, provides a narrow band amplifier with a variable centre-frequency guaranteed to be centred on ω_m and having an easily varied band width. This sounds too good to be true! It is achieved using so-called lock-in detection (Figure 3.12). This is the same as the zero-frequency heterodyne detection scheme described in Section 3.2.4. The main amplification is performed by the amplifier of gain A. This is done at the modulation frequency ω_m. Following this the signal is multiplied by the 'reference signal' $\cos\omega_m t$. The output of the multiplier comprises the required DC signal plus the double frequency term. The purpose of the low pass filter is two-fold. Firstly it removes the double frequency term, but secondly, the cut-off frequency of the filter determines the bandwidth of the amplifier circuit. Thus we see that the centre-frequency is automatically set by the frequency of the reference signal, and the bandwidth by the cut-off frequency.

As mentioned above, when using a Robinson oscillator it is possible to vary the frequency of the tank circuit instead of changing the magnetic field. Now in the lock-in method there is both a sweep through the line and the small oscillatory modulation superimposed on this. The oscillatory modulation is ideally provided as frequency modulation using a varicap diode.

3.5.4 Modulation distortion

In the lock-in detection of CW NMR the signal size was seen to be proportional to the amplitude of the modulation signal b_m. Note, however, that result was valid only in the limit of small b_m – specifically, in the limit of $b_m \ll \Delta B$ where ΔB is the width of the resonance. Also, although we were working in terms of field modulation, it is equally possible, using a Robinson oscillator, to work in terms of frequency modulation. In fact, for the analysis we will now present, it is rather more intuitive to think in terms of frequency modulation. Since the signal increases in size with the magnitude of the modulation it is of interest to see how large the modulation can be, and what distorting effects this can have on the resultant signal. The signal from the tank circuit will be

$$S(t) = \chi''[\omega_p + \Delta\cos(\omega_m t)], \tag{3.16}$$

where ω_p is the current position along the resonance in the absence of the frequency

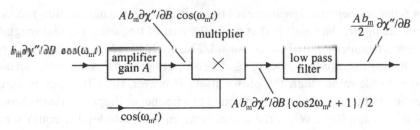

Figure 3.12 Lock-in detection.

modulation; this is the position of the field sweep. Δ is the depth of the modulation and ω_m is the frequency of the modulation.

The response, Equation (3.16), is non-linear. The 'input' is the cosine at frequency ω_m, but the output will comprise the fundamental at ω_m together with harmonics at multiples of this frequency. Here the lock-in detector can be used to advantage since its effect is to extract the component of $S(t)$ at the fundamental (modulation) frequency ω_m. This Fourier component may be found by multiplying by $\pi^{-1}\cos(\omega_m t)$ and integrating over a cycle of the oscillation. This can be done by putting $\omega_m t = x$ and integrating from 0 to 2π. That is

$$I_\omega = \frac{1}{\pi} \int_0^{2\pi} \chi''(\omega_p + \Delta\cos x)\cos x\, dx. \tag{3.17}$$

This is best treated by expanding $\chi''(\omega)$ as a Fourier transform:

$$\chi''(\omega) = \int_{-\infty}^{\infty} f(t)\exp(i\omega t)dt \tag{3.18}$$

so that

$$\chi''(\omega_p + \Delta\cos x) = \int_{-\infty}^{\infty} f(t)\exp[i(\omega_p + \Delta\cos x)t]dt.$$

It then follows that I_ω can be expressed, from Equation (3.17), as

$$I_\omega = \frac{1}{\pi} \int_0^{2\pi} dx \int_{-\infty}^{\infty} dt f(t)\exp[i(\omega_p + \Delta\cos x)t]\cos x,$$

which may be rearranged to give

$$I_\omega = \frac{1}{\pi} \int_0^{2\pi} \cos x \, \exp(it\Delta\cos x)dx \int_{-\infty}^{\infty} f(t)\exp(i\omega_p t)dt.$$

Now the first integral gives a Bessel function (Gradshteyn and Ryzhik, 1965, p. 743)

$$\int_0^{2\pi} \cos x \, \exp(it\Delta\cos x)dx = 2\pi i J_1(\Delta t)$$

and from this we then obtain I_ω:

$$I_\omega = 2i \int_{-\infty}^{\infty} J_1(\Delta t)f(t)\exp(i\omega_p t)dt. \tag{3.19}$$

In the small Δ limit the Bessel function may be expanded to first order: $J_1(\Delta t) \sim \Delta t/2 + \dots$ so that

$$I_\omega \sim i \int_{-\infty}^{\infty} \Delta t f(t)\exp(i\omega_p t)dt.$$

Since

$$it\exp(i\omega_p t) = \frac{\partial}{\partial\omega_p}\exp(i\omega_p t) \tag{3.20}$$

if follows that

$$I_\omega \sim \Delta\frac{\partial}{\partial\omega_p} \int_{-\infty}^{\infty} f(t)\exp(i\omega_p t)dt$$

and, as the integral here is $\chi''(\omega)$ (Equation (3.18)), we obtain in the small Δ limit the linear behaviour, as found from the qualitative discussion of the previous section,

$$I_\omega \sim \Delta\frac{\partial\chi''(\omega)}{\partial\omega_p}; \tag{3.21}$$

the signal is proportional to the magnitude of the modulation and to the derivative of the line profile function.

We now return to the full expression for the signal I_ω, Equation (3.19), to examine the meaning of this result. We slightly reorganise the equation as

$$I_\omega = \Delta \int_{-\infty}^{\infty} \frac{2}{\Delta t} J_1(\Delta t) itf(t)\exp(i\omega_p t)dt.$$

This is Δ times the Fourier transform of the product of $(2/\Delta t)J(\Delta t)$ and $itf(t)$, and this is Δ times the *convolution* of the Fourier transform of $(2/\Delta t)J(\Delta t)$ with the Fourier transform of $itf(t)$. Now the Fourier transform of $itf(t)$ is the derivative of $\chi''(\omega)$, and the Fourier transform of $(2/\Delta t)J(\Delta t)$ is a semicircle (Gradshteyn and Ryzhik, 1965, p. 743):

$$2 \int_{-\infty}^{\infty} \frac{J_1(\Delta t)}{\Delta t} \exp(i\omega t)dt = \frac{4}{\Delta}\left[1 - \left(\frac{\omega}{\Delta}\right)^2\right]^{\frac{1}{2}}.$$

Thus we see that the effect of the modulation of the magnitude Δ on the lock-in detected signal is to convolute the expected derivative signal with a semicircle of radius Δ.

We conclude that while the signal size will increase with increasing modulation depth, the limit on this will be set by the acceptable distortion of the signal.

4

Classical view of relaxation

4.1 Transverse relaxation in solids

4.1.1 Local magnetic fields

In a magnetic resonance experiment the specimen under investigation is placed in a static magnetic field $\mathbf{B_0}$. The spins then precess about this field at the Larmor frequency $\omega_0 = \gamma B_0$. So if the spins start from some coherent initial state, produced for example by a 90° pulse, then they should continue to precess all together in phase, but this is not observed to happen. In practice the precessing magnetisation decays away over a certain time scale. The explanation of this decay is one of the main topics of the present chapter.

If each nucleus experienced precisely the same magnetic field then clearly there could be no transverse relaxation; all spins would continue precessing coherently. However, each precessing nucleus is a magnetic dipole μ and as such, as well as responding to a magnetic field, it is the *source* of a field. This dipolar magnetic field is seen by other spins in its vicinity. Thus each spin sees not only the static field of the magnet, it also sees the (albeit much smaller) dipole fields of its neighbours. The field experienced by the *i*th spin, $\mathbf{B}_i$ is then given by

$$\mathbf{B}_i = \mathbf{B}_0 + \mathbf{b}_i, \tag{4.1}$$

where the dipole field $\mathbf{b}_i$ is

$$\mathbf{b}_i = \frac{\mu_0}{4\pi} \sum_j \left[\frac{\mu}{r_{ij}^3} - \frac{(\mu \cdot \mathbf{r}_{ij})\mathbf{r}_{ij}}{r_{ij}^5} \right] \tag{4.2}$$

and *j* labels the other dipoles in the system.

There is, however, a further reason why the magnetic field will vary from spin to spin. No magnet is perfect; the applied magnetic field will vary over the volume of the sample. These 'local fields', of whatever origin, are all the cause of transverse relaxation. When each spin is in a slightly different magnetic field it precesses at a slightly different rate. So although they may start by precessing in phase, after a certain time the coherence is lost through destructive interference.

The two main contributions to the field variations: neighbours' dipole fields and the effect of a 'poor' (inhomogeneous) magnet, are rather different. The magnet contribution is simply an instrumentational problem whereas the dipolar part represents some intrinsic property of the specimen under investigation. We shall see in Section 4.4 how the technique of *spin echoes* can distinguish between these and that one can usually recover the magnetisation lost through a bad magnet. For the purposes of this section, however, it is sufficient to accept that field variations, of whatever origin, exist in any system. Henceforth we shall use the symbol B_0 to denote the *mean* magnetic field over the specimen, and b_i to denote the deviation from this mean value at the ith spin site.

We should note that in this discussion we have already made a significant approximation. We are regarding each spin independently; each one sees a composite field due to its neighbours together with the external field. This is a *mean field* or *single particle* approximation. In reality a spin not only sees and reacts to its neighbours, but also its behaviour influences the behaviour of the neighbours and by this mechanism spins react back on themselves. These effects, although often not important, can only be accounted for in a many-body treatment of the system. This is done in a quantum approach to relaxation in the following chapter.

We shall make one further simplification before developing the consequences of our elementary model of a magnetic system. It is the component of b_i which is parallel to the applied B_0 which causes variations in the frequency of precession about B_0. Furthermore, since a spin's neighbours are also precessing about B_0, the transverse components of their dipolar fields will have a rapid oscillatory dependence (at ω_0) which will average to zero. For these reasons we shall, for the present, consider only local fields parallel to B_0. These models were first considered by Anderson and Weiss (1953) and Anderson (1954). A readable account is given by Kubo (1961).

4.1.2 Solution of equations of motion

By convention the B_0 field is taken to point along the z axis. The field seen by the jth spin is then

$$\mathbf{B}_j = (B_0 + b_j)\mathbf{k} \tag{4.3}$$

and the equation of motion for this magnetic moment is then

$$\begin{aligned}
\dot{\boldsymbol{\mu}}_j &= \gamma \boldsymbol{\mu}_j \times \mathbf{B}_j \\
&= \gamma \boldsymbol{\mu}_j \times (B_0 + b_j)\mathbf{k}.
\end{aligned} \tag{4.4}$$

We see immediately from the cross product that the z component of $\dot{\boldsymbol{\mu}}_j$ is zero so that this model is certainly incapable of treating longitudinal relaxation. This is related to the neglect of the transverse components of the local fields, and clearly it is a limitation of the model. The fuller treatment overcoming these difficulties is taken up in the next chapter in the quantum mechanical context. An alternative classical approach is adopted in Appendix E, where both transverse and longitudinal relaxation are treated, all components of the local field being considered. For the present, however, remember that our aim is a discussion of the essential features of transverse relaxation.

Since we are then only treating transverse relaxation, it proves expedient to adopt the complex notation incorporating the x and y components of μ_j. As in Section 2.6.2 we define the complex μ as

$$\mu = \mu_x + i\mu_y. \tag{4.5}$$

(Here the subscripts are Cartesian components, not particle labels.) In terms of the complex μ the equation of motion becomes

$$\dot{\mu}_j(t) = -i\gamma(B_0 + b_j)\mu_j(t). \tag{4.6}$$

At this stage we specifically (but temporarily) restrict ourselves to consideration of a solid: an immobile assembly of atoms. Under such conditions the local fields b_i are constant and integration of Equation (4.6) is trivial. We may write the solution immediately:

$$\mu_j(t) = \exp[-i\gamma(B_0 + b_j)t]\mu_j(0) \tag{4.7}$$

and we see, as expected, that each spin precesses at its own rate $\omega_j = \gamma(B_0 + b_j)$. Over the whole specimen, then, there is a spread of precession frequencies.

The magnetisation is the total magnetic moment per unit volume:

$$\mathbf{M} = \frac{1}{V} \sum_{j=1}^{N} \mu_j, \tag{4.8}$$

where the sum is over the N particles in the system. So adopting the complex notation for magnetisation we have similarly

$$M = \frac{1}{V} \sum_{j=1}^{N} \mu_j$$

$$= \frac{1}{V} \sum_{j=1}^{N} \exp[-i\gamma(B_0 + b_j)t]\mu_j(0).$$

We may factorise out the initial magnetisation

$$M(0) = \frac{1}{V} \sum_{j=1}^{N} \mu_j(0) \tag{4.9}$$

on the assumption that the initial value of the magnetic moment $\mu_j(0)$ is independent of its local field b_j. If we also factorise the exponential, the magnetisation expression becomes

$$M(t) = \exp(-i\gamma B_0 t)\frac{1}{N}\sum_{j=1}^{N}\exp(-i\gamma b_j t)M(0).\qquad(4.10)$$

The term $\exp(-i\gamma B_0 t)$ represents the precession of the magnetisation at the Larmor frequency. The summation term is the superposition of oscillations at different frequencies. It is the destructive interference in this sum which is responsible for the decay of the precessing magnetisation. Finally the $M(0)$ term is the initial magnitude of the transverse magnetisation.

Figure 4.1 shows the typical behaviour which follows from this result. The oscillations are at an angular frequency of $\omega_0 = \gamma B_0$ and the shape of the relaxation 'envelope' is given by

$$\frac{1}{N}\sum_{j=1}^{N}\exp(-i\gamma b_j t),$$

which we write succinctly as

$$\langle\exp(-i\gamma bt)\rangle$$

to indicate explicitly that the sum over particles divided by the number of particles gives the mean or average of the various oscillations.

4.1.3 Relaxation times

As we have stated, it is the spread of the individual precession frequencies which causes the destructive interference and the resultant decay of magnetisation. The time scale for this decay is the reciprocal of the spread of frequencies. This is a general property of the superposition of sinusoidal waves and it is essentially the same (see Appendix A) as the Uncertainty Principle of quantum mechanics.

We mentioned in Section 2.2.5 that the characteristic time for the decay of transverse magnetisation, the spin–spin relaxation time, is given the symbol T_2. We thus conclude so far that

$$T_2 \approx 1/(\text{spread in frequencies } \gamma b_j),\qquad(4.11)$$

but how do we quantify this spread of frequencies? We know that the mean of b_j, which we can denote by $\langle b\rangle$, is zero; recall that B_0 was redefined to ensure this. We then measure the spread in b as the RMS value: the mean square deviation from the mean, which is $\langle b^2\rangle^{\frac{1}{2}}$. From this the spin–spin relaxation time may be expressed as

$$\frac{1}{T_2} = \gamma\langle b^2\rangle^{\frac{1}{2}}.\qquad(4.12)$$

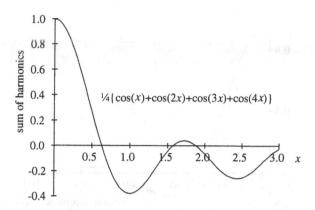

Figure 4.1 Example of destructive interference.

Let us now evaluate the order of magnitude for such relaxation times. The magnitude of the dipolar field a distance r from a magnetic moment μ is given approximately by

$$b \approx \frac{\mu_0}{4\pi} \frac{\mu}{r^3},$$

which for protons at a distance of 2 Å gives a field of some 2×10^{-4} T. The r^{-3} dependence implies a rapid fall-off with separation so that the effect of near neighbours will be dominant. So assuming all orientations of spins are equally probable the RMS value of b will have a similar magnitude:

$$\langle b^2 \rangle^{\frac{1}{2}} \approx 2 \times 10^{-4} \, \text{T}. \tag{4.13}$$

Since the magnetogyric ratio of the proton (Table 1.1) is given by $2\pi \times 42.6 \times 10^6$ radians/(s T) (42.6×10^6 Hz/T), the estimate for T_2 is then approximately 20 μs.

The transverse relaxation of the fluorine nuclear spins in calcium fluoride has been studied in great detail. This substance has a number of features which make it attractive for such work. Only the fluorine atoms carry a nuclear spin, which conveniently has magnitude $\frac{1}{2}$, and these atoms form a simple cubic lattice. Furthermore, the fluorine magnetogyric ratio, at $2\pi \times 40.06 \times 10^6$ radians/(s T) (40.06×10^6 Hz/T) is only slightly smaller than that of protons, ensuring reasonably larger signals. Following the same argument as above, one would expect the fluorine T_2 similarly to be approximately 20 μs. Figure 4.2 shows a typical transverse decay envelope for a crystal of CaF_2. A time scale of 20 μs seems quite appropriate.

4.1.4 Shape of the FID

We now turn to consider the *shape* of the relaxation or, to be precise, the shape of the relaxation envelope. This is given by the expression

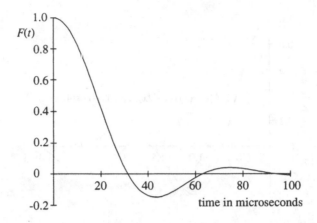

Figure 4.2 FID of calcium fluoride.

$$F(t) = \langle \exp(-i\gamma bt) \rangle, \qquad (4.14)$$

where b is the 'local field' seen by a representative spin. And since γb is the local precession frequency, ω_{loc}, we can write $F(t)$ as

$$F(t) = \langle \exp(-i\omega_{loc}t) \rangle. \qquad (4.15)$$

Let us look at this average in a slightly different, more suggestive, manner and introduce a probability density function for the distribution of local frequencies. We shall write $F(t)$ as

$$F(t) = \int f(\omega)\exp(-i\gamma\omega t)d\omega, \qquad (4.16)$$

where $f(\omega)d\omega$ is the fraction of spins experiencing local frequencies between ω and $\omega + d\omega$. Thus $f(\omega)$ is the distribution of local frequencies, and we see that the FID, $F(t)$, is precisely the Fourier transform of the local frequency distribution. Recalling the Fourier transform relation between the FID and the CW absorption lineshape (Section 2.6.3), we see that in this case the lineshape is the same as the local frequency distribution function.

4.2 Motion and transverse relaxation

4.2.1 Averaging of the local fields

The discussion in the previous sections applied to the study of solids. We made the assumption that the local field experienced by a spin was constant in time. It was that

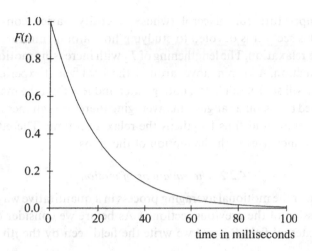

Figure 4.3 Proton transverse relaxation in glycerol.

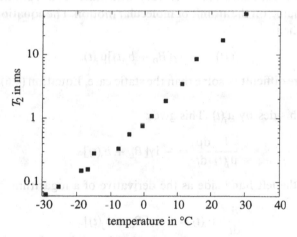

Figure 4.4 Temperature dependence of T_2 in glycerol.

fact which facilitated the immediate integration of Equation (4.6) to show that the FID was related to a sum of different sine waves.

Recall also that we estimated the magnitude of the spin–spin relaxation time T_2 in a typical solid and we obtained a value of about 20 µs. In Figure 4.3 we show a typical proton FID for glycerol which would be observed with a homogeneous $\mathbf{B}_0$ field. We see that the relaxation time is some 20 ms, which is very much greater than that in a solid (as would be observed, for instance, in frozen glycerol). It is clear then that motion has an important effect on T_2, and, to demonstrate this, in Figure 4.4 we show the variation of T_2 with temperature for glycerol (whose viscosity varies considerably with temperature). This section is devoted to studying how atomic motion influences the rate of transverse relaxation. The lengthening of T_2 with increasing motion has a simple qualitative explanation. As a spin moves around, the local field it experiences will vary. In time each spin will sample all fields, both greater and less than the mean $\mathbf{B}_0$, so each will have precessed by a similar angle. This averaging therefore enhances the coherence of the precessing spins and thus lengthens the relaxation time. The efficiency of the averaging process increases with the motion of the spins.

4.2.2 The equations of motion

Let us now consider the motional averaging process in a quantitative way by extending the model discussion of the previous section(s). As before we consider only the fields parallel to the external field $\mathbf{B}_0$, and we write the field seen by the jth spin as

$$\mathbf{B}_j = (B_0 + b_j)\mathbf{k}. \tag{4.17}$$

Now we allow the local fields, the b_j, to vary with time. This variation will be in a random way depending on the atomic or molecular motion. The equation of motion for the complex μ is then

$$\dot{\mu}_j(t) = -i\gamma[B_0 + b_j(t)]\mu_j(t). \tag{4.18}$$

This is slightly more difficult to solve than the static case, Equation (4.6), so we shall go through the steps.

First divide both sides by $\mu_j(t)$. This gives:

$$\frac{1}{\mu_j(t)}\frac{d\mu_j}{dt} = -i\gamma[B_0 + b_j(t)]$$

and we recognise the left hand side as the derivative of a logarithm

$$\frac{d}{dt}\ln\mu_j(t) = -i\gamma[B_0 + b_j(t)].$$

This can be integrated immediately, giving

$$\ln\mu_j(t) = -i\gamma\int_0^t [B_0 + b_j(\tau)]d\tau + \ln\mu_j(0),$$

where the second term on the right hand side is the constant of integration. So taking the exponential, and since B_0 is a constant, we obtain

$$\mu_j(t) = \exp(-i\gamma B_0 t)\exp\left[-i\gamma\int_0^t b_j(\tau)d\tau\right]\mu_j(0). \tag{4.19}$$

This gives the evolution of a given spin in terms of the magnetic fields it experiences. Performing the sum over all spins in the system and factorising out the initial magnetisation as in Equation (4.9), the complex transverse magnetisation is given by

$$M(t) = \exp(-i\gamma B_0 t)\left\{\frac{1}{N}\sum_{j=1}^{N}\exp\left[-i\gamma\int_0^t b_j(\tau)d\tau\right]\right\}M(0). \qquad (4.20)$$

Again we recognise the first term as the Larmor precession part; the second, bracketed term is the relaxation part: the FID envelope; and the final term $M(0)$ is the initial transverse magnetisation.

Our interest focuses on the relaxation part, the FID envelope. As when treating the simpler solid model, we shall denote the average over the spins by angle brackets, giving the relaxation function $F(t)$ as

$$F(t) = \left\langle\exp\left[-i\gamma\int_0^t b(\tau)d\tau\right]\right\rangle. \qquad (4.21)$$

To take this further we note that the integral

$$-\gamma\int_0^t b(\tau)d\tau$$

is the total angle through which the jth spin has precessed in time t: the accumulated phase angle. Denoting this by $\phi_j(t)$,

$$\phi_j(t) = -\gamma\int_0^t b(\tau)d\tau, \qquad (4.22)$$

we see that the relaxation function $F(t)$ becomes

$$F(t) = \langle\exp[i\phi(t)]\rangle. \qquad (4.23)$$

Again, following the procedure adopted for the solid model, we write the average as an integral over a probability distribution function

$$F(t) = \int P[\phi(t)]\exp[i\phi(t)]d\phi. \qquad (4.24)$$

Here $P[\phi(t)]d\phi$ is the probability that a spin will have accumulated a phase angle of between ϕ and $\phi + d\phi$ in a time t.

Once again then everything depends on the nature of the probability function, but now it is the probability distribution of *phase angles*. This is not exactly the same as the distribution of local frequencies. They are only the same in the absence of motion as then $\phi_j \propto \omega_j$ since $\phi_j = \omega_j t$.

4.2.3 Gaussian distribution of phases

To make further progress it is necessary to make some assumptions about the probability function $P(\phi)$. Now the phase angle ϕ is made up from a lot of small contributions; as a spin travels around it samples many different fields for short periods of time. Under such conditions we can appeal to the Central Limit Theorem of probability theory and approximate $P(\phi)$ by a Gaussian function. Since the mean of ϕ is zero, $P(\phi)$ is specified in terms of one parameter, its mean square width $\langle \phi^2 \rangle$. The normalised probability is then given by

$$P(\phi) = \frac{1}{(2\pi \langle \phi^2 \rangle)^{\frac{1}{2}}} \exp\left(-\frac{\phi^2}{2\langle \phi^2 \rangle} \right). \tag{4.25}$$

We can then perform the integral for $F(t)$,

$$F(t) = \frac{1}{(2\pi \langle \phi^2 \rangle)^{\frac{1}{2}}} \int_{-\infty}^{\infty} \exp\left(-\frac{\phi^2}{2\langle \phi^2 \rangle} \right) \exp(i\phi) d\phi, \tag{4.26}$$

which is evaluated by completing the square of the argument of the exponentials. The result is

$$F(t) = \exp[-\langle \phi^2(t) \rangle /2]. \tag{4.27}$$

Box 4.1 Gaussian averaging – another view

We have seen that if $\phi(t)$ has a Gaussian probability distribution then the expression for the FID envelope

$$F(t) = \langle \exp[i\phi(t)] \rangle$$

is equal to the more tractable (as we shall see) form

$$F(t) = \exp[-\tfrac{1}{2}\langle \phi^2(t) \rangle], \tag{4.28}$$

since everything then depends only on the mean square value of ϕ and not on its various other features.

Furthermore it should be evident that if $P(\phi)$ is *approximately* Gaussian then $F(t)$ is given *approximately* by the more tractable Equation (4.28). This result is of considerable importance in NMR and we shall therefore complement its derivation with a view from a somewhat different standpoint.

Let us take Equation (4.23) for $F(t)$, expand the exponential and perform the average term by term:

$$F(t) = \langle 1 \rangle + i\langle \phi(t) \rangle - \langle \phi^2(t) \rangle /2 - i\langle \phi^3(t) \rangle /3! + \dots. \tag{4.29}$$

All the odd terms vanish since $\phi(t)$, being random, is equally likely to be positive or negative and the average of 1 is 1. The series is then

$$F(t) = 1 - \tfrac{1}{2}\langle\phi^2(t)\rangle + \dots.$$

Now this can be regarded as the leading part of exponential expansion,

$$F(t) \approx \exp[-\tfrac{1}{2}\langle\phi^2(t)\rangle],$$

which is precisely the result obtained above by the Gaussian averaging procedure! Up to order ϕ^2 it is clear that Equation (4.28) and Equation (4.29) are equivalent. If we examine the term in ϕ^4 we have

Equation (4.29): $\langle\phi^4(t)\rangle/4!$
Equation (4.28): $\langle\phi^2(t)\rangle^2/8$.

We can see from this that the Gaussian averaging procedure is equivalent to approximating the higher order averages $\langle\phi^n(t)\rangle$ in terms of powers of the second order term $\langle\phi^2\rangle$.

It remains then to study the width of the distribution function, the mean square value $\langle\phi^2(t)\rangle$. In particular we must investigate how it varies with time. We have an expression for the evolution of the phase angle of a representative spin, Equation (4.22),

$$\phi(t) = -\gamma \int_0^t b(\tau)d\tau$$

and $\phi^2(t)$ is simply the square of this. It is convenient to write the square as

$$\phi^2(t) = \gamma^2 \left[\int_0^t b(\tau)d\tau \right]\left[\int_0^t b(\tau)d\tau \right]$$

and then to remove the brackets by labelling the integration variable τ_1 in the first bracket and τ_2 in the second. We can then write

$$\phi^2(t) = \gamma^2 \int_0^t d\tau_1 \int_0^t d\tau_2 b(\tau_1)b(\tau_2).$$

The average can now be performed, giving

$$\langle\phi^2(t)\rangle = \gamma^2 \int_0^t d\tau_1 \int_0^t d\tau_2 \langle b(\tau_1)b(\tau_2)\rangle. \tag{4.30}$$

4.2.4 The autocorrelation function

The average $\langle b(\tau_1)b(\tau_2)\rangle$ is known as the *autocorrelation function* of the random variable $b(\tau)$. In Appendix D we discuss some of the important properties of random variables and their correlation functions. For the present purposes we only need note the property of *stationarity*. For systems in equilibrium the autocorrelation function depends on the τ only through the time difference $\tau_2 - \tau_1$. Such random functions are called *stationary* random functions.

In this case we may then write

$$\gamma^2\langle b(\tau_1)b(\tau_2)\rangle = G(\tau_2 - \tau_1), \tag{4.31}$$

where $G(\tau)$ is the local frequency autocorrelation function. We then have

$$\langle\phi^2(t)\rangle = \int_0^t d\tau_1 \int_0^t d\tau_2 G(\tau_2 - \tau_1)$$

and now one of the integrations may be performed through a change of variables. Transforming to

$$\tau = \tau_2 - \tau_1,$$
$$T = \tau_2 + \tau_1,$$

since G depends only on τ we can integrate over T to obtain

$$\langle\phi^2(t)\rangle = 2\int_0^t (t - \tau)G(\tau)d\tau. \tag{4.32}$$

The expression for the FID envelope, Equation (4.27), can then finally be written as

$$F(t) = \exp\left[-\int_0^t (t - \tau)G(\tau)d\tau\right]. \tag{4.33}$$

This equation is the end result of this section. The derivation was based upon various assumptions about the system. We shall see in the next section that quite a lot of physics is contained in this expression, and that it does indeed reflect the behaviour of real systems under a variety of circumstances.

4.3 Consequences of the model

4.3.1 The correlation time

We discussed in the last section, in a qualitative way, how the more rapid the motion the more efficient the process of averaging away the local magnetic fields: the relaxation

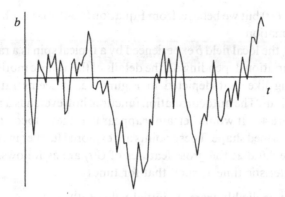

Figure 4.5 Random variation of the local magnetic field.

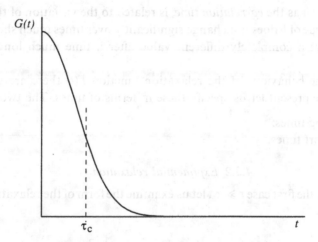

Figure 4.6 Autocorrelation function for random field.

time T_2 increases. We then derived the key equation for the relaxation of transverse magnetisation, $F(t)$,

$$F(t) = \exp\left[-\int_0^t (t - \tau)G(\tau)\mathrm{d}\tau \right]. \tag{4.34}$$

We asserted that this equation contains much physics, and the least we expect is a quantitative support for the above-mentioned qualitative ideas. This will be the aim of the present section.

Our interest has been transferred from the transverse relaxation function $F(t)$ to the local field (really frequency) autocorrelation function $G(t)$. Ultimately, of course, we want to know about $F(t)$ but we believe, from Equation (4.34), that the behaviour of $G(t)$ will yield that information.

As we have stated, the local field h experienced by a typical spin is a random function of time, its precise variation depending on the details of the atomic motion. We expect a behaviour something like that depicted in Figure 4.5. It seems rather difficult to quantify such behaviour. The autocorrelation function, however, has a much smoother evolution, as in Figure 4.6. It would certainly appear that the autocorrelation function has a more easily described shape. These points are explored further in Appendix B. We observe from Figure 4.6 that the gross features of $G(t)$ are as follows.

There is a characteristic time τ_c such that for times

$t \ll \tau_c$: $G(t)$ varies negligibly from its initial value $G(0)$
$t \gg \tau_c$: $G(t)$ has decayed to zero.

The time τ_c, known as the *correlation time*, is related to the variation of the local field $b(t)$. The magnitude of b does not change significantly over times much shorter than τ_c, while it will have a completely different value after a time much longer than the correlation time.

In studying the behaviour of the relaxation function $F(t)$ there are two cases to consider. For the present let us specify these in terms of times. The two cases are:

(a) $t \gg \tau_c$: long times;
(b) $t \ll \tau_c$: short times.

4.3.2 Exponential relaxation

Considering now the first case $t \gg \tau_c$, let us examine the form of the relaxation function, Equation (4.33):

$$F(t) = \exp\left[-\int_0^t (t - \tau)G(\tau)\mathrm{d}\tau \right].$$

For times t greater than the correlation time τ_c we know that the correlation function has decayed away to zero. The argument of the integral is then zero, so that no harm is done if the upper limit of the integral is extended to infinity. If this is done then the expression for $F(t)$ simplifies:

$$F(t) = \exp\left[-\int_0^\infty (t - \tau)G(\tau)\mathrm{d}\tau \right]$$

$$= \left\{ \exp\left[-t \int_0^\infty G(\tau)\mathrm{d}\tau \right] \right\}\left\{ \exp\left[+\int_0^\infty \tau G(\tau)\mathrm{d}\tau \right] \right\}. \tag{4.35}$$

The second term is simply a constant and the first term describes an exponential decay, the integral giving the inverse relaxation time. Thus we may write

$$F(t) \propto \exp(-t/T_2),$$ (4.36)

where

$$\frac{1}{T_2} = \int_0^\infty G(\tau)d\tau.$$ (4.37)

We have then two conclusions. Firstly when $t \gg \tau_c$ the relaxation is exponential, and secondly we have the expression for the relaxation rate T_2^{-1} as the area under the autocorrelation function.

Let us consider further the expression for T_2. So far we do not have a precise definition for the correlation time τ_c. We have only specified behaviour for times much less and much greater than τ_c. We understand that essentially τ_c refers to the time scale over which $G(\tau)$ decays appreciably. This is some measure of the width of the correlation function. We can then formulate a precise definition of the width as the area divided by the (initial) height. In this way we are led to the definition

$$\tau_c = \frac{1}{G(0)} \int_0^\infty G(\tau)d\tau.$$ (4.38)

This may be incorporated into the expression for T_2 giving

$$\frac{1}{T_2} = G(0)\tau_c.$$ (4.39)

Now $G(0)$, the initial value of the autocorrelation function, has a special meaning; we have encountered it before, in another guise. From the definition of $G(t)$ we have

$$G(0) = \gamma^2\langle b(0)b(0)\rangle$$
$$= \gamma^2\langle b^2\rangle \text{ or } \langle \omega_{loc}^2\rangle,$$ (4.40)

the mean square value of the local frequencies. It was the square root of this quantity which we found to be the inverse relaxation time when considering immobile spins. We are now able to write the expression for T_2 as

$$\frac{1}{T_2} = \langle \omega_{loc}^2\rangle\tau_c.$$ (4.41)

There are two factors which determine the relaxation time. There is the mean square strength of the local fields (this was the only thing that affected the T_2 of a rigid lattice of spins) and, as expected, the motion now influences the relaxation through the dependence on the correlation time τ_c. Turning again to Figure 4.4 showing the variation of T_2 in glycerol as a function of temperature, we see that the increase in T_2 with temperature indicates how the correlation time is decreasing: the motion is getting faster. We shall return to an evaluation of $\langle \omega_{loc}^2 \rangle$ and τ_c and hence T_2' for various systems in Chapter 7. We turn now to the second case, case (b) which applies when $t \ll \tau_c$.

4.3.3 Gaussian relaxation

For short times such that $t \ll \tau_c$ we know that the autocorrelation function $G(t)$ will have changed very little from its initial value. Thus in the expression for the FID, Equation (4.33),

$$F(t) = \exp\left[-\int_0^t (t - \tau)G(\tau)d\tau \right]$$

we may replace $G(\tau)$ by its initial value whereupon it may be taken out of the integral

$$F(t) = \exp\left[-G(0)\int_0^t (t - \tau)d\tau \right].$$

This integral may be evaluated, giving

$$F(t) = \exp[-G(0)t^2/2]. \tag{4.42}$$

We have then a Gaussian relaxation which decays on a time scale of the order of $G(0)^{-1/2}$. In other words, T_2 (which does not have a precise definition here) is given by

$$\frac{1}{T_2} \approx [G(0)]^{\frac{1}{2}}. \tag{4.43}$$

Recall that $G(0) = \langle \omega_{loc}^2 \rangle$, so we then obtain

$$\frac{1}{T_2} \approx \langle \omega_{loc}^2 \rangle^{\frac{1}{2}}, \tag{4.44}$$

which is precisely the result obtained in Section 4.1.3 for the relaxation time for an assembly of immobile spins.

4.3.4 Condition for exponential or Gaussian decay

We have seen thus far that the very early part of the relaxation is Gaussian while the very late part is exponential, and we have found the relevant time scales for these processes. However, what we really want to know is: what is the behaviour of most of the relaxation? Is it mostly Gaussian, mostly exponential, or a mixture of both? These

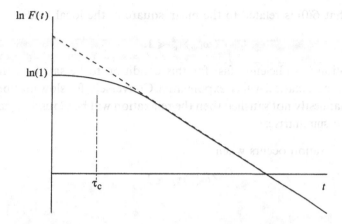

Figure 4.7 Transition between Gaussian and exponential relaxation.

questions are best answered by appeal to Equation (4.35) which is true for long times $t \gg \tau_c$,

$$F(t) = \left\{ \exp\left[-t \int_0^\infty G(\tau) d\tau \right] \right\} \left\{ \exp\left[+\int_0^\infty \tau G(\tau) d\tau \right] \right\}.$$

The first term represents exponential decay and the second term is a constant. Recalling that $F(0) = 1$ we can draw the true behaviour of the logarithm of $F(t)$ as in Figure 4.7. We wish to find the condition that most of the relaxation is exponential. From Figure 4.7 we see that this will be so if the $t = 0$ value of the projected back exponential curve approximates the true value of unity. That is, the integral representing the 'overshoot' should be very small:

$$\int_0^\infty \tau G(\tau) d\tau \ll 1.$$

Now recalling the general property of the correlation function, namely that $G(t) \approx G(0)$ for $t \ll \tau_c$ while $G(t) \approx 0$ for $t \gg \tau_c$, we see that this integral is approximately equal to

$$\int_0^{\tau_c} \tau G(\tau) d\tau,$$

which to within a factor of 2 is equal to $G(0)\tau_c^2$. We conclude then that the condition for most of the relaxation to be exponential is that

$$G(0)\tau_c^2 \ll 1.$$

or, recalling that $G(0)$ is related to the mean square of the local fields

$$\langle \omega_{loc}^2 \rangle \tau_c^2 \ll 1.$$

When the motion is sufficiently fast for this condition to be satisfied, then we can conclude that the relaxation will be exponential. Conversely, for slow motion, when the condition is manifestly not satisfied then the relaxation will be Gaussian (according to this model). To summarise:

Exponential relaxation occurs when

$$\langle \omega_{loc}^2 \rangle \tau_c^2 \ll 1$$

and then

$$T_2^{-1} = \langle \omega_{loc}^2 \rangle \tau_c.$$

Gaussian relaxation occurs when

$$\langle \omega_{loc}^2 \rangle \tau_c^2 \gg 1$$

and then

$$T_2^{-1} \approx \langle \omega_{loc}^2 \rangle^{\frac{1}{2}}.$$

We can regard $\langle \omega_{loc}^2 \rangle$ as the mean square dephasing frequency, so then $\langle \omega_{loc}^2 \rangle^{-\frac{1}{2}}$ is the average dephasing time in the local fields. The slow motion case is then indicated when the motion is slow compared with the dephasing time in the local field. In other words, the local fields then appear static just as in the solid case treated in Section 4.1.2.

Both models seem to have given the same relaxation time, but here we appear also to have predicted the shape of the relaxation. This is because we have *put in* the Gaussian distribution function for $P(\phi)$. In fact this is probably not valid in the case of no motion, since then the Central Limit Theorem does not come to our assistance. We see this in the FID for calcium fluoride in Figure 4.2. While the relaxation time has the correct order of magnitude, the shape of the relaxation is definitely not Gaussian. These questions will be taken up in Chapter 6 when we shall have the full machinery of quantum mechanics at our disposal.

4.4 Spin echoes

4.4.1 Recovery of lost magnetisation

In Section 4.1 we promised to show how one can recover the transverse magnetisation lost due to the dephasing in an inhomogeneous magnet. The key to this is the use of spin echoes; we follow the treatment of Cowan (1977). Imagine the spins precessing after a 90° pulse. Those spins in a slightly greater field will precess that much faster, and those in a slightly smaller field will precess that much slower. Thus after sufficient time the spins will have dephased and the precessing magnetisation is lost.

However, consider what would happen if a time t after the 90° pulse we were able to reverse the motion of the spins. The spins in the greater field will have precessed that much further. They will now travel in the opposite direction at their faster rate. While the spins in the lower field, which have travelled that much less, will now travel backwards at their slower rate. After a further time t they will come back into phase. The transverse magnetisation will grow causing what is known as a spin echo.

To achieve this the motion of the spins must be reversed. How can this be done? The answer is to use a 180° pulse, as we shall see in the next section.

4.4.2 Effect of a 180° pulse

We return to the model of a collection of magnetic moments in a magnetic field. The magnetic field consists of the constant $\mathbf{B}_0$ together with a further contribution $\mathbf{b}_j$ which varies from spin to spin. We imagine the $\mathbf{b}_j$ to arise from variations in the strength of the applied magnetic field over the volume of the specimen. Furthermore we consider the motion to be sufficiently slow, $\langle \omega_{loc}^2 \rangle \tau_0^2 \gg 1$, that we can neglect (in this section) the time variation of the $\mathbf{b}_j$. Let us follow the course of a typical experiment.

At time $t = 0$ we apply the usual 90° pulse to our system to initialise the transverse relaxation. As we know, the result of this is to tip the equilibrium magnetisation in the z direction over into the transverse plane. After a time t the complex magnetisation will be given by Equation (4.10)

$$M(t) = \exp(-i\omega_0 t)\frac{1}{N}\sum_{j=1}^{N}\exp(-i\gamma b_j t)M(0).$$

We now continue our discussion in the frame rotating at the Larmor frequency. We do this because, as we saw in Section 2.5.4, the rotation effects of the RF pulses occur in the rotating frame. The result of transforming to the rotating frame is to remove the $\exp(-i\omega_0 t)$ term from the relaxation function so that $M(t)$ appears as a stationary but shrinking vector. In the rotating frame then

$$M(t) = \frac{1}{N}\sum_{j=1}^{N}\exp(-i\gamma b_j t)M(0). \tag{4.45}$$

The effect of a 180° pulse is to rotate all the spins through this angle about an axis in the transverse plane. The actual direction of the axis is unimportant, but it proves convenient to take the x axis. The result of a 180° rotation about the x axis is to transform the components of the magnetisation vector $\mathbf{M}$ as follows:

$$\left.\begin{array}{l} M_x \rightarrow M_x, \\ M_y \rightarrow -M_y, \\ M_z \rightarrow -M_z. \end{array}\right\} \tag{4.46}$$

So considering the complex transverse magnetisation, the effect of a 180° pulse is to transform M into its complex conjugate M^*

$$M \to M^*. \tag{4.47}$$

4.4.3 Formation of a spin echo

At time t we shall apply a 180° pulse. Immediately after this the magnetisation is given by the complex conjugate of Equation (4.45),

$$M(t) = \frac{1}{N} \sum_{j=1}^{N} \exp(+i\gamma b_j t) M^*(0). \tag{4.48}$$

The effect of subsequent precession is still to multiply the complex magnetic moments by $\exp(-i\tau b_j t)$ so that at time τ after the 180° pulse we have

$$M(t + \tau) = \frac{1}{N} \sum_{j=1}^{N} \exp(-i\gamma b_j \tau) \exp(+i\gamma b_j t) M^*(0)$$

$$= \frac{1}{N} \sum_{j=1}^{N} \exp[-i\gamma b_j(\tau - t)] M^*(0). \tag{4.49}$$

We observe that the relaxation process has been reversed. In particular, when $t = \tau$ we have

$$M(2t) = M^*(0); \tag{4.50}$$

the initial magnetisation has been fully recovered.

If the initial magnetisation was created with a 90° pulse about the x axis so that it pointed along the y axis then $M(0)$ is pure imaginary, so that the echo is along the $-y$ axis. Conversely, if the initial magnetisation was created with a 90° pulse about the y axis so that it pointed along the x axis then $M(0)$ is pure real so that the echo is in the same direction as in Figure 4.1.

The way the echo starts to grow from zero is the opposite of the initial FID, and following the echo peak the dephasing once again is similar to the initial relaxation following the 90° pulse. The relaxation process is the same, so we can conclude in general that the shape of the spin echo is like two back-to-back FIDs.

4.4.4 T_2 and T_2^*

In practice the full height of the FID is not recovered in a spin echo. There are two reasons for this. Firstly, we neglected the effects of motion in the field inhomogeneity. We shall see in Section 4.5.4 that this is usually permissible. Secondly, the spin-echo technique of this section does not recover magnetisation lost through the local fields of dipolar origin. Since the 180° pulse reverses the spins it reverses the sources of dipolar

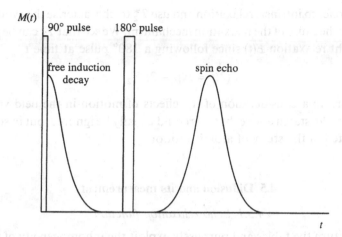

Figure 4.8 Formation of a spin echo.

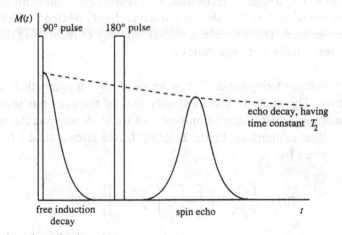

Figure 4.9 Relaxation of spin echoes.

fields as well. Thus in the products of **μ** and **b** in the equations of motion, as both are reversed, the result is unchanged. The relaxation proceeds as before, unaffected by the 180° pulse. This is a considerable advantage. In this way the spin-echo technique can distinguish between the dipolar and 'bad magnet' origins of local fields. Because of this there is a small change in terminology that we shall introduce. The resultant spin-echo relaxation is shown in Figure 4.9.

The dipolar relaxation is an intrinsic property of the system being studied. Conversely the bad magnet relaxation is usually simply an instrumentational problem, which can often be overcome through the use of spin echoes. We henceforth reserve the symbol T_2 to refer to intrinsic relaxation and use T_2^* to characterise the evolution of the FID whatever the cause of the relaxation mechanism. We see that T_2 can be found from the echo height relaxation $E(t)$ since following a 180° pulse at time t

$$E(2t) = \exp(-2t/T_2). \tag{4.51}$$

We now turn to a consideration of the effects of motion in the field variations of external origin. We stated above that this could usually be ignored, but in some cases it can be exploited in the study of atomic motion.

4.5 Diffusion and its measurement

4.5.1 Echo relaxation function

We shall now turn the tables and purposely exploit the inhomogeneity of the applied magnetic field. We shall see that when the motion of the spins is sufficiently fast the full height of the echo will no longer be recovered. This relaxation of the echo height is the subject of the present section and we shall see that it is closely related to the diffusion of the spins in the specimen. This provides a convenient way of measuring the diffusion coefficient in gases, liquids and even solids.

We are thus here considering the deliberate application of a magnetic field gradient over our specimen. The analysis follows closely that of the previous section. Now, however, we incorporate the effects of motion and we shall see that the recovery of magnetisation is then incomplete. In the rotating frame after a time t the complex magnetisation is given by

$$M(t) = \frac{1}{N} \sum_{j=1}^{N} \exp\left\{-\left[i\gamma \int_0^t b_j(\tau)d\tau\right]\right\} M(0). \tag{4.52}$$

Let us now consider a 180° pulse about the x axis at time t; we take the complex conjugate of M. Immediately following the pulse we have

$$M(t) = \frac{1}{N} \sum_{j=1}^{N} \exp\left\{+\left[i\gamma \int_0^t b_j(\tau)d\tau\right]\right\} M^*(0)$$

and after a further time t the echo will appear with a height (normalised to $E(0) = 1$)

$$E(2t) = \frac{1}{N} \sum_{j=1}^{N} \exp\left\{-\left[i\gamma \int_t^{2t} b_j(\tau)d\tau\right]\right\} \exp\left\{+\left[i\gamma \int_0^t b_j(\tau)d\tau\right]\right\}. \tag{4.53}$$

We see that now because of the variation of the $b_j(t)$ there is no longer full cancellation of the 'positive time' and the 'negative time' components. The above expression can be simplified:

$$E(2t) = \frac{1}{N} \sum_{j=1}^{N} \exp\left\{ -\left[i\gamma \int_{t}^{2t} b_j(\tau)d\tau - \int_{0}^{t} b_j(\tau)d\tau \right] \right\}$$

and using angle brackets to indicate the average over particles we can write $E(2t)$ as

$$E(2t) = \left\langle \exp\left\{ -\left[i\gamma \int_{t}^{2t} b_j(\tau)d\tau - \int_{0}^{t} b_j(\tau)d\tau \right] \right\} \right\rangle. \tag{4.54}$$

We observe that this is very similar to Equation (4.23), where now we have a resultant phase angle $\phi_j(t)$ given by

$$\phi_j(2t) = -\gamma \left[\int_{t}^{2t} b_j(\tau)d\tau - \int_{0}^{t} b_j(\tau)d\tau \right]. \tag{4.55}$$

So we can use the results of Section 4.2.2 where we saw that, assuming ϕ to be a Gaussian random variable, the relaxation function could be expressed as

$$\exp(-\langle \phi^2 \rangle/2),$$

i.e.

$$E(2t) = \exp\left\{ -\frac{\gamma^2}{2} \left\langle \left[\int_{t}^{2t} b_j(\tau)d\tau - \int_{0}^{t} b_j(\tau)d\tau \right]^2 \right\rangle \right\}$$

or

$$E(t) = \exp\left\{ -\frac{\gamma^2}{2} \left\langle \left[\int_{t/2}^{t} b_j(\tau)d\tau - \int_{0}^{t/2} b_j(\tau)d\tau \right]^2 \right\rangle \right\}. \tag{4.56}$$

This is best evaluated using the identity

$$(a - b)^2/2 = a^2 + b^2 - (a + b)^2/2.$$

Following the discussion of Section 4.2.4, and introducing the local field autocorrelation function $G(t)$ in a similar way, we obtain the expression for the echo relaxation function

$$E(t) = \exp\left\{-\int_0^t (t-\tau)[G(\tau/2) - G(\tau)]\mathrm{d}\tau\right\}. \tag{4.57}$$

It must be emphasised that this expression applies for spins relaxing through motion in an externally impressed field inhomogeneity. It does not apply to internuclear dipole fields for which both spins and field are reversed by a 180° pulse.

4.5.2 Diffusion autocorrelation function

We shall now see how Equation (4.57) may be applied to the problem of studying atomic or molecular diffusion. We shall find the benefits of actually applying an inhomogeneous magnetic field.

In practice one applies a magnetic field gradient G together with the constant $\mathbf{B}_0$ field. The mathematics of a linear gradient turn out to be a little difficult to handle. By imagining a sinusoidal field variation we can make a more direct connection with the phenomenon of diffusion while at the same time we will simplify the calculations. It is then a simple matter to take the appropriate limit at the end of the calculation.

It is clear, from the shape of the sine curve, that a portion which is small compared with the wavelength is approximately linear. Thus when the wavelength is significantly greater than the specimen dimensions this corresponds to a constant field gradient over the sample. Let us take a magnetic field pointing in the z direction, which varies with z in a sinusoidal manner:

$$b(z) = A\sin(qz). \tag{4.58}$$

The magnitude of the field gradient in the z direction is then

$$\frac{\partial b(z)}{\partial z} = Aq\cos(qz),$$

whose mean square value G^2 is given by

$$G^2 = A^2 q^2 / 2. \tag{4.59}$$

In terms of G we can therefore write the field variation as

$$b(z) = \frac{\sqrt{2}G}{q}\sin(qz) \tag{4.60}$$

and the experimentally realisable linear gradient then corresponds to the limit $q \to 0$.

Our general aim in this section is to calculate the form of the echo relaxation function $E(t)$ for the case of particle diffusion in a magnetic field gradient, whereupon we shall see the dependence on the diffusion coefficient. To this end we must evaluate the appropriate autocorrelation function $G(t)$, since from that Equation (4.57) will give $E(t)$. Recall the definition of the correlation function, Equation (4.31):

$$G(t) = \gamma^2 \langle b(0)b(t) \rangle.$$

For the present case the variation of b with position is sinusoidal. The time dependence of the b experienced by the particles is determined by their motion

$$b(t) = b[z(t)] = \frac{\sqrt{2G}}{q} \sin[qz(t)], \tag{4.61}$$

where $z(t)$ is the z coordinate of our representative particle at time t. The correlation function can then be written as

$$G(t) = \frac{2\gamma^2 G^2}{q^2} \langle \sin[qz(0)]\sin[qz(t)] \rangle. \tag{4.62}$$

What is important here is the distance travelled in a time t, $[z(t) - z(0)]$, which we shall denote by $\Delta z(t)$. Let us therefore expand the $\sin[qz(t)]$ as

$$\sin[qz(t)] = \sin[qz(0)]\cos[q\Delta z(t)] + \cos[qz(0)]\sin[q\Delta z(t)],$$

so that

$$G(t) = \frac{2\gamma^2 G^2}{q^2} \{ \langle \sin^2[qz(0)]\cos[q\Delta z(t)] \rangle + \langle \sin[qz(0)]\cos[qz(0)]\sin[q\Delta z(t)] \rangle \}.$$

Now the initial position $z(0)$ is in no way correlated with the distance $\Delta z(t)$ travelled in time t. Therefore the averages of the $\sin^2$ and the $\sin \times \cos$ may be performed independently of the latter factor, $\cos(q\Delta z(t))$. The average of $\sin^2$ is $\frac{1}{2}$ and since $\sin x \cos x = \sin(2x)/2$ the average of that term is zero. Thus we have:

$$G(t) = \frac{\gamma^2 G^2}{q^2} \langle \cos[q\Delta z(t)] \rangle. \tag{4.63}$$

We are already familiar with averages of this kind. It is similar to that performed in the calculation of the FID relaxation function $F(t)$ presented earlier in this chapter. Now if the particle motion is diffusive then the probability distribution for $\Delta z(t)$ is Gaussian, and the average is then the same as that performed in Section 4.2.3. We thus immediately write the average as

$$\langle \cos[q\Delta z(t)] \rangle = \exp\{ - [q^2 \langle \Delta z^2(t) \rangle / 2] \}. \tag{4.64}$$

Using random walk arguments, as demonstrated by Reif (1965), for $\langle \Delta z^2(t) \rangle$, or by direct solution of the diffusion equation, one finds

$$\langle \Delta z^2(t) \rangle = 2Dt, \tag{4.65}$$

where D is the diffusion coefficient in the z direction. So the correlation function in this case is finally

$$G(t) = \frac{\gamma^2 G^2}{q^2} \exp(-q^2 D t), \qquad (4.66)$$

an exponentially relaxing function.

4.5.3 Cubic echo decay

Finally, we are in a position to find the echo relaxation function $E(t)$ through the substitution of the exponential expression for $G(t)$, Equation (4.66), into Equation (4.57). We obtain:

$$E(t) = \exp\left\{ -\int_0^t (t - \tau) \frac{\gamma^2 G^2}{q^2} [\exp(-q^2 D\tau/2) - \exp(-q^2 D\tau)] d\tau \right\}.$$

Upon integration one finds

$$E(t) = \exp\left\{ -\frac{\gamma^2 G^2}{q^6 D^2} [4\exp(-q^2 D t/2) - \exp(-q^2 D t) - 3 + q^2 D t] \right\}. \qquad (4.67)$$

For the case of the linear gradient we then take the limit $q \to 0$ whereupon we obtain

$$E(t) = \exp(-\gamma^2 G^2 D t^3/12). \qquad (4.68)$$

This is our required result, showing how the echo height varies with the diffusion coefficient. Note the cubic variation with time (Torrey, 1952). This is a somewhat unusual form for a relaxation function, and whenever it is observed in NMR measurements it is regarded as an indication of diffusive motion in a field gradient. A more detailed discussion of diffusion measurement is given by Kärger *et al.* (1988).

If we know the value of an imposed field gradient then this method allows us to obtain the value for the diffusion coefficient. Alternatively, given a fluid of known diffusion coefficient we can calibrate the field gradient. Of course, one has an independent check on the magnitude of the gradient from the shape and duration of the FID, as considered in Problem 4.2.

4.5.4 Rapid diffusion

It is instructive to examine the $q \to 0$ limit in a little more detail since it is this which results in the characteristic cubic decay of the echo. The limit may be effected through an expansion of the exponentials in Equation (4.67); it is the arguments of the exponentials which must be small quantities. We are therefore requiring that

$$q^2 Dt \ll 1 \qquad (4.69)$$

and whenever this condition is satisfied the echo relaxation will be observed to be cubic as in Equation (4.68).

Physically this condition is that in a typical measurement time t a representative particle must diffuse a distance much less than the period of the sinusoidal field variation. In other words, for the duration of the measurement the field any spin sees is essentially a uniform gradient. There could therefore be many periods of the sinusoid over the specimen, and yet the relaxation still be cubic. Note that the dimensions of the specimen have not yet entered into the consideration.

If, however, the diffusion coefficient is very large, such as in a gas, then a particle could possibly traverse the sample chamber many times during its dephasing. Here the finite size of the specimen is of paramount importance and this case must be paid special attention. It is possible to model this circumstance, a finite sized specimen in a uniform gradient, by a sinusoidal field variation whose period is some four times the specimen dimension a (in the z direction):

$$q \approx \pi/2a.$$

In this case the condition for cubic decay, Equation (4.69), may well not be satisfied. Then Equation (4.67) will give an indication of the form of the echo relaxation. In particular for rapid diffusion, when

$$q^2 Dt \gg 1$$

(the opposite of the cubic decay condition), Equation (4.67) gives

$$E(t) \approx \exp\left(-\frac{\gamma^2 G^2}{q^4 D} t \right), \qquad (4.70)$$

i.e. exponential relaxation with a time constant T_2 given by

$$\frac{1}{T_2} = \frac{\gamma^2 G^2}{q^4 D}. \qquad (4.71)$$

This is, in fact, a representation of the familiar motional averaging expression for T_2 as given in Equation (4.41):

$$\frac{1}{T_2} = \langle \omega_{loc}^2 \rangle \tau_c$$

since the mean square local frequency variation and the time to traverse such a fluctuation are essentially

$$\left.\begin{aligned} \langle \omega_{loc}^2 \rangle &\approx \frac{\gamma^2 G^2}{q^2}, \\[2mm] \tau_c &\approx \frac{1}{q^2 D}. \end{aligned}\right\} \qquad (4.72)$$

This result, that for sufficiently fast diffusion the relaxation is exponential, is to be expected since if the spins traverse the extent of the sample cell many times during a measurement then the fields they see become completely random and the particle motion simply averages away these field variations.

We learn from this that the spin-echo technique does not always recover fully the transverse magnetisation lost because of an inhomogeneous magnetic field. It will do so when the particles move only a fraction of the specimen size.

4.6 Measuring relaxation times

In the following section we shall discuss some of the ways that the relaxation times of NMR may be measured using pulse techniques. In many respects this subject matter should be part of the previous chapter, which was devoted to detection methods. However, it could not be covered until the discussion of spin echoes in the present chapter. The validity of the methods to be described for measuring relaxation times is in no way reliant on 'classical' methodology – in that respect they do not form a cognate part of the present chapter. Nevertheless these are all based on the use of spin echoes, which involve dephasing and rephasing of magnetic moments precessing in inhomogeneous magnetic fields, and such phenomena are described quite adequately in classical terms. With that justification the present section on further experimental matters is placed at this point in the book.

4.6.1 Measurement of T_2^*

In Section 4.4.4 we explained that T_2^* was the characteristic time associated with the decay of the FID, usually due to the inhomogeneity of the B_0 field. The simplest way, therefore, of measuring this relaxation time is to apply a 90° pulse to create transverse magnetisation and to record its subsequent free precession, shown in Figure 4.10. A curve can be fitted through the data, from which T_2^* may be obtained.

Alternatively, and as explained in Section 4.4.3, a spin echo has the form of two back-to-back FIDs. Therefore it is possible to find T_2^* from an echo following a 90°–180°–*echo* sequence. This is more convenient if the signal is being captured electronically since the reception process is occurring far away from the high voltage RF pulses which can have the effect of temporarily overloading the sensitive measuring equipment.

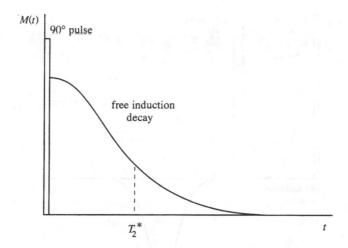

Figure 4.10 Measurement of T_2^*.

It should be pointed out that T_2^* is determined primarily by the magnet inhomogene-ity (one could, for instance, be characterising the field profile prior to a diffusion measurement). Generally there is no reason for the field inhomogeneity to have a Lorentzian distribution, so one is unlikely to have a precisely exponential relaxation. Thus T_2^* will not be well defined, and this time constant is usually used simply as a rough measure of the magnetisation lifetime.

4.6.2 Measurement of T_2

From the discussions of the previous sections of this chapter we know that T_2 is the time constant associated with the *intrinsic* decay of transverse magnetisation, irrespec-tive of the homogeneity of the magnetic field. We have seen that the decay due to the field inhomogeneity can be reversed by a 180° pulse. The echo height in a 90°–t–180°–t–echo sequence will have had a time $2t$ for relaxation to occur so that if the relaxation is exponential the echo height will have decayed to $\exp(-2t/T_2)$ of its initial magnitude. Figure 4.9 shows this.

In measurements such as these the height of the echo gives a single data point. In order to establish the value of T_2 it is necessary to capture a number of data points by repeating the 90°–t–180°–t–echo sequence with different values of the time delay t. In doing this it will be necessary to wait between each sequence for a time of the order of a few T_1 for the equilibrium M_z to be established. There is, however a more economical way of making such measurements.

Recall the fundamental spin echo result expressed in Equation (4.47). This treated the echo which was formed in a 90°–t–180°–t–echo sequence. We deliberately chose the 180° refocusing pulse to be along the x axis (in the rotating frame) so that the effect of

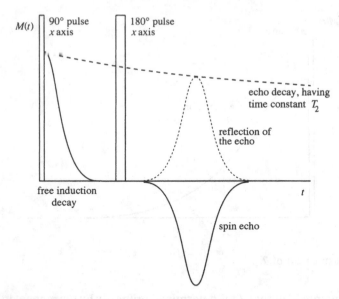

Figure 4.11 Spin echo when the 90° pulse and the 180° pulse are colinear.

this pulse on the complex transverse magnetisation $M = M_x + iM_y$ was simply one of the complex conjugation. We saw that if $M(0)$ was the initial magnetisation, then following a 90°–t–180°–t–*echo* sequence, the height of the echo was $M^*(0)$, reduced by the effects of T_2 relaxation. This meant that if the initiating 90° pulse was applied along the y axis then since $M(0)$ is real (the x direction), the echo would also be real. In other words the echo would also be pointing in the x direction. If, on the other hand, the initiating 90° pulse were applied along the x axis then $M(0)$ would be pointing along the y direction; it would be pure imaginary. In this case the echo, proportional to the complex conjugate of the magnetisation, would be pointing in the direction *opposite* to the initial magnetisation: in the $-y$ direction. This is shown in Figure 4.11. In fact this was how the original spin echoes were observed since the early pulsed NMR spectrometer systems did not have the facility for applying pulses along different directions.

The main purpose of this section is to discuss the effect of applying multiple 180° pulses. It should be apparent, particularly since the latter half of an echo is just like an FID, that it should be possible to recover the magnetisation of an echo again and again, by the repeated application of 180° pulses. This scheme was originally proposed by Carr and Purcell in 1954, but at that early stage, they were working with colinear pulses. Therefore since the signal was in the y direction then each subsequent echo, being the complex conjugate of the previous, pointed in the opposite direction. An echo sequence produced by the Carr–Purcell technique is shown in Figure 4.12.

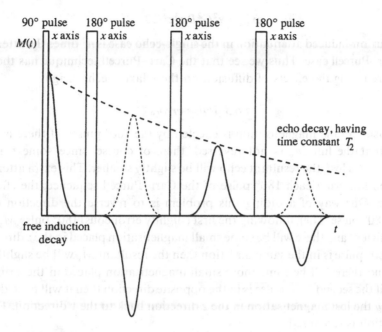

Figure 4.12 Spin echoes produced by the Carr–Purcell method.

It can be appreciated that the Carr–Purcell method results in a considerable saving of time, as compared with the use of separate $90°-t-180°-t-echo$ sequences for different t. There is a further advantage when one considers the effect of motion of the spins. We saw in Section 4.5 that diffusive motion of the spins in an inhomogeneous field (the same inhomogeneity which causes the T_2^* relaxation) contributes an $\exp(-\gamma^2 G^2 D t^3/12)$ term to the relaxation of the echo height. We compare a single echo, produced by a $90°-t/2-180°-t/2-echo$ sequence, occurring at time t, with the nth echo of a Carr–Purcell sequence, also occurring at time t. In other words the time interval in the Carr–Purcell sequence $\tau/2$ is $t/2n$. In both cases the T_2 relaxation will have caused a reduction of the echo height over the time duration t. However, considering the effects of diffusion, we know for the single echo sequence that the extra reduction of the echo height will be

$$\exp(-\gamma^2 G^2 D t^3/12),$$

but in the multiple pulse case the relaxation is reinitiated following each 180° pulse. In this case the extra reduction in the echo height will be

$$\exp(-\gamma^2 G^2 D n \tau^3/12)$$

and since $\tau = t/n$ it follows that the attenuation in the single echo case may be written as

$$\exp(-\gamma^2 G^2 D n^3 \tau^3 / 12).$$

So the diffusion-induced attenuation in the single-echo case is n^2 times the attenuation in the Carr–Purcell case. Thus we see that the Carr–Purcell technique has the added benefit of reducing the effects of diffusion on the relaxing echo height.

4.6.3 Pulse errors

The refocusing of the magnetisation is effected by the 180° pulses. If there is a small error, so that the full M_y is not reversed, then, of course, since some transverse magnetisation is lost, the resultant echo will be slightly smaller. This extra attenuation will happen following each 180° pulse of the Carr–Purcell sequence; the effects are cumulative. One way of avoiding this problem is to reverse the direction of each *alternate* 180° pulse. Then following the first (slightly erroneous) 180° pulse M_y will be slightly deficient and there will be some small magnetisation placed in the z direction. If the next 180° pulse is in the same direction then the resultant M_y will be slightly more deficient and there will be some more small magnetisation placed in the z direction. However, if the second 180° pulse is in the opposite direction then it will have the effect of *returning* the lost magnetisation in the z direction back to the y direction. Thus the magnetisation is recovered.

A pictorial demonstration of this result is given in the Carr and Purcell (1954) paper. To complement that we give here an analytic treatment. If we denote the magnetisation vector in the rotating frame by the column vector

$$\mathbf{M} = \begin{pmatrix} M_x \\ M_y \\ M_z \end{pmatrix},$$

then the effect of a 180° pulse about the x axis, which is 'deficient' by an angle α, is given by the matrix

$$\mathbf{R}(180) = \begin{pmatrix} 1 & 0 & 0 \\ 0 & -\cos\alpha & -\sin\alpha \\ 0 & \sin\alpha & -\cos\alpha \end{pmatrix}.$$

The matrix representing the corresponding rotation about the $-x$ axis, $\mathbf{R}(-180)$ is obtained by changing the sign of α. Let us consider a spin whose local field differs from the Larmor frequency by Δ. Its time evolution in the rotating frame, giving its dephasing, will be given by the evolution operator

$$\mathbf{U}(t) = \begin{pmatrix} \cos(\Delta t) & \sin(\Delta t) & 0 \\ -\sin(\Delta t) & \cos(\Delta t) & 0 \\ 0 & 0 & 1 \end{pmatrix}.$$

The magnetisation of the first echo, at time $2t$, is given by

$$\mathbf{E}_1 = \mathbf{U}(t)\mathbf{R}(180)\mathbf{U}(t)\mathbf{M}(0).$$

In the Carr–Purcell sequence the magnetisation starts in the y direction as it is created by a 90° pulse about the x axis. Then $\mathbf{M}(0)$ only has a component in the y direction, $M_y(0)$. Then the components of the first echo are, to leading order in the error angle α,

$$\mathbf{E}_1 = \begin{pmatrix} \dfrac{\alpha^2}{4}\sin(2\Delta t) \\[2mm] -1 + \dfrac{\alpha^2}{4}[1 + \cos(2\Delta t)] \\[2mm] \alpha\cos(\Delta t) \end{pmatrix} M_y(0) \quad \text{when} \quad \mathbf{M}(0) = \begin{pmatrix} 0 \\ M_y(0) \\ 0 \end{pmatrix}.$$

The $\sin(2\Delta)$ term goes to zero when averaged over all spins. Thus the echo points in the $-y$ direction, and it is attenuated by a term in α^2. We see that there is a small magnetisation in the z direction as well.

Now we consider the application of the 180° pulse in the opposite direction, to produce the second echo. The magnetisation at the point of the second echo is

$$\mathbf{E}_2 = \mathbf{U}(t)\mathbf{R}(-180)\mathbf{U}(t)\mathbf{E}_1.$$

The components of $\mathbf{E}_2$ are, to leading order in α,

$$\mathbf{E}_2 = \begin{pmatrix} \dfrac{\alpha^2}{2}\sin(2\Delta t) \\[2mm] 1 - \dfrac{\alpha^4}{16}[1 - \cos(4\Delta t)] \\[2mm] \alpha^3\cos(\Delta t)\sin^2(\Delta t) \end{pmatrix} M_y(0) \quad \text{when} \quad \mathbf{M}(0) = \begin{pmatrix} 0 \\ M_y(0) \\ 0 \end{pmatrix}.$$

We observe here again that the x component will average to zero. So the second echo points in the $+y$ direction, as expected. The important point, however, is that the reduction in height due to the error in pulse length is proportional to the *fourth* power of α. If this calculation is done with the second pulse applied in the *same* direction then the magnitude of the echo is reduced by a term proportional to the *square* of α, as in the first echo.

An alternative procedure, to compensate for errors in pulse lengths, was suggested by Meiboom and Gill (1958). They were the ones to propose that the 180° pulse should be applied perpendicular to the initiating 90° pulse. In this case the initial magnetisation is along the x direction and the echoes all point in this same direction. The (slightly erroneous) 180° pulse is now applied along the direction of the magnetisation. Its effect is thus to reverse the sign of just the dephased part of the magnetisation. When the pulse is deficient the dephased part of the magnetisation is not rotated quite far enough. Just

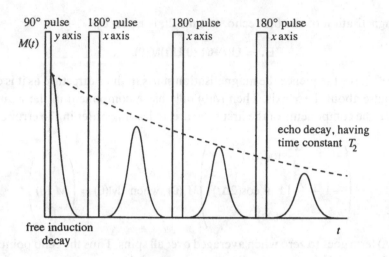

Figure 4.13 Carr–Purcell–Meiboom–Gill sequence.

as in the original arguments on spin echoes, this dephased part of the magnetisation subsequently does not need to be rotated quite so far to return it to the correct position. Thus in this case there is no cumulative effect when the 180° pulses are all in the same direction, perpendicular to the direction of the 90° pulse. This is known as the Carr–Purcell–Meiboom–Gill sequence. An example is shown in Figure 4.13.

A pictorial demonstration of this result is given in the Meiboom and Gill (1958) paper. To complement this we give here an analytic treatment analogous to that for the Carr–Purcell case above. As before, the magnetisation of the first echo is given by

$$\mathbf{E}_1 = \mathbf{U}(t)\mathbf{R}(180)\mathbf{U}(t)\mathbf{M}(0).$$

Now in the Meiboom–Gill modified sequence the magnetisation starts in the x direction as it is created by a 90° pulse about the y axis and $\mathbf{M}(0)$ only has a component in the x direction, $M_x(0)$. Then the components of the first echo are, to leading order in α,

$$\mathbf{E}_1 = \begin{pmatrix} 1 - \dfrac{\alpha^2}{4}[1 - \cos(2\Delta t)] \\ -\dfrac{\alpha^2}{4}\sin(2\Delta t) \\ -\alpha\sin(\Delta t) \end{pmatrix} M_x(0) \text{ when } \mathbf{M}(0) = \begin{pmatrix} M_x(0) \\ 0 \\ 0 \end{pmatrix}.$$

The y and z terms go to zero when averaged over all spins. Thus the echo points in the $+x$ direction, and it is attenuated by a term in α^2.

Now we consider the application of the second 180° pulse, which is in the same direction, to produce the second echo. The magnetisation at the point of the second echo is

$$\mathbf{E}_2 = \mathbf{U}(t)\mathbf{R}(180)\mathbf{U}(t)\mathbf{E}_1.$$

The components of $\mathbf{E}_2$ are, to leading order in α,

$$\mathbf{E}_2 = \begin{pmatrix} 1 - \dfrac{\alpha^4}{16}[1 - \cos(4\Delta t)] \\ \dfrac{\alpha^2}{2}\sin(2\Delta t) \\ -\alpha^3\sin(\Delta t) \end{pmatrix} M_x(0) \text{ when } \mathbf{M}(0) = \begin{pmatrix} M_x(0) \\ 0 \\ 0 \end{pmatrix}.$$

We observe here that the y and z components will average to zero. So the second echo points in the $+x$ direction, as expected. The important point, however, is that the reduction in height due to the error in pulse length is proportional to the *fourth* power of α. If this calculation is done with the second pulse applied in the *opposite* direction then the magnitude of the echo is reduced by a term proportional to the *square* of α, as in the first echo.

The above analytic treatments show how the pulse reversal in the Carr–Purcell case and the different axes of the initiating 90° pulse in the Meiboom–Gill case, both have the effect of removing the main effects of the pulse error so that first remaining error term in the echo is of the order of α^4.

We should emphasise that when we specify the axis of a pulse, this is in the rotating frame. Thus when the axis of rotation is shifted through 90° this is implemented by applying a 90° phase shift to the RF sinusoid of the pulse.

4.6.4 Measurement of T_1

In studying T_1 it is the longitudinal component of the magnetisation which must be observed. This can be disturbed from its equilibrium state with an initial pulse and then the magnetisation will return to equilibrium. Ordinarily it is not possible to observe M_z directly, and it must be rotated into the transverse plane with a 90° pulse in order to be detected. Then one can use a 180° pulse to create an echo proportional to the magnetisation for measurement. As explained in Section 4.6.1 it is convenient to measure an echo away from the high voltages of the RF pulses.

This scheme is shown in Figure 4.14 and the echo height is found as follows. The initiation pulse acts on the equilibrium magnetisation in the z direction, M_0, and results in a non-equilibrium z component of magnetisation which we denote by $M(0)$. There may be some transverse magnetisation as well, which will decay with a time constant T_2^* and which will be ignored. If the initiation pulse has the duration for a 180° rotation, then M_0 will have been inverted and there will be no transverse magnetisation. On the

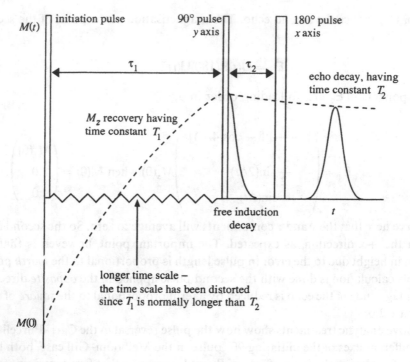

Figure 4.14 Measurement of T_1 by spin echo method.

other hand, if the sequence is initiated by a 90° pulse, then $M(0)$ will be zero and there will be an observable free induction decay. For the present we will not specify the size of the initiation pulse; we simply start with a non-equilibrium $M(0)$.

The z component of magnetisation will return towards equilibrium according to Equation (2.10). This will happen for the time τ_1 so that it has the magnitude

$$M_z = M_0 - [M_0 - M(0)]\exp(-\tau_1/T_1).$$

At this stage the 90° pulse will rotate this M_z into the transverse plane. There will be an FID, so the magnetisation *could* be measured by extrapolating to the initial height of the FID. Alternatively, the measurement can be made with a spin echo. At a time τ_2 after the 90° pulse a 180° pulse is applied and after a further time τ_2 the echo will appear. During the time interval of $2\tau_2$ the transverse magnetisation will have suffered some T_2 relaxation, so the height of the echo E will be

$$E = \{M_0 - [M_0 - M(0)]\exp(-\tau_1/T_1)\}\exp(-2\tau_2/T_2). \qquad (4.73)$$

This equation shows that for the measurement of T_1 the time interval is kept fixed at a time short compared with T_2 but long compared with T_2^* so that there is negligible relaxation but the transverse magnetisation has decayed away before the 180° pulse is applied. The time interval τ_1 is varied so that E is recorded as a function of τ_1. Then T_1 is found by fitting the data to Equation (4.73).

If the initiation pulse is a good and homogeneous 180° pulse so that it accurately reverses the equilibrium magnetisation then we have $M(0) = -M_0$. Then, if it is known that the T_1 relaxation is truly exponential, the echo signal will pass through zero for the value of τ_1 which makes E of Equation (4.73) equal to zero, i.e.

$$0 = M_0[1 - 2\exp(-\tau_1/T_1)],$$

or

$$T_1 = \tau_1 \ln 2.$$

Thus it is possible to estimate T_1 by varying τ_1 until the echo vanishes.

The general case, in which a curve is fitted through a set of captured data, gives a more reliable result. In such cases a 180° initiation pulse gives a greater signal range, but a 90° initiation pulse (if exact) will set $M(0)$ to zero, thus simplifying the equation to which the data must be fitted.

4.6.5 Techniques when T_1 is long

If one is using a 180°–τ_1–90°–τ_2–180°–τ_2–*echo* sequence to study T_1 then it is necessary to wait a time of the order of $5T_1$ for the z component of magnetisation to recover to within 1% of its equilibrium value before starting the next sequence. If T_1 is very long (it can be minutes, or even hours in a gas), then this pulse sequence becomes impracticable. In this case it is preferable to initiate the sequence with a 90° pulse, starting from zero magnetisation, since then the longitudinal magnetisation before the initiation pulse is unimportant. This means that each pulse sequence can start immediately following the previous echo; it is no longer necessary to wait $5T_1$.

Frequently the 90°–τ_1–90°–τ_2–180°–τ_2–*echo* sequence can be prone to systematic error when repeated in rapid succession because of the inhomogeneity of the B_1 field of the initiation pulse. In practice the magnetisation is fully destroyed only in a limited region of the specimen. Elsewhere the effective pulse angle will be greater or less than 90° and there will be some resultant M_z. In these regions the magnetisation will depend on the state preceding the initiation pulse. As a result the measured T_1 value will be less than the true relaxation time.

The fundamental requirement of this procedure is to destroy all longitudinal magnetisation at the start of the relaxation process. This may be achieved through the use of a number of 90° pulses applied in succession. These pulses are applied rapidly compared with T_1 so there is negligible 'blurring' of the time origin, but a time greater than T_2^* separates the pulses so that there is no transverse magnetisation present to be

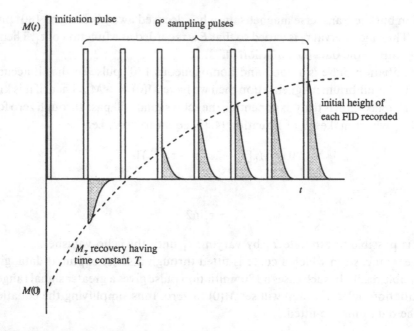

Figure 4.15 Measurement of T_1 by the small-tip sampling method.

rotated back into the longitudinal direction. In this way all longitudinal magnetisation is destroyed so that it can relax to equilibrium from a well-defined zero value.

In all methods (discussed in this book) for studying T_1 it is necessary to tip the longitudinal magnetisation into the transverse plane in order to measure it, and the application of the 90° pulse destroys the longitudinal magnetisation in the process. However, if there is sufficient magnetisation available to provide a large signal then it is possible to *sample* M_z without destroying it, through the application of a small-angle pulse. Then T_1 may be found in a single sequence, in a time of the order T_1. A pulse of angle θ will give a transverse signal of $\sin\theta$ times the value of M_z, and at the same time it reduces M_z to $\cos\theta$ times its previous value. So the idea is to choose a tip angle θ which is large enough to provide a measurable signal, but small enough so that negligible reduction of M_z occurs.

A practical implementation of this procedure is shown in Figure 4.15. An initiation pulse of perhaps 180° is followed by a number of equally-spaced sampling pulses of angle θ and the amplitude of the FID signal is recorded following each pulse. These amplitudes then relax over the time scale of T_1. Following n sampling pulses the longitudinal magnetisation will have suffered a reduction, due to the pulses, of $(\cos\theta)^n$. Thus application of 10 sampling pulses of 2° tip angle will introduce an error of 0.6% while with pulses of 4° tip angle the error increases to 2.4%.

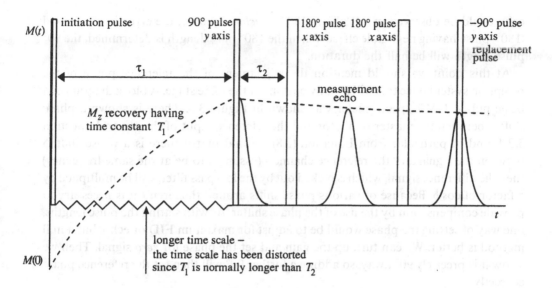

Figure 4.16 Measurement of T_1 by the measurement–replacement method.

A third, and rather ingenious, solution to the problem of measuring very long T_1s is shown in Figure 4.16. The sequence starts just as in the straightforward method of the previous section. The magnetisation proceeds to the formation of the echo, whose height is recorded. But after this a further 180° pulse is applied, to give a second echo, and precisely at the instant of the second echo, a $-90°$ pulse is applied, whose effect is to return the transverse magnetisation to the z direction. Then a further period of longitudinal relaxation can proceed, after which the magnetisation is moved back into the transverse plane, measured and replaced again. This scheme can be continued until the longitudinal magnetisation has decayed. Clearly there is a requirement for T_2 to be long so that negligible magnetisation is lost while the magnetisation is in the transverse plane.

4.6.6 Accurate adjustment of the spectrometer

We have mentioned the importance of ensuring the lengths of the RF pulses are set correctly. It would appear that the best way of setting the 90° pulse would be to adjust it so as to maximise the size of the FID. If the amplitude of the RF pulse is guaranteed to be constant for the duration of the pulse then things can be set from the length of the 180° pulse. And this is much easier to set precisely since there should be *no* free induction decay following the 180° pulse. Thus the pulse length can be set to make the FID very small, then the spectrometer gain can be increased significantly and the remnant signal set to zero by finely adjusting the duration of the pulse. Alternatively the echo could be observed in a '180°'–*t*–'180°'–*t*–*echo* sequence, the error in the second '180°' pulse having negligible effect. When the 180° pulse length is determined, the 90° pulse length will be half the duration.

At this point we should mention the adjustment of the reference phase of the reception system since this will usually be done at the same stage. A block diagram of a basic pulsed NMR spectrometer was shown in Figure 3.3. There is shown a phase shifter between the master oscillator and the reference input of the mixer. In Section 3.2.4, and in particular from Equation (3.8) we find that if there is a phase shift ϕ between the signal and the reference channels (assumed to be at the same frequency) then the difference signal, which is picked out by the low pass filter, will be multiplied by a factor of $\cos\phi$. Because of various phase shifts around the circuit it is necessary to provide compensation by the use of the phase shifter. As with setting the pulse lengths, one way of setting the phase would be to adjust for maximum FID or echo, but a null method is better. We can turn up the gain and set the phase for *zero* signal. Then we know it is precisely 90° away, so adding a further 90° will then set the reference phase correctly.

It is sometimes observed that the quadrature signal cannot be set exactly to zero in this way. From the discussion at the end of Section 2.6.3 we see that the quadrature signal will be proportional to the sine transform of the lineshape function, and this will only be zero when the shape function is symmetric about ω_0. The non-vanishing of the quadrature signal is thus due to an asymmetry in the distribution of the B_0 field over the specimen. This can have its origin in the magnetic properties and asymmetric shape of the specimen holder. Further discussion of this matter, which can have significant consequences in metrological applications, is contained in the paper by the author (Cowan, 1996).

4.6.7 *Data processing and curve fitting*

In times gone by, data points were recorded by hand and plotted on graph paper. For an exponential relaxation $\exp(-t/T_2)$ the points may be plotted on semilog paper, or the logarithm of the signal may be plotted on linear paper. In both cases the data points are expected to lie on a straight line whose slope gives the relaxation time. There is the temptation, when calculating the relaxation times by computer, to use a linear regression procedure on the logarithm of the signal heights, but this is *wrong*. Recalling a manual fit using graph paper, when the error bars are placed on the data it is seen that they increase in size as the signal heights become smaller; this is because we are taking the logarithm. Now the common least-squares procedure for fitting a straight line gives the best estimate of the curve parameters, giving *equal weight* to all data points. This is clearly incorrect.

When fitting on paper 'by eye' one automatically accounts for this by taking less account of the later data. But the computer does not do this when using the linear regression algorithm. The correct procedure is to do a full non-linear fit to the curve. However, non-linear curve fitting is an iterative procedure and it can take longer to perform. Also, non-linear fitting algorithms usually require initial 'guess values' for the fit parameters. This can be done using the incorrect linear fit to the transformed data since the linear fit is done by formula directly, without iteration. The reader is referred to the book by Press, Teukolsky, Vetterling and Flannery (1992) for an extensive treatment of such numerical procedures.

Fitting T_1 data by hand can present further problems. The magnetisation does not relax to zero, so the data cannot be plotted directly. Instead, it is necessary to determine the equilibrium M_0 and subtract this from each data value. In practice, M_0 is determined by making the magnetisation measurement without having applied the initiation pulse; τ_1 is effectively infinite. However, whatever error there is in this measurement is then added to each data point in the plot, thus the importance of an accurate determination of the equilibrium magnetisation.

This is not a problem when a non-linear fitting procedure is performed by computer. However, in this case *three* fit parameters must be determined: T_1, $M(0)$ and M_0, rather than the two parameters T_2 and $M(0)$ in the transverse case. Unfortunately both the time for convergence and the likely error on the answers are rapidly increasing functions of the number of fit parameters.

5

Quantum treatment of relaxation

5.1 Introduction

5.1.1 Relaxation and resonance

In the simplest pulsed NMR experiments a system is studied as it returns to equilibrium following some initial disturbance. In the CW case it is the Fourier transform of the relaxing (transverse) magnetisation which is observed as resonance absorption. The purpose of this chapter is to treat such relaxation processes from quantum mechanical first principles. It is likely that the novice reader will encounter concepts in this chapter which are conceptually difficult. For this reason a gentle, pedagogically-oriented approach has been adopted which the experienced reader might find somewhat laboured. The intention in this introduction to relaxation is to concentrate on the *physics* of the process in terms of familiar concepts and without the encumbrance of new techniques.

5.1.2 The equilibrium state

The use of the word *state* in connection with equilibrium is rather different from that in quantum mechanics. In discussing the approach to equilibrium we need concepts from both statistical mechanics and quantum mechanics, which sometimes use different terminology. The quantum mechanical state of a system is described by a wavefunction and the Schrödinger equation for the wavefunction gives the time evolution of that state in terms of the system Hamiltonian. The eigenstates of the Hamiltonian are states of fixed energy and these have the property of being time independent. Statistical mechanics tells us that when a system is in thermal equilibrium it may be found in any of its energy eigenstates, with a probability

$$P \propto \exp(-E/kT),$$

the Boltzmann factor, where E is the energy of the eigenstate, k is Boltzmann's constant and T is the absolute temperature. Thus an equilibrium state is actually made up from a large number of quantum states.

5.1.3 The present approach

The additional feature which we are using in this chapter is the averaging over a probability distribution function. It is this which adds the statistical element to quantum mechanics, giving the discipline of quantum statistical mechanics. There is well-established machinery to deal with this, transferring interest from the conventional wavefunctions of quantum mechanics to what is known as the density operator. We have chosen, however, in this chapter to eschew explicit use of the density operator. The quantum mechanics of relaxation will be discussed in conventional elementary terms, taking thermal averages as required. The cognisant will recognise that in essence we are introducing the density operator through the 'back door' but the aim here is to present the physics of relaxation in terms of familiar methods and concepts. A detailed treatment of the density operator follows in Chapter 9, where many of the results of this chapter are discussed and generalised.

A further point should be made concerning the present approach, for readers already familiar with the density operator treatment of relaxation. Our approach uses the Heisenberg picture of dynamical evolution; the time dependence is contained in the quantum mechanical operators representing the (magnetisation) variables of interest and this time dependence is described by the conventional Heisenberg equation of motion. The density operator, to the extent that it does appear, simply provides an operator description of the Boltzmann distribution of states of an equilibrium system. Conversely, the conventional density operator treatment of relaxation uses the Schrödinger picture of dynamical evolution. Here it is the density operator itself which varies with time – the operators representing the variables of interest are constant – but in this there is a paradox. The fundamental equation of motion for the density operator leads to a reversible time evolution. So how can such a density operator exhibit the irreversible relaxation behaviour which we know happens in practice? The answer will be found in Chapter 9. The merit of our present approach is that such difficulties are circumvented; in the Heisenberg picture irreversibility appears naturally upon taking a macroscopic average (magnetisation) over the microscopic variables (magnetic moments). The two approaches are further contrasted in the discussion at the end of Section 9.1.7.

5.2 Expectation values of quantum operators

5.2.1 Expectation values: statement of the problem

We noted in Section 2.3 that if a quantum system is in a state described by a wavefunction $\psi_s(x)$ then the expectation value A of an operator $\mathscr{A}$ is given by an expression of the form

$$A = \langle \mathscr{A} \rangle = \int \psi_s^*(x) \mathscr{A} \psi_s(x) \mathrm{d}x. \tag{5.1}$$

This is simply the *ss* matrix element of the operator $\mathscr{A}$,

$$A = \mathscr{A}_{ss}.$$

To find the value of an observable quantity we see that we must first know which state the system is in; we must know its wavefunction. For simple systems this may be found by solving the Schrödinger equation subject to the relevant boundary conditions. For complex systems this may be prohibitively difficult, particularly if there are 10^{23} or so variables to consider. Under such circumstances an alternative approach is required. It is indeed fortunate that an appropriate method is available.

5.2.2 *Systems in equilibrium*

Let us consider initially a system in thermal equilibrium. For such a system we accept the impossibility of knowing its precise quantum state, but what we do know is the probability of finding the system in any one of its stationary states: its energy eigenstates.

A system in thermal equilibrium is characterised by a temperature T and the probability that the systems is found in the energy state s is given by the Boltzmann factor

$$P_s = \text{const} \times \exp(-E_s/kT),$$

where E_s is the energy eigenvalue of the eigenstate s. The constant is found by normalisation since

$$\sum_{\text{all states}} P_s = 1.$$

Thus the probability P_s is

$$P_s = \frac{\exp(-E_s/kT)}{\sum_s \exp(-E_s/kT)}.$$

We recognise the denominator as the partition function for the system.

The observed value of an operator $\mathscr{A}$ will be the average over the set of possible states s:

$$A = \sum_s \mathscr{A}_{ss} P_s$$

or

$$A = \frac{\sum_s \mathscr{A}_{ss} \exp(-\beta E_s)}{\sum_s \exp(-\beta E_s)}, \qquad (5.2)$$

where β is an inverse measure of the temperature,

$$\beta = 1/kT.$$

Our current aim is to cast this expression for A, Equation (5.2), into a form more amenable to calculation. Now the energy E_s of the state s may be regarded as the ss matrix element of the Hamiltonian or energy operator $\mathscr{H}$ and, of course, $\mathscr{H}$ is diagonal in this representation since the states considered are energy eigenstates.

$$E_s = \mathscr{H}_{ss}.$$

We are particularly interested in the quantity

$$\exp(-\beta E_s) = \exp(-\beta \mathscr{H}_{ss}).$$

However since $\mathscr{H}$ is here diagonal we have the important result that

$$\exp(-\beta E_s) = |\exp(-\beta \mathscr{H})|_{ss},$$

where the quantity on the right hand side is the ss matrix element of the operator $\exp(-\beta \mathscr{H})$. A function of an operator may be understood in terms of its power series expansion, and the above result may be verified directly from the exponential series. Similarly, since $\exp(-\beta \mathscr{H})$ also is diagonal we have the result that

$$\mathscr{A}_{ss} |\exp(-\beta \mathscr{H}|_{ss} = |\mathscr{A}(\exp - \beta \mathscr{H})|_{ss}.$$

We are now in a position to write the expression for the observed value A as

$$A = \frac{\sum_s |\mathscr{A} \exp(-\beta \mathscr{H})|_{ss}}{\sum_s |\exp(-\beta \mathscr{H})|_{ss}}.$$

Both the numerator and denominator are seen to be the sum of the diagonal matrix elements of operators. Now the diagonal sum of a matrix, known as the trace, is a rather special quantity. In particular, its value is independent of the representation used. Thus the matrix does not *have* to be diagonal. So in evaluating the expression for A, which is now written purely in terms of the operators $\mathscr{A}$ and $\mathscr{H}$:

$$A = \mathrm{Tr}\{\mathscr{A} \exp(-\beta \mathscr{H})\} / \mathrm{Tr}\{\exp(-\beta \mathscr{H})\}, \qquad (5.3)$$

any set of states can be used; in order to find the expectation value of an observable we no longer need to find the eigenstates of the system. The symbol Tr represents the trace of its matrix argument.

It is important to appreciate the significance of this result. As already stated it is prohibitively difficult to obtain the stationary states of a macroscopic system. The expression for the thermal averages in terms of traces of operators obviates the need to find them. Any convenient basis set of states may be used; the evaluation of expectation values for systems in thermal equilibrium thus becomes a *relatively* simple problem.

Box 5.1 Curie's Law for systems with interactions

As a simple demonstration of the above result we shall derive Curie's Law, Equation (2.7), for a complex, interacting system. We use Equation (5.3) to find the expectation value of the magnetisation **M**. Let us assume a magnetic field **B** applied in the z direction, so we need to find the expectation value of

$$M_z = \frac{1}{V}\sum \mu_z = \frac{\gamma\hbar}{V}\sum_i I_z^i,$$

which we do with the aid of Equation (5.3):

$$\langle M_z \rangle = \frac{\hbar\gamma}{V}\frac{\sum_i \mathrm{Tr}\{I_z^i \exp(-\beta\mathcal{H})\}}{\mathrm{Tr}\{1\}}.$$

In the spirit of the high temperature/low polarisation limit we expand the exponential to leading order in β; we will see this is the first order term. We write

$$\langle M_z \rangle = \frac{\hbar\gamma}{V}\frac{\sum_i \mathrm{Tr}\{I_z^i(1 - \beta\mathcal{H})\}}{\mathrm{Tr}\{1\}},$$

whereupon we see the first term vanishes since the trace of I_z vanishes. Thus we have

$$\langle M_z \rangle = -\frac{\hbar\gamma\beta}{V}\frac{\sum_i \mathrm{Tr}\{I_z^i\mathcal{H}\}}{\mathrm{Tr}\{1\}},$$

the factor β indicating, immediately, the $1/T$ behaviour of the magnetisation. The Hamiltonian for our system will comprise the Zeeman term, the interaction with the external magnetic field, plus a term presenting the interspin interactions. We shall express this as

$$\mathcal{H} = -\hbar\gamma B_z I_z + \mathcal{H}'$$

so that the expression for the magnetisation is then

$$\langle M_z \rangle = \frac{\hbar^2 \gamma^2 \beta B_z}{V} \frac{\sum_i \text{Tr}\{I_z^i I_z\}}{\text{Tr}\{1\}} - \frac{\hbar \gamma \beta}{V} \frac{\sum_i \text{Tr}\{I_z^i \mathcal{H}'\}}{\text{Tr}\{1\}}.$$

If the interaction Hamiltonian is bilinear in the spin operators then the trace in the last term will be over *three* such operators and therefore zero. In this case we are left with the first term.

The I_z is made up of a sum over all I_z^j, but only the $i = j$ term will be non-zero. Since all spins are equivalent, the sum is the same as N times the trace over a single spin, so that

$$\langle M_z \rangle = N_v \hbar^2 \gamma^2 \beta B_z \frac{\text{Tr}\{(I_z^i)^2\}}{\text{Tr}\{1\}},$$

where N_v is the number of spins per unit volume.

The numerator trace is easily evaluated for spin $\frac{1}{2}$. Since I_z is diagonal, with elements $+\frac{1}{2}$ and $-\frac{1}{2}$, then I_z^2 is also diagonal with elements $+\frac{1}{4}$ and $+\frac{1}{4}$, the trace of which is $\frac{1}{2}$. The denominator trace is trivial; we have a 2×2 unit matrix, so its trace must be 2. The result for spin $\frac{1}{2}$ is then

$$\langle M_z \rangle = \frac{N_v \hbar^2 \gamma^2}{4kT} B_z,$$

which we see is identical to the result of Chapter 2, Equation (2.7). In the generalisation to spin of arbitrary magnitude I the traces are given by $\text{Tr}\{I_z^2\} = I(I + 1)(2I + 1)/3$ and $\text{Tr}\{1\} = I(I + 1)$, as discussed in Problem 5.2. The result is then that in the *high temperature/low polarisation limit*

$$\langle M_z \rangle = \frac{N_v \hbar^2 \gamma^2 I(I + 1)}{3kT} B_z,$$

the standard Curie Law result.

5.2.3 Time independence of equilibrium quantities

It is instructive at this point to consider how the expectation value of a quantity varies with time for a system in thermal equilibrium. Of course, we understand equilibrium as that state reached by a system when everything has settled down and there is no further change. Thus intuitively we expect no time dependence for equilibrium expectation values. In this section we shall see how this result follows from the formalism thus far developed.

We choose to work on the Heisenberg picture of quantum mechanics where time dependence resides in the operators and not in the wavefunctions (as in the Schrödinger picture). The time evolution of an operator $\mathscr{A}$ is described by the Heisenberg equation of motion:

$$\frac{\mathrm{d}\mathscr{A}}{\mathrm{d}t} = -\frac{\mathrm{i}}{\hbar}[\mathscr{A},\mathscr{H}].$$

The bracket indicates the commutator of the operators and $\mathscr{H}$ is the Hamiltonian for the system. For an isolated system $\mathscr{H}$ is constant in time and the equation of motion may then be integrated to give

$$\mathscr{A}(t) = \exp\left(\frac{\mathrm{i}}{\hbar}\mathscr{H}t\right)\mathscr{A}\exp\left(-\frac{\mathrm{i}}{\hbar}\mathscr{H}t\right).$$

Here $\mathscr{A} = \mathscr{A}(0)$, the Schrödinger representation of the operator. The above result may be verified directly by differentiation.

The expectation value $A(t)$ is given by Equation (5.3), which now reads

$$A(t) = \mathrm{Tr}\{\mathscr{A}(t)\exp(-\beta\mathscr{H})\}/\mathrm{Tr}\{\exp(-\beta\mathscr{H})\}$$

$$= \frac{\mathrm{Tr}\left\{\exp\left(\frac{\mathrm{i}}{\hbar}\mathscr{H}t\right)\mathscr{A}\exp\left(-\frac{\mathrm{i}}{\hbar}\mathscr{H}t\right)\exp(-\beta\mathscr{H})\right\}}{\mathrm{Tr}\{\exp(-\beta\mathscr{H})\}}.$$

Now the trace has the important property that the operators within it may be permuted cyclically. Thus the first factor $\exp[(\mathrm{i}/\hbar)\mathscr{H}t]$ may be moved round to be placed after the final $\exp(-\beta\mathscr{H})$ giving

$$\mathrm{Tr}\left\{\mathscr{A}\exp\left(-\frac{\mathrm{i}}{\hbar}\mathscr{H}t\right)\exp\left(-\beta\mathscr{H}\right)\exp\left(\frac{\mathrm{i}}{\hbar}\mathscr{H}t\right)\right\}.$$

The last two operators commute with each other as they are both power series in the same operator $\mathscr{H}$, so we can swap these over whereupon the time evolution operators cancel since

$$\exp\left(-\frac{\mathrm{i}}{\hbar}\mathscr{H}t\right)\exp\left(\frac{\mathrm{i}}{\hbar}\mathscr{H}t\right) = 1.$$

This gives the result

$$A(t) = \mathrm{Tr}\{\mathscr{A}\exp(-\beta\mathscr{H})\}/\mathrm{Tr}\{\exp(-\beta\mathscr{H})\}.$$

All time dependence has vanished; the quantity A is constant in time,

$$A(t) = A(0).$$

We see then, as expected, that for systems in thermal equilibrium, the expectation values or observable quantities remain constant. Mathematically this followed from the fact that the Hamiltonian generating the time evolution was the same as the Hamiltonian that specified the equilibrium state in the Boltzmann factor.

5.2.4 Systems not in equilibrium

We have seen that there is no time dependence in the equilibrium state. But what we are really interested in is the time evolution of a non-equilibrium state. For such a system we know that there *will* be time dependence, and that the system will evolve towards equilibrium; this is precisely what we observe in a pulsed NMR experiment.

In the simplest pulsed NMR experiment we watch the magnetisation relax from some initially created non-equilibrium state back to the steady state value of

$$M_0 = (\chi_0/\mu_0)B_0.$$

That is, we observed M_z relaxing to M_0 – longitudinal relaxation – and we observe M_x and M_y relaxing to zero – transverse relaxation. In this and the following sections we shall derive quantum mechanical expressions to describe such relaxation processes. For transverse relaxation this is the quantum analogue of much of the material in Chapter 4, but we will also now be treating longitudinal relaxation in a similar fashion.

We saw in the previous section how to calculate quantities for a system in equilibrium and we demonstrated explicitly that such expectation values were constant in time. This followed because the Hamiltonian that caused the initial Boltzmann distribution of states was the same as that which generated the time evolution. These Hamiltonians need not be the same; consider the following hypothetical experiment to study longitudinal relaxation.

Imagine that we have applied a magnetic field B' to our system way back in the past. If this field has been applied for a sufficiently long time then the system will be in equilibrium with a magnetisation $M' = (\chi_0/\mu_0)B'$ (linearity and isotropy being assumed). Now at time $t = 0$ let us quickly change the field to a new value B''. Then, of course, the system will relax to its new equilibrium state.

We can say that initially the system had a Hamiltonian $\mathcal{H}''$ which at time $t = 0$ was changed to a different one. So in the language of the precious section, in an expression such as

$$A(t) = \mathrm{Tr}\{\mathcal{A}(t)\exp(-\beta\mathcal{H})\}/\mathrm{Tr}\{\exp(-\beta\mathcal{H})\}$$

since the initial Boltzmann distribution of energy states was determined by the original

$\mathcal{H}'$, this is the Hamiltonian to use in the $\exp(-\beta\mathcal{H})$. However, for times $t > 0$ the evolution of $A(t)$ is generated by the new Hamiltonian $\mathcal{H}''$. This then is the Hamiltonian to use in the $\exp[(i/\hbar)\mathcal{H}t]$ expressions.

Thus the time dependence of the observable A is given in this case by

$$A(t) = \frac{Tr\left\{\exp\left(\frac{i}{\hbar}\mathcal{H}''t\right)\mathcal{A}\exp\left(-\frac{i}{\hbar}\mathcal{H}''t\right)\exp(-\beta\mathcal{H}')\right\}}{Tr\{\exp(-\beta\mathcal{H})\}}. \tag{5.4}$$

In general the Hamiltonians $\mathcal{H}'$ and $\mathcal{H}''$ will not commute. Under these circumstances the procedure performed in the previous section cannot be carried out; we may no longer swap round the operators and thus cancel the time dependence. In this case then there will be a variation in A as it approaches it new equilibrium value.

In essence what we have done in this section is to regard a non-equilibrium state simply as an equilibrium state of a different or 'fictitious' Hamiltonian. The initial state relates to the fictitious Hamiltonian and the subsequent time evolution is generated by the real system Hamiltonian. Clearly not every non-equilibrium state may be described in this way in terms of a fictitious Hamiltonian, but where this is possible it does admit considerable simplification.

5.3 Behaviour of magnetisation in pulsed NMR

5.3.1 The Hamiltonian

We now continue with the principal task of this chapter which is to obtain expressions for the relaxation of the magnetisation from the non-equilibrium states created by 90° and 180° RF pulses. In quantum theory a system and its evolution are specified by the Hamiltonian operator. Therefore we first consider the general form of the Hamiltonian for an assembly of nuclear spins. We shall write

$$\mathcal{H} = \mathcal{H}_z + \mathcal{H}_m + \mathcal{H}_d. \tag{5.5}$$

Here $\mathcal{H}_z$ is the Zeeman part of the Hamiltonian, the interaction between the individual magnetic moments and the applied static magnetic field. The magnetic energy density is given by $-\mathbf{B}_0\cdot\mathbf{M}$, thus the Zeeman Hamiltonian operator is

$$\mathcal{H}_z = -B_0\gamma\hbar I_z = -\hbar\omega_0 I_z, \tag{5.6}$$

where I_z is the operator for the total z component of the spin angular momentum for the system. The Hamiltonian $\mathcal{H}_m$ generates any motion of the particles. It contains the potential and kinetic energy of the particles. In many cases we shall see that motion may be treated in a classical fashion, when the precise form of $\mathcal{H}_m$ becomes unimportant. Finally, $\mathcal{H}_d$ represents the interspin interactions. This is frequently the internuclear dipole–dipole interaction – hence the subscript. However, it could also include a quadrupole or a hyperfine interaction.

5.3.2 Some approximations

At this stage we make the approximation (except where otherwise explicitly stated) that the static magnetic field B_0 is very much greater than the 'local fields' associated with the interaction Hamiltonian $\mathcal{H}_d$. This is no restriction for the majority of NMR experiments where B_0 is invariably much greater than the few gauss (10^{-4} T) of the dipolar or other interactions. The consequence of this assumption is that the spin energy states are then essentially the equally spaced eigenstates of the Zeeman Hamiltonian, the effect of $\mathcal{H}_d$ being to cause small splittings of these energy levels. In the Boltzmann distribution of energy states for the system, the operator $\exp(-\beta\mathcal{H})$, we may therefore ignore the operator $\mathcal{H}_d$ as its effect is negligible when compared with $\mathcal{H}_Z$.

We shall also, for the present, ignore any motion Hamiltonian in the Boltzmann distribution. We also further make the 'high temperature' or 'low polarisation' approximation that we used in Section 2.1 and again in Box 5.1 when discussing Curie's Law. In the present context this means making the approximation

$$\exp(-\beta\mathcal{H}) \approx 1 - \beta\mathcal{H}. \tag{5.7}$$

These assumptions and their validity will be discussed further in later sections. At this stage they simplify considerably and thus expedite the calculations in this section.

The relaxation of the quantity A, in Equation (5.4), can now be written as

$$A(t) = \text{Tr}\{\mathcal{A}(t)(1 - \beta\mathcal{H}')\}/\text{Tr}\{1\},$$

where $\mathcal{H}'$ is the fictitious Hamiltonian related to the initial non-equilibrium state concerned. The denominator is the trace of the unit operator. It takes this simple form because the trace of $\mathcal{H}'$, essentially a spin angular momentum operator, is zero since the trace of I_z is zero. This is clear since such operators have cancelling positive and negative eigenvalues. Also, as we are interested in the behaviour of magnetisation: M_x, M_y and M_z, whose traces similarly are zero, we assume $\text{Tr}\{\mathcal{A}\} = 0$ whereupon we have

$$A(t) = -\beta\text{Tr}\{\mathcal{A}(t)\mathcal{H}'\}/\text{Tr}\{1\}. \tag{5.8}$$

Notice the pre-factor $\beta = 1/kT$, in conformity with Curie's Law.

5.3.3 90° and 180° pulses

In pulsed NMR experiments the non-equilibrium states of interest are created using bursts of magnetic field oscillating at the Larmor frequency in the transverse plane. In this way the equilibrium magnetisation, which was pointing in the z direction, can be rotated to the x direction (a 90° pulse) or the $-z$ direction (a 180° pulse). A more realistic view of the creation of non-equilibrium states with the RF pulses is given in Problem 5.3. In this section we shall adopt the spirit of Section 5.2.4., i.e. we will regard these states as the equilibrium configuration of a fictitious Hamiltonian.

This is easy. To create the magnetisation along the x axis we simply 'require the $\mathbf{B}_0$ field to have been applied along the x axis'. The fictitious Zeeman Hamiltonian for a 90° pulse is then by analogy with Equation (5.6),

$$\mathscr{H}_{90} = -B_0\gamma\hbar I_x = -\hbar\omega_0 I_x. \tag{5.9}$$

Similarly, to create the magnetisation along the $-z$ direction, i.e. to reverse it, we 'require the $\mathbf{B}_0$ field to have been pointing along the $-z$ axis'. The fictitious Hamiltonian for a 180° pulse is then

$$\mathscr{H}_{180} = +B_0\gamma\hbar I_z = +\hbar\omega_0 I_z. \tag{5.10}$$

These then are the Hamiltonians to use in Equation (5.8) when we want to study the relaxation of magnetisation following 90° and 180° pulses. Incidentally, we note that if we had not neglected the motion Hamiltonian in the Boltzmann distribution, it would have remained unchanged by these fictitious magnetic fields.

The demonstration of the effects of a 90° and 180° pulse from quantum mechanical first principles, obtaining the above results, is treated in Problem 5.3.

5.4 The relaxation functions of NMR

5.4.1 Relaxation of longitudinal and transverse magnetisation

In an experiment to study transverse relaxation we first create the transverse magnetisation by applying a 90° pulse, then we observe the transverse magnetisation, say M_x. In this case we have then (Equation (5.8))

$$M_x(t) = -\beta\mathrm{Tr}\{M_x(t)\mathscr{H}_{90}\}/\mathrm{Tr}\{1\}$$

or, using the established form for $\mathscr{H}_{90}$,

$$M_x(t) = -\beta\hbar\omega_0\mathrm{Tr}\{M_x(t)I_x\}/\mathrm{Tr}\{1\}.$$

Now M_x is the magnetic moment (x component) per unit volume,

$$M_x = \frac{1}{V}\sum_i \mu_x^i = \frac{\gamma\hbar}{V}I_x$$

and so

$$M_x(t) = \frac{\gamma\beta\hbar^2\omega_0}{V}\mathrm{Tr}\{I_x(t)I_x\}/\mathrm{Tr}\{1\}. \tag{5.11}$$

To study longitudinal relaxation a 180° pulse is applied to reverse the magnetisation. We are then interested in how M_z relaxes back to its equilibrium value.

$$M_z(t) = -\beta \text{Tr}\{M_z(t)\mathcal{H}_{180}\}/\text{Tr}\{1\}$$

or

$$M_z(t) = -\frac{\gamma\beta\hbar^2\omega_0}{V}\text{Tr}\{I_z(t)I_z\}/\text{Tr}\{1\}. \tag{5.12}$$

Of course we cannot (simply) observe the magnetisation in the longitudinal direction, we usually tip it into the transverse plane with a 90° pulse in order to observe it.

Box 5.2 *M_z does not really relax to zero*

Equation (5.12) implies that the z component of magnetisation will relax to zero following a disturbance. Certainly in the high temperature limit we are considering the equilibrium magnetisation will be small, but it is non-zero and given by Curie's Law, as we have seen. We can see that the magnetisation does indeed relax to the equilibrium, Curie Law, value by taking the high temperature limit somewhat more carefully. This will also justify our adoption of Equation (5.12) as containing the essential details of the relaxation process.

Let us start with Equation (5.4) for the relaxation of a quantity from an initial non-equilibrium state:

$$A(t) = \frac{\text{Tr}\{\mathcal{A}(t)\exp(-\beta\mathcal{H}')\}}{\text{Tr}\{\exp(-\beta\mathcal{H}')\}}.$$

Here $\mathcal{H}'$ is the fictitious Hamiltonian responsible for the production of the initial state. We use Equation (5.10) for the effective Hamiltonian for a 180° pulse, reexpressed in the following way.

$$\mathcal{H}_{180} = +\hbar\omega I_z = \mathcal{H}_{eq} + 2\hbar\omega_0 I_z,$$

where the final term presents the *deviation* from the equilibrium state.

The expression for the expectation value of I_z is then

$$\langle I_z(t)\rangle = \frac{\text{Tr}\{I_z(t)\exp(-\beta\mathcal{H}_{eq})\exp(+2\beta\hbar\omega_0 I_z)\}}{\text{Tr}\{\exp(-\beta\mathcal{H}_{eq})\exp(+2\beta\hbar\omega_0 I_z)\}},$$

where the exponentials have been factored since the arguments commute. We now make the approximation that the deviation from equilibrium is small. In other words, the creation of a non-equilibrium I_z has negligible effect on the system. In that case we may expand the second exponential in both the numerator and the denominator. We shall expand to first order in the numerator and to zeroth order in the denominator. This gives

$$\langle I_z(t) \rangle = \frac{\text{Tr}\{I_z(t)\exp(-\beta\mathcal{H}_{eq})\}}{\text{Tr}\{\exp(-\beta\mathcal{H}_{eq})\}} + 2\beta\hbar\omega_0 \frac{\text{Tr}\{I_z(t)I_z\exp(-\beta\mathcal{H}_{eq})\}}{\text{Tr}\{\exp(-\beta\mathcal{H}_{eq})\}},$$

where, as we saw in Section (5.2.3), the first term is the equilibrium value of the magnetisation. The second term contains the time dependence. This represents the relaxation. In the high temperature limit the exponentials would be replaced by the unit operator and Equation (5.12) would be recovered.

In conclusion the constant, equilibrium value of the magnetisation is present, and in the high temperature limit, Equation (5.12) does indeed describe the relaxation process correctly.

5.4.2 Complex form for transverse relaxation

We saw in Section 2.6.2 how the precessing magnetisation was conveniently treated in terms of the complex quantity

$$M = M_x + iM_y.$$

The main advantage was that the precession then involved multiplication by $\exp(i\omega t)$ rather than a messy mixture of $\cos(\omega t)$ and $\sin(\omega t)$ which is required when using separate M_x and M_y. This advantage obviously carries over to the present quantum case, but now we have the further advantage that the operators

$$I_+ = I_x + iI_y,$$
$$I_- = I_x - iI_y$$

are the spin raising and lowering operators of quantum mechanics. These have rather special properties as the reader is probably well aware, and any good quantum mechanics text book will show.

For these reasons we shall obtain an expression for the transverse complex magnetisation in terms of these operators. We saw that M_x was given by

$$M_x(t) = \frac{\gamma\beta\hbar^2\omega_0}{V}\text{Tr}\{I_x(t)I_x(0)\}/\text{Tr}\{1\}.$$

The first operator in the trace, $I_x(t)$, is what we are observing. The second operator was previously written as I_x. This is, of course, the zero time value of the Heisenberg operator $I_x(t)$; thus here we write it as $I_x(0)$ to show this explicitly and to enhance the symmetry of the expression. This second term is I_x because the 90° pulse rotated the equilibrium I_z into I_x.

The expression for the behaviour of the y component of the magnetisation, after the same 90° pulse, will then be

$$M_y(t) = \frac{\gamma \beta \hbar^2 \omega_0}{V} \mathrm{Tr}\{I_y(t)I_x(0)\}/\mathrm{Tr}\{1\}.$$

We are interested in the complex $M = M_x + iM_y$ which then can be written as

$$M(t) = \frac{\gamma \beta \hbar^2 \omega_0}{V} \mathrm{Tr}\{I_+(t)I_x(0)\}/\mathrm{Tr}\{1\}.$$

We shall show that this equation can be expressed in the more convenient form

$$M(t) = \frac{\gamma \beta \hbar^2 \omega_0}{2V} \mathrm{Tr}\{I_+(t)I_-(0)\}/\mathrm{Tr}\{1\}.$$

The proof is as follows.

The new expression for $M(t)$ is given essentially by $\mathrm{Tr}\{I_+(t)I_x(0)\}$, so let us consider the effect of adding the other part to it, $i\mathrm{Tr}\{I_+(t)I_y(0)\}$. Now

$$i\mathrm{Tr}\{I_+(t)I_y(0)\} = i\mathrm{Tr}\{I_x(t)I_y(0)\} - \mathrm{Tr}\{I_y(t)I_y(0)\}.$$

We know that the behaviour of the system is unchanged by a rotation about the z axis, so the above expressions will be the same if we make the transformations I_y to I_x and I_x to $-I_y$. Thus our extra part above is equal to

$$-i\mathrm{Tr}\{I_y(t)I_x(0)\} - \mathrm{Tr}\{I_x(t)I_x(0)\} = -\mathrm{Tr}\{I_+(t)I_x(0)\}.$$

So if we subtract this expression from $\mathrm{Tr}\{I_+(t)I_x(0)\}$ we obtain

$$\mathrm{Tr}\{I_+(t)I_-(0)\} = 2\mathrm{Tr}\{I_-(t)I_x(0)\}.$$

The expression thus takes the required form

$$M(t) = \frac{\gamma \beta \hbar^2 \omega_0}{2V} \mathrm{Tr}\{I_+(t)I_-(0)\}/\mathrm{Tr}\{1\}. \tag{5.13}$$

5.4.3 *The interaction picture*

The most rapid part of the motion of the transverse magnetisation is the Larmor precession. Superimposed on this is the actual relaxation. It would seem advantageous then to separate these motions in the expression for the magnetisation. This we shall do.

We are considering an assembly of particles, the Hamiltonian of which is

$$\mathcal{H} = \mathcal{H}_z + \mathcal{H}_m + \mathcal{H}_d.$$

It is this Hamiltonian then which generates the time evolution for the system. The equation of motion for $I_+(t)$ is then

$$\frac{d}{dt}I_+(t) = \frac{i}{\hbar}[(\mathcal{H}_z + \mathcal{H}_m + \mathcal{H}_d), I_+(t)],$$

which has the solution

$$I_+(t) = \exp\left[+\frac{i}{\hbar}(\mathcal{H}_z + \mathcal{H}_m + \mathcal{H}_d)t\right]I_+(0)\exp\left[-\frac{i}{\hbar}(\mathcal{H}_z + \mathcal{H}_m + \mathcal{H}_d)t\right]. \qquad (5.14)$$

If $\mathcal{H}_d$ were zero then we would have only simple Larmor precession. It is instructive to demonstrate this explicitly. The equation of motion for $I_+(t)$ is then

$$\frac{d}{dt}I_+(t) = \frac{i}{\hbar}[(\mathcal{H}_z + \mathcal{H}_m), I_+(t)].$$

Now the motion Hamiltonian commutes with the spin operators such as I_+. This is clearly so since they operate in different spaces; $\mathcal{H}_m$ operates in the real space of the particle coordinates while I_+ operates in spin space. Since $\mathcal{H}_m$ commutes with I_+, this simplifies the above equation to

$$\frac{d}{dt}I_+(t) = \frac{i}{\hbar}[\mathcal{H}_z, I_+(t)],$$

but since $\mathcal{H}_z$ is given by

$$\mathcal{H}_z = \hbar\omega_0 I_z$$

the equation of motion becomes

$$\frac{d}{dt}I_+ = i\omega_0[I_z, I_+].$$

This commutator is easily seen to be given by

$$[I_z, I_+] = I_+,$$

so the equation of motion is then

$$\frac{d}{dt}I_+ = i\omega_0 I_+.$$

On integration, the solution is found to be

$$I_+(t) = \exp(i\omega_0 t)I_+(0).$$

Substituting this into the expression for the (complex) transverse magnetisation $M(t)$

we find that

$$M(t) = \frac{\gamma\beta\hbar^2\omega_0}{2V}\exp(i\omega_0 t)\mathrm{Tr}\{I_+I_-\}/\mathrm{Tr}\{1\}.$$

We see that the time dependence is all in the oscillatory factor $\exp(i\omega_0 t)$; there is Larmor precession but no relaxation.

The essence of what we have just done is to demonstrate the result

$$\exp\left(\frac{i}{\hbar}\mathscr{H}t\right)I_+\exp\left(-\frac{i}{\hbar}\mathscr{H}t\right) = \exp(i\omega_0 t)I_+,$$

an indication that $\mathscr{H}_z$ generates rotations about the z axis.

Box 5.3 Some comments

Working in the $+,-,z$ representation has a number of calculational advantages. The matrix representations for the raising and lowering operators (for spin $\frac{1}{2}$) are

$$I_+ = \begin{pmatrix} 0 & 1 \\ 0 & 0 \end{pmatrix}, I = \begin{pmatrix} 0 & 0 \\ 1 & 0 \end{pmatrix},$$

so their product is simply

$$I_+I_- = \begin{pmatrix} 1 & 0 \\ 0 & 0 \end{pmatrix},$$

the trace of which is easily evaluated!

Furthermore, it is only when taking the complex combination of I_x and I_y that the rotation of the Larmor precession falls out as a complex factor $\exp(i\omega t)$. In the x,y,z representation the precession mixes the I_x and I_y with cos and sin factors; this makes for a more clumsy description.

We turn now to the main subject of this particular section: the interaction picture. In Chapter 2 we have already discussed the main pictures of quantum mechanics: the Heisenberg picture and the Schrödinger picture. Recall that in the Heisenberg picture the time evolution of a system is contained in the operators whereas in the Schrödinger picture it is the wavefunctions that vary with time. Clearly both pictures give the same results for the evolution of observables.

The interaction picture, sometimes referred to as the Dirac picture, is a picture intermediate between those of Heisenberg and Schrödinger. In this picture the time evolution is shared between the operators and the wavefunctions. It is particularly valuable when the Hamiltonian operator generating the time evolution splits naturally into two parts.

Consider some operator I that in the Heisenberg picture varies as

$$I(t) = \exp(i\mathscr{H}t/\hbar)I(0)\exp(-i\mathscr{H}t/\hbar).$$

We may introduce an evolution operator $U(t) = \exp(i\mathscr{H}t/\hbar)$ so that the time dependence of I may be expressed as

$$I(t) = U(t)I(0)U^{-1}(t).$$

All information about the time variation of the system is contained in $U(t)$ so attention may be focused on that operator. Now let us express the Hamiltonian as the sum of two parts

$$\mathscr{H} = \mathscr{H}_a + \mathscr{H}_b.$$

The evolution operator is then

$$U(t) = \exp[i(\mathscr{H}_a + \mathscr{H}_b)t/\hbar]$$

and one might be tempted to factorise the exponential, to express $U(t)$ as

$$\exp(i\mathscr{H}_a t/\hbar)\exp(i\mathscr{H}_b t/\hbar).$$

In general this is *wrong*. Whereas such factorisation holds true for scalars it does not for operators. The only exception is when the operators $\mathscr{H}_a$ and $\mathscr{H}_b$ commute with each other. The machinery of the interaction picture is closely related to the way in which such evolution operators *can* be factorised, albeit not in the above simple form.

To proceed with the discussion, let us temporarily ignore the motion Hamiltonian to keep things simple; it may be added in later. We are concerned then with the evolution operator

$$U(t) = \exp[i(\mathscr{H}_d + \mathscr{H}_z)t/\hbar].$$

We saw at the start of this section that in the absence of the dipolar interaction the transverse magnetisation displays simple Larmor precession: the evolution produced by $\exp(i\mathscr{H}_z t/\hbar)$ while the effect of including the influence of $\mathscr{H}_d$ is to cause relaxation. That is, the main part of the motion is still precession, but superimposed upon this will be the relaxation.

Considering now the evolution operator for this system, it would clearly be advantageous to factorise out the precession term. In other words we would like a factorisation of the form

$$\exp[i(\mathscr{H}_d + \mathscr{H}_z)t/\hbar] = S(t)\exp(i\mathscr{H}_z t/\hbar). \tag{5.15}$$

Clearly the operator $S(t)$ will not have the simple form $\exp(i\mathscr{H}_d t/\hbar)$, but we can regard

Equation (5.15) as the *definition* of $S(t)$. This problem is taken up in Appendix C where the form for $S(t)$ is obtained.

For the present let us not concern ourselves about the precise expression for $S(t)$, but simply explore the consequences of such a factorisation of $U(t)$. We start from Equation (5.13) for the decay of the complex transverse magnetisation:

$$M(t) = \frac{\gamma\beta\hbar^2\omega_0}{2V} \mathrm{Tr}\{I_+(t)I_-(0)\}/\mathrm{Tr}\{1\}.$$

The time evolution of I_+ may be written using the evolution operator

$$I_+(t) = U(t)I_+(0)U^{-1}(t).$$

Now we use the factorised expression for $U(t)$ and its hermitian conjugate

$$U(t) = S(t)\exp(i\mathcal{H}_zt/\hbar),$$
$$U^+(t) = \exp(-i\mathcal{H}_zt/\hbar)S^+(t).$$

(Since $U(t)$ and $S(t)$ are unitary operators their hermitian conjugate is equal to their inverse.) The above expressions enable us to write $I_+(t)$ as

$$I_+(t) = S(t)\exp(i\mathcal{H}_zt/\hbar)I_+(0)\exp(-i\mathcal{H}_zt/\hbar)S^+(t),$$

and we immediately recognise the Larmor precession in the middle of this expression:

$$\exp(i\mathcal{H}_zt/\hbar)I_+(0)\exp(-i\mathcal{H}_zt/\hbar) = \exp(i\omega_0t)I_+(0).$$

The consequence of this is that the expression for transverse relaxation may now be written:

$$M(t) = \frac{\gamma\beta\hbar^2\omega_0}{2V} \frac{\mathrm{Tr}\{S(t)I_+(0)S^+(t)I_-(0)\}}{\mathrm{Tr}\{1\}}\exp(i\omega_0t).$$

Performing a similar procedure on the expression for longitudinal relaxation (for consistency) we obtain in an analogous way

$$M_z(t) = \frac{\gamma\beta\hbar^2\omega_0}{2V} \frac{\mathrm{Tr}\{S(t)I_z(0)S^+(t)I_z(0)\}}{\mathrm{Tr}\{1\}}.$$

If we keep our wits about us we could write

$$S(t)I_+(0)S^+(t) = I_+(t),$$

where the time dependence of *this* $I(t)$ does not have the time evolution generated by the full Hamiltonian, but only the partial evolution of $S(t)$; the precession has been factorised out.

It is in this sense that the designation *interaction picture* is appropriate. The effect of the 'interesting' part has been isolated. Also one sees clearly that the transformation to this interaction picture is equivalent to going into a reference frame rotating at the Larmor frequency.

Remembering that when we are using this interaction picture evolution for the operators, we have the relaxation expressions

$$M(t) = \frac{\gamma \beta \hbar^2 \omega_0}{2V} \frac{\text{Tr}\{S(t)I_+(0)S^+(t)I_-(0)\}}{\text{Tr}\{1\}} \exp(i\omega_0 t),$$

$$M_z(t) = \frac{\gamma \beta \hbar^2 \omega_0}{2V} \frac{\text{Tr}\{S(t)I_z(0)S^+(t)I_z(0)\}}{\text{Tr}\{1\}}.$$

(5.16)

Box 5.4 Comment

Clearly going into the interaction picture has had no effect on the expression for $M_z(t)$. In other words rotation of I_z about the z axis achieves nothing. In this interaction picture the $I_z(t)$ is the same as the old $I_z(t)$ in the previous Heisenberg picture. We have performed the same transformation in both the longitudinal and the transverse cases to preserve as much symmetry in the treatment as possible. For both cases all time evolution is contained in the (as yet undetermined) operator $S(t)$.

5.4.4 *The relaxation functions*

For transverse relaxation the effect of transforming to the interaction picture was to make explicit the Larmor precession. Having factorised out the $\exp(i\omega_0 t)$ it follows that the remaining time evolution, that in the interaction picture $\text{Tr}\{I_+(t)I_-(0)\}$, is simply the decay of the length of the transverse magnetisation vector. In other words the oscillatory part has been taken out leaving the smooth decay to zero of the envelope.

Ordinarily it is the smooth decay of this envelope which is observed in an experiment through the use of phase sensitive detection or diode demodulation. Apart from the convenience of subsequently processing the signals with 'AF' rather than 'RF' electronics, such demodulation is necessary to reduce the noise bandwidth of the spectrometer. For all these reasons we focus attention on this relatively slow varying function, and we are therefore led to the introduction of what we shall refer to as the fundamental relaxation functions of NMR. These functions are normalised to unity at zero time. The transverse relaxation function is denoted by $F(t)$ (the F indicates Free induction decay), and it is given by

$$F(t) = \text{Tr}\{I_+(t)I_-(0)\}/\text{Tr}\{I_+I_-\};$$

(5.17)

the longitudinal relaxation function $L(t)$ is given similarly by

$$L(t) = \text{Tr}\{I_z(t)I_z(0)\}/\text{Tr}\{I_zI_z\}.$$ (5.18)

Remember that both expressions are using the interaction picture evolution. The actual transverse and longitudinal magnetisation are related to these functions through

$$M(t) = \frac{\gamma\beta\hbar^2\omega_0}{2V} \frac{\text{Tr}\{I_+I_-\}}{\text{Tr}\{1\}} \exp(i\omega_0 t)F(t),$$

(5.19)

$$M_z(t) = \frac{\gamma\beta\hbar^2\omega_0}{2V} \frac{\text{Tr}\{I_z^2\}}{\text{Tr}\{1\}} L(t).$$

On evaluation of the traces, as in Box 5.1, we see that the initial magnetisation corresponds to the Curie Law value, as expected for the transverse magnetisation following a 90° pulse and the longitudinal magnetisation following a 180° pulse.

The two following chapters of this book will be devoted mainly to considering the shape and extent of these relaxation functions $F(t)$ and $L(t)$ for a variety of systems. As a preliminary, however, we must first consider the nature of the Hamiltonian that causes the relaxation: that which we have labelled as $\mathcal{H}_d$.

5.5 The dipolar and other interactions

5.5.1 The dipole–dipole interaction

Fundamental to NMR are the nuclear magnetic dipoles which process in the applied magnetic fields, but a magnetic dipole has a dual manifestation. It responds to a magnetic field, experiencing a torque which results in the Larmor precession. This is the passive feature of a magnetic dipole. There is, however, an active aspect of a dipole. This is the second feature: a dipole is also a source of magnetic fields. These dipolar fields are seen by the other nuclear dipoles. This is the most common internuclear interaction in NMR.

A magnetic moment μ will produce, at a displacement r from itself, a magnetic field B given by

$$B = \frac{\mu_0}{4\pi}\left[\frac{\mu}{r^3} - 3\frac{(\mu \cdot r)r}{r^5}\right].$$ (5.20)

Let us consider briefly the effects of this interaction. To understand the nature of the dipole field it is necessary to take account of the fact that the dipoles producing the fields are precessing at the Larmor frequency. The component of μ parallel to the static field B_0 will produce a static dipolar field, but the perpendicular component of the dipole moment will produce fields rotating (components oscillating) at the Larmor frequency.

If the dipole responding to this field has the same magnetogyric ratio as that producing it then it will experience an on-resonance oscillating field which can cause it to change its orientation. If, however, its magnetogyric ratio is different this will not happen. It is clear then that dipole interactions between 'like' and 'unlike' spins will be somewhat different. This will be further emphasised when we consider the quantum description of the dipolar interaction.

There is also a particular feature that arises because the dipoles are both causing and responding to the fields. We discussed previously in Section 2.5 how a spin experiencing transverse fields fluctuating at the Larmor frequency can have the z component of its orientation altered; random variations thus cause longitudinal relaxation. Now, both dipoles participating in the interaction can simultaneously alter their longitudinal components. This happens for field fluctuations at double the Larmor frequency. Double frequency fluctuations then will also contribute towards longitudinal relaxation.

The direct approach in considering the effects of the dipole local fields would be to consider the torque exerted by one dipole on another and in this way attempt to solve the equations of motion. We shall not do this. Instead we go immediately for a quantum mechanical treatment. This puts the emphasis on the energy operator for the interaction, its Hamiltonian. The two dipoles in the pairwise interaction are then treated on an equal basis; the distinction between which one causes the field and which one responds to it is removed. In fact this is not uniquely a quantum feature, it would also follow from a classical Hamiltonian treatment. There is, however, a particular advantage of the quantum viewpoint. It gives a physical picture of the interaction in terms of transitions between spin states or spin flips. It is then possible to classify the different sorts of processes that are a result of the dipolar interaction.

5.5.2 *The dipolar Hamiltonian*

Two magnetic moments μ_1 and μ_2 separated by a displacement $\mathbf{r}$ will have an energy of interaction E given by

$$E = \frac{\mu_0}{4\pi}\left[\frac{\mu_1 \cdot \mu_2}{r^3} - 3\frac{(\mu_1 \cdot \mathbf{r})(\mu_2 \cdot \mathbf{r})}{r^5}\right], \tag{5.21}$$

which may, of course, be interpreted as the scalar product of one magnetic moment with the dipole field produced by the other.

The magnetic moments are related to the nuclear spins by

$$\mu_1 = \gamma_1 \hbar \mathbf{I}^1,$$
$$\mu_2 = \gamma_2 \hbar \mathbf{I}^2,$$

so that the Hamiltonian operator for the dipolar interaction between spins 1 and 2 takes the form

$$\mathcal{H}_{12} = \frac{\mu_0}{4\pi}\gamma_1\gamma_2\hbar^2\left[\frac{\mathbf{I}^1\cdot\mathbf{I}^2}{r^3} - 3\frac{(\mathbf{I}^1\cdot\mathbf{r})(\mathbf{I}^2\cdot\mathbf{r})}{r^5}\right], \tag{5.22}$$

where $\mathbf{I}^1$ and $\mathbf{I}^2$ are the nuclear spin operators. At this stage we allow for the general case of spins 1 and 2 being different through the labelling of their magnetogyric ratios γ_1 and γ_2.

The general form of this interaction is that it is bilinear in the spin operators $\mathbf{I}^1$ and $\mathbf{I}^2$: i.e. the Hamiltonian has the general form

$$\mathcal{H}_{12} = \sum_{\alpha,\beta=x,y,z} A^{\alpha\beta}I_\alpha^1 I_\beta^2.$$

Once again it proves convenient to use the spin raising and lowering operators I_+ and I_-. In other words instead of expressing the dipolar Hamiltonian in terms of the I_x, I_y and I_z of each member of the pair, we shall use the basis set I_+, I_z, I_-. Then the expression for the bilinear interaction may be written as the sum of the nine terms:

$$\begin{aligned}\mathcal{H}_{12} = \quad & A^{++}I_+^1 I_+^2 + A^{+z}I_+^1 I_z^2 + A^{z+}I_z^1 I_+^2 \\ & + A^{zz}I_z^1 I_z^2 + A^{+-}I_+^1 I_-^2 + A^{-+}I_-^1 I_+^2 \\ & + A^{z-}I_z^1 I_-^2 + A^{-z}I_-^1 I_z^2 + A^{--}I_-^1 I_-^2.\end{aligned}$$

When expressed in this way, and considering spin $\frac{1}{2}$ for simplicity, each part of the Hamiltonian has an immediate interpretation. The term with $I_+^1 I_+^2$ indicates that the spin 1 and the spin 2 are both flipped up. That with $I_+^1 I_-^2$ indicates that the spin 1 is flipped up while the spin 2 is flipped down. The term with $I_z^1 I_z^2$ indicates that neither spin is flipped, and the term with $I_+^1 I_z^2$ indicates that the spin 1 is flipped up while the spin 2 remains unchanged. This picture of the effect of the interaction is helpful in understanding what is going on particularly when combined with such concepts as energy conservation.

We have mentioned previously that in the usual sort of NMR experiments there is a large static magnetic field applied. The energy of a magnetic moment in this field, the Zeeman energy, is much larger than the energy of interaction such as the dipole–dipole energy. To zeroth order the energy levels of the spins are then simply the equally spaced Zeeman levels. The effect of the dipolar interaction may then be regarded as inducing transitions between these states. The above expansion of the dipole Hamiltonian then indicates the transitions induced by each term. This is shown schematically in Figure 5.1 for two (unlike) spins $\frac{1}{2}$.

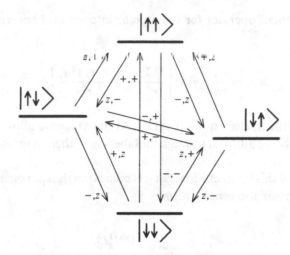

Figure 5.1 Transitions between Zeeman levels induced by components of the dipolar interaction.

5.5.3 Like spins

In this section we consider the dipolar Hamiltonian for like spins. Clearly the mathematical expression for the Hamiltonian is the same as that for unlike spins; the former is simply a special case of the latter. However, as discussed in Section 5.5.1, some of the effects can be different in the like case and this will lead to a different classification of the various terms in the Hamiltonian. For example, whereas the term in $I_z^1 I_z^2$ always conserves Zeeman energy, a term such as $I_+^1 I_-^2$ only conserves energy for like spins; the energy lost by one spin flipping up must be balanced by that gained when the other flips down. It is the energy conserving terms that dominate transverse relaxation while the non-conserving terms are responsible for longitudinal relaxation.

Another way of looking at this is to appreciate that for like spins it is only the net number of spins flipped in an interaction that is of importance. This number can range between $+2$ for the term in $I_+^1 I_+^2$ to -2 for the term in $I_-^1 I_-^2$. Designating this number by m the like-spin dipole Hamiltonian may be written

$$\mathcal{H}_d = \sum_{m=-2}^{2} D_m, \tag{5.23}$$

where the D_m can be related to the expansion terms in the previous section. Thus, for instance, the contribution to D_0 from spins number 1 and 2 will be given by

$$A^{zz} I_z^1 I_z^2 + A^{+-} I_+^1 I_2^2 + A^{-+} I_-^1 I_+^2.$$

The natural normalisation for such spin operators is provided by the basis set of spin flip operators T_{ij}^m, tensor operators of order two, which are given by

$$T_{ij}^0 = \mathbf{I}^i \cdot \mathbf{I}^j - 3I_z^i I_z^j,$$

$$T_{ij}^1 = \left(\frac{3}{2}\right)^{\frac{1}{2}}(I_z^i I_+^j + I_+^i I_z^j) = -(T_{ij}^{-1})^+,$$

$$T_{ij}^2 = -\left(\frac{3}{2}\right)^{\frac{1}{2}} I_+^i I_+^j = (T_{ij}^{-2})^+.$$

(5.24)

It is particularly convenient to use these operators since they satisfy the angular momentum commutation relations

$$[I_z, T_{ij}^m] = mT_{ij}^m,$$
$$[I_\pm, T_{ij}^m] = [6 - m(m \pm 1)]^{\frac{1}{2}} T_{ij}^{m \pm 1},$$

(5.25)

which are very useful in performing calculations.

The corresponding spatial functions that go with the T_{ij}^m to make the dipole components D_m are the spherical harmonics of order two: $Y_2^m(\Omega_{ij})$. Here the argument Ω_{ij} represents the orientation of the vector $\mathbf{r}_{ij}$ joining the ith and the jth spins. In other words Ω_{ij} is the angular part (θ_{ij}, ϕ_{ij}) of the vector $\mathbf{r}_{ij} = (r_{ij}, \theta_{ij}, \phi_{ij})$.

The full expression for the D_m is then given by

$$D_m = \frac{\mu_0 \hbar^2 \gamma^2}{4\pi \sqrt{5}} \sum_{i,j} (-1)^m \frac{Y_2^{-m}(\Omega_{ij})}{r_{ij}^3} T_{ij}^m.$$

(5.26)

This seemingly complicated formula is, of course, nothing more than the original expression for the dipole interaction, Equation (5.22), (restricted to like spins and summed over all spin pairs) but written in a form much more suited to calculation. It is immediately apparent for instance from Equation (5.25) that the $m = 0$ component D_0 commutes with I_z, and therefore with $\mathcal{H}_Z$; this is the term that conserves Zeeman energy.

5.5.4 Electron–nucleus interactions

Thus far we have considered the interaction of nuclear magnetic moments with externally applied magnetic fields and with themselves. We turn now to a consideration of the influence of the electrons in the specimen on the nuclear magnetic moment. The electron–nuclear interaction can take a variety of different forms, but very generally, the influence of electrons has two different aspects: electric and magnetic.

Orbiting electrons can produce magnetic fields in the vicinity of the nuclei. This can cause a small shift of the resonance frequency and because the electron orbits will be determined by the nature of the chemical bonds they may be participating in, this effect is called a *chemical shift*.

At first sight it might be thought that the *electric* interaction would be of no consequence for nuclear *magnetic* behaviour. That it is otherwise due to the Wigner–Eckart theorem of quantum mechanics. These ideas are taken up in the next sub-section.

5.5.5 Quadrupole interactions: classical picture

A nucleus has structure; it is not simply a point charge. One consequence of this structure is that its electric charge will be distributed over the volume in some, as yet unspecified, way. Electronic electric fields will interact with the nuclear electric charge. In particular, spatial variations of the electric field will couple to the charge distribution and this may be described in terms of the nuclear electric multipole moments. If $\rho(\mathbf{r})$ describes the charge distribution of the nucleus and $V(\mathbf{r})$ is the electric potential at $\mathbf{r}$ then the energy of this interaction is given by the integral over the nuclear volume:

$$E = \int \rho(\mathbf{r}) V(\mathbf{r}) \mathrm{d}^3 r.$$

Expanding $V(\mathbf{r})$ about the centre of the nucleus, the interaction energy takes the form of a series:

$$E = \int \rho(\mathbf{r}) \left[V_0 + \sum_{\alpha=x,y,z} \left(\frac{\partial V}{\partial r_\alpha} \right)_0 r_\alpha + \tfrac{1}{2} \sum_{\alpha,\beta=x,y,z} \left(\frac{\partial^2 V}{\partial r_\alpha \partial r_\beta} \right)_0 r_\alpha r_\beta + \dots \right] \mathrm{d}^3 r.$$

In the first term, which may be written $V_0 \int \rho(\mathbf{r}) \mathrm{d}^3\mathbf{r}$, the integral gives the total electric charge of the nucleus. This term is thus the electrostatic interaction of the charge and the electric potential at its centre; for a point charge this would be the only term.

We can express the second term as

$$\sum_{\alpha=x,y,z} \left(\frac{\partial V}{\partial r_\alpha} \right)_0 \int \rho(\mathbf{r}) r_\alpha \mathrm{d}^3 r.$$

The integral here represents the electric dipole moment of the nucleus and we see that it couples with the first spatial derivative of the electric potential, i.e. the electric field. Nuclei are not observed to have electric dipole moments. There are quantum mechanical arguments why this should be so, related to the fact that the nuclear ground state should be non-degenerate and therefore of definite parity. Since $\rho(\mathbf{r})$ is proportional to the squared magnitude of the wavefunction, this must then be an even function of r. All integrals involving odd powers of r_α therefore vanish, and this includes that for the dipole moment.

The third term in the energy series is

$$\frac{1}{2} \sum_{\alpha,\beta=x,y,z} \left(\frac{\partial^2 V}{\partial r_\alpha \partial r_\beta}\right)_0 \int \rho(\mathbf{r}) r_\alpha r_\beta \mathrm{d}^3 r.$$

The integral is closely related to the nuclear electric quadrupole tensor and the derivative represents nothing more than the various components of the electric field gradient at the nucleus. This is the term of interest to us. We see that there is an interaction between the electric quadrupole moment and the electric field gradients it experiences.

From this expression we immediately note two things. If the field gradients arise from an environment with cubic symmetry then the off-diagonal elements of the derivatives of V vanish while the three diagonal elements are equal. However, the sum of the diagonal elements must be zero by Laplace's equation, so all elements are zero; there can be no quadrupolar interaction in a cubic crystal. Secondly, a spin $\frac{1}{2}$ has spherical symmetry; it cannot have a quadrupole moment. Thus only spins having $I \geq 1$ can participate in quadrupole effects.

Our intention, in this and the following sub-section, is to cast the expression for the quadrupole interaction into the language of quantum mechanics; the electrostatic interaction energy will become the quadrupole Hamiltonian operator. At the same time, for it to be of any use to us, the result must in some way involve the nuclear spin operators. Before embarking on this, however, there are a number of simplifications which can be made to the classical energy of interaction.

The conventional definition of the electric quadrupole moment tensor is

$$Q_{\alpha\beta} = \int \rho(\mathbf{r})(3r_\alpha r_\beta - \delta_{\alpha\beta} r^2)\mathrm{d}^3 r, \tag{5.27}$$

where $\delta_{\alpha\beta} = 1$ if $\alpha = \beta$ and zero otherwise. The addition of this second term into the definition brings out the symmetry of the charge distribution, and the trace of the quadrupole tensor vanishes. With this definition of the quadrupole tensor, its energy of interaction with the magnetic field gradients may be written

$$\frac{1}{6} \sum_{\alpha,\beta=x,y,z} \left(\frac{\partial^2 V}{\partial r_\alpha \partial r_\beta}\right)_0 Q_{\alpha\beta} + \frac{1}{2} \sum_{\alpha=x,y,z} \left(\frac{\partial^2 V}{\partial r_\alpha^2}\right) \int \rho(\mathbf{r}) r^2 \mathrm{d}^3 r.$$

Since the field gradients arise from charges external to the nucleus, the second derivatives of the potential obey Laplace's equation, $\nabla^2 V = 0$, and the second term vanishes and thus the energy of the quadrupolar interaction is simply

$$E_Q = \frac{1}{6} \sum_{\alpha,\beta=x,y,z} \left(\frac{\partial^2 V}{\partial r_\alpha \partial r_\beta}\right)_0 Q_{\alpha\beta}. \tag{5.28}$$

The quadrupole tensor has nine elements. From its definition, Equation (5.27), we see that it is symmetric, i.e $Q_{\alpha\beta} = Q_{\beta\alpha}$. Thus it has only three independent off-diagonal elements. Since the tensor has zero trace there are only two independent diagonal elements. In all, then, there are five independent quantities. Now it takes three variables to specify the orientation of the quadrupole, leaving two parameters intrinsic to the quadrupole itself. But these can be reduced to one, as the following argument demonstrates.

Within the nucleus the various constituent charges will be precessing rapidly about the direction of its magnetic moment. The time scale of this is extremely short; so far as external interactions with the quadrupole are concerned we can take the time average. Thus $Q_{\alpha\beta}$ may be taken to have cylindrical symmetry, with its axis parallel to its spin angular momentum. If one of the α, β, γ axes is chosen to be in the direction of the spin then $Q_{\alpha\beta}$ will be diagonal and if the spin is taken to be in the z direction, then $Q_{xx} = Q_{yy}$. Since the trace of this tensor is zero, we are able to express all its elements in terms of a single scalar Q, which we define by

$$Q = \frac{1}{e} \int \rho(\mathbf{r})(3\tilde{z}^2 - r^2) \mathrm{d}^3 r,$$

where $\tilde{z}$ is along the direction of the spin angular momentum. We have divided out the elementary charge e so that Q has the dimensions of $(\text{length})^2$. This is essentially the diagonal element of the quadrupole tensor in the direction of the spin axis. This quantity is loosely referred to as the quadrupole moment. Thus the single parameter Q together with the spin vector are sufficient to describe the quadrupole tensor.

5.5.6 *Quadrupole interactions: quantum expressions*

In making the transition to quantum mechanics the quadrupole tensor now becomes an operator while the energy of the quadrupole interaction becomes the quadrupole Hamiltonian

$$\mathscr{H}_Q = \frac{1}{6} \sum_{\alpha,\beta=x,y,z} \left(\frac{\partial^2 V}{\partial r_\alpha \partial r_\beta}\right)_0 Q_{\alpha\beta}. \tag{5.29}$$

The quadrupole tensor operator $Q_{\alpha\beta}$ here is the quantum mechanical analogue of the classical quantity defined in Equation (5.27). The form of this operator is found with the aid of the Wigner–Eckart Theorem. We shall not delve into the technicalities of this theorem but the basic idea is that if two tensor operators have the same symmetry then they must be proportional to one another. Thus in going over to quantum mechanics the traceless tensor $3r_\alpha r_\beta - \delta_{\alpha\beta}r^2$ in Eq. (5.27) may be replaced by an operator proportional to $[3(I_\alpha I_\beta + I_\beta I_\alpha)/2] - I(I + 1)\delta_{\alpha\beta}$. Observe that the classical product $r_\alpha r_\beta$ has gone to the symmetrised product of spin operators, and that $I(I + 1)$ is the square of the magnitude of the spin. The result for the operator $Q_{\alpha\beta}$ is

$$Q_{\alpha\beta} = eQ \left(\frac{3(I_\alpha I_\beta - I_\beta I_\alpha)/2 - \delta_{\alpha\beta} I(I+1)}{6I(2I-1)} \right). \tag{5.30}$$

Spin operators have appeared here for the first time in the discussion of quadrupolar effects. It was the Wigner–Eckart Theorem which led to the introduction of spin operators and it is because of their presence that the nuclear electric quadrupole moment can be relevant to NMR.

As when treating the internuclear dipolar interaction, it is convenient (and suggestive) to express the spin operators appearing in the quadrupole operator in terms of the raising and lowering, or spin flip, operators I_+ and I_-. Although this interaction involves a single nuclear spin we see that since Equation (5.30) involves products of two spin operators the change in the quantum number m can range from -2 to $+2$. It is, of course, perfectly acceptable to have $\Delta m = 2$ for a single spin since here we are dealing with spins I greater than or equal to 1.

By analogy with the dipolar case and the Equation (5.23) we shall write the quadrupole interaction in a form which emphasises change in I_z of the various terms:

$$\mathcal{H}_Q = \sum_{m=-2}^{2} \mathcal{Q}_m. \tag{5.31}$$

The individual quadrupole Hamiltonian components are given by

$$\mathcal{Q}_m = \frac{eQ}{2I(2I-1)}(-1)^m V^{-m} T^m, \tag{5.32}$$

where the T^m are the basis set of spin flip operators as defined in Equation (5.24), but here with $i = j$. The V^m are linear combinations of the various components of the electric field gradient:

$$V^0 = -2V_{zz},$$

$$V^{\pm 1} = \frac{2}{3}(V_{zx} \pm iV_{zy}),$$

$$V^{\pm 2} = \frac{4}{3}\left[\frac{1}{2}(V_{xx} - V_{yy}) \pm iV_{xy} \right], \tag{5.33}$$

where $V_{\alpha\beta}$ are the various derivatives of the electric potential: the elements of the electric field gradient tensor,

$$V_{\alpha\beta} = \frac{\partial^2 V}{\partial r_\alpha \partial r_\beta} = -\frac{\partial E_\alpha}{\partial r_\beta} = -\frac{\partial E_\beta}{\partial r_\alpha}. \tag{5.34}$$

In some substances the quadrupolar splittings can be particularly large. For the covalently bonded solid halogens the frequencies are given in Table 5.1. These are the

Table 5.1. *Quadrupole splittings for*
some solid halogens.

Nucleus	F in MHz
Chlorine	80
Bromine	500
Iodine	2000

frequencies at which *pure quadrupole resonance* will be observed in the absence of a magnetic field. We see that with quadrupoles' interaction there are two distinct régimes. At high applied magnetic fields, or when the quadrupole splitting is small, the quadrupole interaction has a small perturbative effect on the Zeeman levels, just as in the dipole case. On the other hand, in a low or zero external field, it is the quadrupole interaction which determines the energy levels.

5.5.7 Chemical shifts: basic ideas

It is observed experimentally that the resonance frequency of a nucleus in a free atom can be slightly different from that in bulk matter. Furthermore, the magnitude of the shift from the free-spin value can be different in different chemical compounds, thus the designation *chemical shift*. To understand the origin of these small shifts one must remember that the precession frequency of the nucleus is determined by the total magnetic field it experiences and there may be fields of origin other than the dominant externally applied B_0 field. The cause of the chemical shift is the small magnetic fields created at the nucleus, because the orbital motion of the electrons may be modified in the presence of the B_0 field. This effect may be regarded as an indirect interaction between the externally applied magnetic field and the nuclear spins, mediated by the electrons.

Motion of an electron will contribute to the system's energy through a term involving the electron kinetic energy. As explained in Appendix D, when considering motion of a charged particle in a magnetic field, it proves expedient to use the *canonical momentum* p defined by

$$p = mv - eA, \tag{5.35}$$

where mv is the product of the particle's mass and its velocity: the conventional momentum, e is the charge of the particle and A is the vector potential of the magnetic field which the charge experiences. The electron kinetic energy T is then given by

$$T = \frac{1}{2}mv^2 = \frac{1}{2m}(p + eA)^2. \tag{5.36}$$

At the position $\mathbf{r}$ of the electron there will be two contributions of relevance to the magnetic field (or the vector potential). There is the externally applied magnetic field whose vector potential may be written as:

$$\mathbf{A_0} = \frac{1}{2}\mathbf{B_0} \times \mathbf{r}$$

and there is the magnetic field produced by the magnetic moment of the nucleus. The vector potential of this nuclear dipole field may be expressed as

$$\mathbf{A_n} = \frac{\mu_0}{4\pi}(\mathbf{\mu} \times \mathbf{r})/r^3$$

and the expression for the electron kinetic energy is

$$T = \frac{1}{2m}(\mathbf{p} + e\mathbf{A_0} + e\mathbf{A_n})^2. \tag{5.37}$$

In multiplying out this expression the various terms represent different types of interaction. The indirect, electron-mediated interaction between the external field and the nuclear moment is given by the term in $\mathbf{A_0} \cdot \mathbf{A_n}$. This gives a contribution to the energy, E', of

$$E' = \frac{e^2}{m}\mathbf{A_0} \cdot \mathbf{A_n}$$

$$= \frac{e^2}{2m}\frac{\mu_0}{4\pi}(\mathbf{B_0} \times \mathbf{r}) \cdot (\mathbf{\mu} \times \mathbf{r})/r^3, \tag{5.38}$$

which must be summed over all electrons in the vicinity. Observe that this expression is bilinear in $\mathbf{\mu}$ and the $\mathbf{B_0}$: i.e the contribution to the energy may be written as

$$E' = \sum_{\alpha,\beta=x,y,z} \mu_\alpha S_{\alpha\beta} B_\beta, \tag{5.39}$$

where the tensor elements $S_{\alpha\beta}$ are given by

$$S_{\alpha\beta} = \frac{e^2}{2m}\frac{\mu_0}{4\pi}\left(\frac{\delta_{\alpha\beta}}{r} - \frac{r_\alpha r_\beta}{r^3}\right),$$

which may be ascertained by direct expansion of the vector products in Equation (5.38).

The interpretation of Equation (5.39) is that the effect may be regarded as an additional field $\mathbf{b}$ at the nucleus, with components

$$b_\alpha = - \sum_{\beta = x,y,z} S_{\alpha\beta} B_\beta. \qquad (5.40)$$

This is an example of *diamagnetism*. What is happening is that when the B_0 field is applied the electron orbits adjust themselves in an attempt to neutralise the magnetic field. This screening field is proportional to, although no necessarily parallel to, the applied magnetic field.

So far our treatment has been entirely classical, but classical mechanics is not sufficient to give the electrons positions **r** over which the expression for the screening tensor $S_{\alpha\beta}$ must be summed. In going over to quantum theory the shift in the energy E' becomes an addition to the Hamiltonian operator, $\mathscr{H}$, and the coordinates, r_α, become position operators in the screening tensor operator. The screening tensor is the expectation value of this evaluated over the electron quantum states. It is fortunate that the effect we are considering is small. This entitles us to use first order perturbation theory which, in this context, means that the expectation values are evaluated over the unperturbed electron states: i.e the states as they are in the absence of the applied magnetic field. The screening tensor is then given by

$$S_{\alpha\beta} = \frac{e^2}{2m} \frac{\mu_0}{4\pi} \sum_{\text{states } s} \left\langle s \left| \left(\frac{\delta_{\alpha\beta}}{r} - \frac{r_\alpha r_\beta}{r^3} \right) \right| s \right\rangle.$$

There is also a paramagnetic contribution to the chemical shift, a consequence of the polarization of the electronic orbits by the applied field. In calculation this arises from the second order perturbation treatment of the energy shift. Since this involves knowledge of the excited electronic states its calculation is much more complicated. However, it will result in an expression similar to Equation (5.39) and hence to a screening tensor and a resultant local field as described by Equation (5.40).

5.5.8 *Isotropic chemical shift*

In fluids the molecular motion will rotate the electron states in a random fashion. The mean effect on the screening tensor may be found since $r_\alpha r_\beta$ will average to zero unless $\alpha = \beta$ and, since $\langle r_\alpha^2 \rangle = \langle r_\beta^2 \rangle = \langle r_\gamma^2 \rangle$, these must each be given by $r^2/3$. The average screening tensor is then isotropic: diagonal with diagonal elements equal. The isotropic screening tensor is given by

$$S_{\alpha\beta}^{(i)} = \delta_{\alpha\beta} \sigma = \frac{1}{3} \text{Tr}\{S_{\alpha\beta}\}.$$

The extra magnetic field is then parallel to $\mathbf{B}_0$ as well as being proportional to it. This results in small shift of the resonance frequency, which we may write as

$$\omega = \gamma B_0 (1 - \sigma),$$

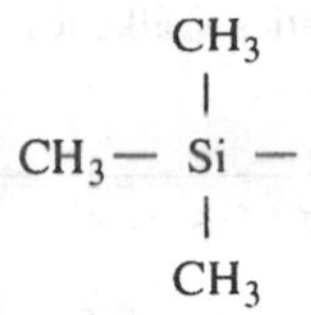

$$CH_3$$
$$|$$
$$CH_3 - \; Si \; - CH_3$$
$$|$$
$$CH_3$$

Figure 5.2 Structure of chemical shift reference standard TMS.

where the dimensionless quantity σ is known as the *isotropic chemical shift*.

A classic example of this is the NMR spectrum of ethanol. The chemical formula for ethanol is CH_3CH_2OH and the proton resonance shows three absorption lines with intensities in the proportion $3:2:1$. The NMR spectrum is able to distinguish the protons in different chemical environments: the three protons in the CH_3 group contribute to one resonance distinct from the resonance of the two protons in the CH_2 group, which in turn is distinct from that of the one proton in the OH group. In this way NMR may be used as an analytical tool in chemistry once a standard reference is established against which chemical shifts may be measured. The accepted proton standard is tetramethylsilane (TMS). This has twelve equivalent protons (Figure 5.2), providing a strong signal, and if the carbon atoms comprise the rarer isotope ^{13}C then TMS may also be used as the ^{13}C chemical shift reference, which is particularly convenient since carbon and hydrogen are the principal constituents of organic compounds.

Chemists measure the proton chemical shifts by comparison with the resonance frequency of protons in TMS. They adopt the symbol δ for 10^6 times the σ defined above: i.e δ measures the chemical shift in parts per million (ppm). There are tables of δ which indicate values for different chemical environments. In general for protons the value of δ varies from 0 to about 10 ppm (Figure 5.3). Since the absolute value of the chemical shift is proportional to applied magnetic field, the resolution is enhanced by working at higher frequencies.

5.5.9 Anisotropic chemical shift

In solids there is no averaging of the screening tensor and its full expression must be used. The screening field is then no longer parallel to the $\mathbf{B}_0$ field but the parallel component will cause a frequency shift. The resonance frequency is given by

$$\omega = \gamma B_0 (1 - S_{zz}).$$

This is telling us that in a crystalline solid the chemical shift will depend on the orientation of the specimen; the three diagonal elements of the screening tensor may be found rather than their average. This potentially greater power of chemical shift analysis in solids is not so easy to exploit. There is usually considerable dipolar

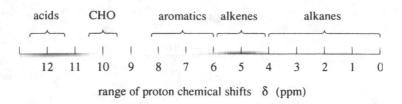

range of proton chemical shifts   δ (ppm)

Figure 5.3 Proton chemical shifts.

broadening of the lines so that the smaller chemical shifts cannot be resolved without special techniques such as spinning the specimen or complex pulse sequences.

In fluids the isotropic screening tensor results in shift of the resonance frequency: the chemical shift. The remainder of the screening tensor, the anisotropic part, $S_{\alpha\beta}{}^{(a)} = S_{\alpha\beta} - S_{\alpha\beta}{}^{(i)}$, will cause field (of zero average) fluctuating by virtue of the molecular motion modulating the orientation of the electronic orbits. There will thus be a fluctuating local magnetic field

$$\mathbf{b}_{loc}(t) = [S_{zx}{}^{(a)}(t)\mathbf{i} + S_{zy}{}^{(a)}(t)\mathbf{j} + S_{zz}{}^{(a)}(t)\mathbf{k}]B_0,$$

a potential relaxation mechanism which may be treated by the methods of the previous chapter.

6

Dipolar lineshape in solids

6.1 Transverse relaxation: rigid lattice lineshape

6.1.1 Introduction

The calculation (or at least the attempt at calculation) of the dipolar-broadened NMR absorption lineshape in solids has been one of the classical problems in the theory of magnetic resonance. Of course, the lineshape is the Fourier transform of the free precession decay so that a calculation of one is equivalent, formally, to a calculation of the other.

The method for performing such calculations was pioneered by Waller in 1932 and Van Vleck in 1948. However, to the present date no fully satisfactory solution has been found, despite the vast number of publications on the subject and the variety of mathematical techniques used. Nor is there likely to be. General expressions for transverse relaxation were given in the previous chapter. Restriction to a rigid lattice solid: the absence of a motion Hamiltonian, results in a considerable simplification of the equations, as we shall see. Nevertheless it is still a many-body problem of considerable complexity.

The various attempts at solving the problem of the transverse relaxation profile in solids have usually been based on the use of certain approximation methods whose validity is justified *a posteriori* by the success (or otherwise) of their results. We shall be examining some of these; none is really satisfactory. Conversely, and it may come as a surprise to discover, the more complicated case of a fluid system often permits approximations to be made which are well justified and with such approximations the resulting equations may be solved. This will be treated in the following chapter, although we have had a foretaste of this in Chapter 4.

Returning to the problem of transverse relaxation in solids, we start with Equation (5.17) for the transverse relaxation function $F(t)$:

$$F(t) = \frac{\text{Tr}\{I_+(t)I_-(0)\}}{\text{Tr}\{I_+I_-\}}, \tag{6.1}$$

157

although here we must be careful about the time dependence of $I_+(t)$. To obtain this time evolution it is easiest to go back to Equation (5.13) for the behaviour of the (complex) transverse magnetisation

$$m(t) \propto \text{Tr}\{I_+(t)I_-(0)\} \tag{6.2}$$

generated by the full Hamiltonian, as in Equation (5.14). In the present case this is

$$I_+(t) = \exp\left[\frac{i}{\hbar}(\mathcal{H}_z + \mathcal{H}_d)t\right]I_+(0)\exp\left[-\frac{i}{\hbar}(\mathcal{H}_z + \mathcal{H}_d)t\right]. \tag{6.3}$$

Now, in general, $\mathcal{H}_z$ and $\mathcal{H}_d$ do not commute so that the expression cannot be factorized simply to make the Larmor precession induced by $\mathcal{H}_z$ explicit. However, as stated previously, we are restricting our considerations to cases where $\mathcal{H}_z$ is very much greater than the dipolar interaction. In other words the applied magnetic field is very much greater than the gauss or so of the dipole fields. Thus $\mathcal{H}_d$ may be regarded as a small perturbation.

In first order perturbation theory it is only the diagonal part of the perturbing Hamiltonian which has any effect. Translated to the present case this means that the dominant effect will come from that part of $\mathcal{H}_d$ which commutes with $\mathcal{H}_z$. Keeping only that contribution will result in considerable simplification since then the exponentials of Equation (6.3) will factorize. If we express $\mathcal{H}_d$ as the sum of the various spin-flip components as in Equation (5.23)

$$\mathcal{H}_d = \sum_{m=-2}^{2} D_m,$$

then the part of $\mathcal{H}_d$ which commutes with $\mathcal{H}_z$, referred to as the 'truncated' dipolar Hamiltonian, is simply D_0, as discussed at the end of Section 5.5.3. Equation (6.3) then reduces to

$$I_+(t) = \exp\left(\frac{i}{\hbar}D_0 t\right)I_+(0)\exp\left(-\frac{i}{\hbar}D_0 t\right)\exp(i\omega_0 t). \tag{6.4}$$

The Larmor precession is indicated by the final exponential. The transverse decay envelope $F(t)$ may then be written as

$$F(t) = \frac{\text{Tr}\left\{\exp\left(\frac{i}{\hbar}D_0 t\right)I_+\exp\left(-\frac{i}{\hbar}D_0 t\right)I_-\right\}}{\text{Tr}\{I_+ I_-\}} \tag{6.5}$$

and it is (essentially) the evaluation of this expression which has taken up the time and energy of so many rigid lattice lineshape devotees for so many years.

6.1.2 A pair of spins

It is the complexity of the calculation which prohibits the analytic derivation of the FID/lineshape of a general NMR specimen. However, solutions may be found in special cases. Thus if the spins occur in isolated pairs then Equation (6.1) may be evaluated exactly. This may also be done for small clusters of spins although the difficulty of the calculation increases rapidly.

The chemical formula of gypsum is $CaSO_4 \cdot 2H_2O$. In crystals of this material the separation between the water molecules is much greater than that between the hydrogen nuclei within a water molecule. The proton resonance is thus determined primarily by a pairwise interaction. To a large extent then, the NMR behaviour may be approximated by that of a pair of spins.

Box 6.1 Comment on traditional calculation method

Calculation of the absorption spectrum for a pair of spins $\frac{1}{2}$ is conventionally effected by an apparently trivial method. The energy eigenvalues for the pair are found, paying particular attention to the splitting of the free-spin Zeeman levels by the dipole–dipole interaction. Application of time dependent perturbation theory (Fermi's Golden Rule) then yields the allowed transitions which can be stimulated by RF irradiation.

The present approach, based on consideration of the time domain, is more direct, making fewer assumptions; it is a calculation of the time evolution from quantum mechanical first principles. Thus in the frequency domain the selection rule $\Delta \omega = \pm 1$ and the requirement that the irradiation field be perpendicular to the B_0 field emerge as results rather than being inputs. The main assumption remaining is that of linearity; the magnitude of the B_1 field must not be too large. It is only then that the CW absorption line is the Fourier transform of the free precession decay, Equation (6.5).

While it is not necessary in a first reading for the student to work through the derivations of the present subsection, the formalism does indicate how an exactly soluble system may be treated in the time domain; in other words it shows how Equation (6.5) may be used in practice.

The trace in Equation (6.5) may be expressed as a sum over some suitable set of states:

$$F(t) \propto \sum_p \langle p | \exp\left(\frac{i}{\hbar} D_0 t\right) I_+ \exp\left(-\frac{i}{\hbar} D_0 t\right) I_- | p \rangle.$$

We have temporarily neglected the denominator, which is just a pure number. In terms of the set of states we can express the products of the operators in terms of their matrix elements:

$$F(t) \propto \sum_{pqrs} \langle p|\exp\left(\frac{i}{\hbar}D_0 t\right)|q\rangle \langle q|I_+|r\rangle \langle r|\exp\left(-\frac{i}{\hbar}D_0 t\right)|s\rangle \langle s|I_-|p\rangle.$$

This would be very complicated to evaluate, particularly the matrix elements of the exponential function of the operators. However, a considerable simplification results if the argument of the exponential is in a diagonal representation; if D_0 is diagonal then so is $\exp(iD_0 t)$. Luckily, since D_0 and $\mathcal{H}_z$ commute, these two operators have a common set of eigenstates.

The eigenstates of $\mathcal{H}_z$ are known. For spin $\frac{1}{2}$ particles these are the 'up' and 'down' states of the individual particles. So for our two spins we have the four states $|\uparrow\uparrow\rangle, |\uparrow\downarrow\rangle$, $|\downarrow\uparrow\rangle$ and $|\downarrow\downarrow\rangle$. These four states are eigenstates of the Zeeman Hamiltonian. But $|\uparrow\downarrow\rangle$ and $|\downarrow\uparrow\rangle$ are degenerate, and individually they are not eigenstates of D_0. We must take the (normalised) symmetric and antisymmetric combinations of the degenerate states:

$$\frac{1}{\sqrt{2}}(|\uparrow\downarrow\rangle + |\downarrow\uparrow\rangle) \text{ and } \frac{1}{\sqrt{2}}(|\uparrow\downarrow\rangle - |\downarrow\uparrow\rangle)$$

together with $|\uparrow\uparrow\rangle$ and $|\uparrow\downarrow\rangle$.

Since, in terms of these states, D_0 is diagonal, then so are the exponential operators. This enables us to write the FID as

$$F(t) \propto \sum_{pq} \langle p|\exp\left(\frac{i}{\hbar}D_0 t\right)|p\rangle \langle p|I_+|q\rangle \langle q|\exp\left(-\frac{i}{\hbar}D_0 t\right)|q\rangle \langle q|I_-|p\rangle.$$

Then denoting

$$\langle p|D_0|p\rangle = \hbar\omega_p, \tag{6.6}$$

we have

$$\langle p|\exp\left(\frac{i}{\hbar}D_0 t\right)|p\rangle = \exp(i\omega_p t), \tag{6.7}$$

so that we obtain

$$F(t) \propto \sum_{pq} \exp[i(\omega_p - \omega_q)t]\langle q|I_-|p\rangle^2, \tag{6.8}$$

which is a sum of sinusoidal oscillations.

The CW absorption will thus consist of a collection of delta functions. The position of each line (measured with respect to the Larmor frequency of the free spins) corresponds to the difference between the eigenfrequencies of states connected by the I_- operator. We note that the selection rule $\Delta m = \pm 1$ is automatically respected.

To see which pairs of states contribute to the absorption lines and the FID, we must examine the matrix elements of the spin lowering operator I_-. Let us denote the four spin states, the simultaneous eigenstates of the Zeeman and the truncated dipolar Hamiltonian, by

$$
\begin{aligned}
|1\rangle &= |\uparrow\uparrow\rangle \\
|0\rangle &= \frac{1}{\sqrt{2}}(|\uparrow\downarrow\rangle + |\downarrow\uparrow\rangle) \\
|-1\rangle &= |\downarrow\downarrow\rangle \\
|s\rangle &= \frac{1}{\sqrt{2}}(|\uparrow\downarrow\rangle - |\downarrow\uparrow\rangle),
\end{aligned}
\hspace{2cm} (6.9)
$$

the first three being the triplet states and the last the singlet state.

The operator I_- is the sum of lowering operators for the two spins being considered

$$
I_- = I_-^1 + I_-^2.
$$

Operating with this on the state $|1\rangle$ we obtain:

$$
\begin{aligned}
I_-|1\rangle &= I_-^1 |\uparrow\uparrow\rangle + I_-^2 |\uparrow\uparrow\rangle \\
&= |\downarrow\uparrow\rangle + |\uparrow\downarrow\rangle \\
&= \sqrt{2}|0\rangle,
\end{aligned}
$$

a multiple of the state $|0\rangle$. Thus we have the matrix element of I_-

$$
\langle 0|I_-|1\rangle = \sqrt{2}. \hspace{2cm} (6.10)
$$

Now apply I_- to the state $|0\rangle$:

$$
\begin{aligned}
I_-|0\rangle &= \frac{1}{\sqrt{2}}(I_-^1 |\uparrow\downarrow\rangle + I_-^1 |\downarrow\uparrow\rangle + I_-^2 |\uparrow\downarrow\rangle + I_-^2 |\downarrow\uparrow\rangle) \\
&= \frac{1}{\sqrt{2}}(|\downarrow\downarrow\rangle + |\downarrow\downarrow\rangle) \\
&= \sqrt{2}|-1\rangle.
\end{aligned}
$$

This is a multiple of the state $|-1\rangle$ so that we have the matrix element

$$
\langle -1|I_-|0\rangle = \sqrt{2}. \hspace{2cm} (6.11)
$$

For the state $|-1\rangle$, the lowering operators are applied to the 'down' states, giving zero: no non-zero matrix elements result. Finally we come to the singlet state. Operating with I_- on this will flip the 'up' spin 'down' giving

$$|\downarrow\downarrow\rangle - |\downarrow\downarrow\rangle = 0,$$

so that there is no contribution from the singlet state.

There are thus two non-zero matrix elements: two allowed transitions, to be considered. We must find the eigenfrequencies of the relevant states. The expression for the truncated dipolar Hamiltonian is found from Equation (5.26):

$$D_0 = \frac{\mu_0}{4\pi} \frac{\hbar^2 \gamma^2}{2r^3} (1 - 3\cos^2\theta)(3I_z^1 I_z^2 - \mathbf{I}^1 \cdot \mathbf{I}^2), \qquad (6.12)$$

where r is the separation of the two spins and θ is the angle between the line joining the spins and the z axis. The eigenvalues of D_0 may be found as the matrix elements taken between the eigenstates. The matrix elements are obtained as above for I_-. In that way one finds:

$$\langle 1|D_0|1\rangle = \hbar\Delta/3,$$
$$\langle 0|D_0|0\rangle = -2\hbar\Delta/3,$$
$$\langle -1|D_0|-1\rangle = \hbar\Delta/3,$$
$$\langle s|D_0|s\rangle = 0,$$

where

$$\Delta = \frac{\mu_0}{4\pi} \frac{3\hbar\gamma^2}{4r^3} (1 - 3\cos^2\theta).$$

The matrix elements are related directly to the precession frequencies of the various sinusoidal components of the transverse magnetisation. The frequencies were given in Equation (6.6), which now reads

$$\omega_1 = \Delta/3,$$
$$\omega_0 = -2\Delta/3,$$
$$\omega_{-1} = \Delta/3,$$
$$\omega_s = 0,$$

and we are now in a position to assemble our results to obtain the expression for the FID. We have (Equation (6.8))

$$F(t) \propto \sum_{p,q} \exp[i(\omega_p - \omega_q)t]\langle q|I_-|p\rangle^2$$

and there are only two non-zero matrix elements of I_-, given by Equations (6.10) and (6.11), corresponding to $p = 1$, $q = 0$ and $p = 0$, $q = -1$. In summary:

$p = 1, q = 0$ $\qquad \langle 0|I_-|1\rangle^2 = 2 \qquad \omega_p - \omega_q = \omega_1 - \omega_0 = \Delta,$
$p = 0, q = -1$ $\quad \langle -1|I_-|0\rangle^2 = 2 \qquad \omega_p - \omega_q = \omega_0 - \omega_{-1} = \Delta.$

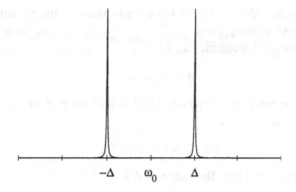

$-\Delta \qquad \omega_0 \qquad \Delta$

Figure 6.1 CW absorption spectrum of a pair of spins $\frac{1}{2}$.

The expression for $F(t)$ is then

$$F(t) \propto 2[\exp(i\Delta t) + \exp(-i\Delta t)],$$

which we observe to be a real function. Thus the FID is an oscillation:

$$F(t) = \cos(\Delta t) \qquad (6.13)$$

and the CW absorption (Figure 6.1) consists of a pair of sharp lines either side of the free-spin resonance, at frequencies

$$\omega = \omega_0 \pm \Delta,$$

i.e.

$$\omega = \omega_0 \pm \frac{\mu_0}{4\pi} \frac{3}{2} \frac{\hbar\gamma^2}{2r^3} (1 - 3\cos^2\theta). \qquad (6.14)$$

Box 6.2 Breadth of spectral lines

The calculation just performed has been for an isolated pair of spins $\frac{1}{2}$. In practice, of course, one is concerned with a *collection* of such spin pairs. In this case there will be interactions between the pairs resulting in a spreading of the energy states and a consequent broadening $\Delta\omega$ of the absorption lines. Correspondingly the FID, given by Equation (6.13), will have a finite lifetime, decaying to zero on a time scale $\Delta t \sim 1/\Delta\omega$.

6.1.3 Pake's doublet

In a powdered specimen all orientations θ will be present. The absorption spectrum will then be a superposition of lines, and the FID the corresponding superposition of cosines. Let us calculate this; it should have application to the proton resonance in powdered or polycrystalline gypsum.

We write the resonance frequency as

$$\omega = \omega_0 + \Delta,$$

where the relation between the frequency offset Δ and the pair orientation θ may be written from Equation (6.14) as

$$\Delta(\theta) = \delta(1 - 3\cos^2\theta). \tag{6.15}$$

Since θ varies between 0 and π, the range of Δ is

$$-2\delta \le \Delta \le \delta.$$

Given the distribution $p(\theta)$ of angles θ we want to find the corresponding distribution of offset frequencies Δ. We denote this distribution by $P(\Delta)$. Just as $p(\theta)d\theta$ is the probability of occurrence of an angle between θ and $\theta + d\theta$, so $P(\Delta)d\Delta$ is the probability of a frequency offset in the corresponding interval Δ to $\Delta + d\Delta$. We write the same interval in terms of either angle or frequency:

$$p(\theta)d\theta = P(\Delta)d\Delta,$$

so that the frequency distribution may be expressed as

$$P(\Delta) = p(\theta)\frac{d\theta}{d\Delta}.$$

In other words, given the distribution for orientations θ and the relation between Δ and θ, we can obtain the distribution for frequency deviations Δ, the CW absorption lineshape.

Differentiating Equation (6.15) we obtain

$$\frac{d\theta}{d\Delta} = 6\delta\cos\theta\sin\theta.$$

The distribution of orientations $p(\theta)$ is proportional to $\sin\theta$. Then considering the normalisation requirement on the probability:

$$\int_0^\pi p(\theta)d\theta = 1,$$

this gives, for the probability

$$p(\theta) = \tfrac{1}{2}\sin\theta.$$

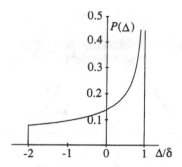

Figure 6.2 Distribution of (positive) frequency offsets for a powdered specimen.

Thus we have:

$$P(\Delta) = \frac{1}{12\delta\cos\theta}$$

$$= \frac{1}{2\delta[12(1 - \Delta/\delta)]^{\frac{1}{2}}}$$

which is shown in Figure 6.2.

This is the distribution of frequency offsets of one of the lines. As Equation (6.14) indicates, for every resonance line there is another symmetrically placed the other side of ω_0. Therefore one must add in the mirror image of the $P(\Delta)$ for the distribution of line pairs. The sum is shown in the dotted line of Figure 6.3. The solid line has been calculated by including a broadening of the constituent resonance lines to account for the interaction between the spin pairs. This solid line is remarkably similar to the observed CW absorption line of powdered gypsum. This lineshape was first calculated and observed by Pake in 1948 and it has thus come to be known as a *Pake doublet*.

The corresponding transverse relaxation $F(t)$ is shown in Figure 6.4. Qualitatively it looks very similar to the FID of calcium fluoride as shown in Figure 4.2. While we observe relaxation of the transverse magnetisation to zero, this does not occur in a monotonic fashion. This is a characteristic feature of most solid systems. Some general aspects of this phenomenon will be discussed at the end of this chapter.

6.1.4 Gypsum monocrystal

In considering the lineshape for a crystal of gypsum there is a complication since there are two water molecules in a unit of gypsum, each having a different orientation. Since these proton pairs have different values of θ, this results in four resonance lines. In practice the interactions between the protons of different water molecules in the crystal

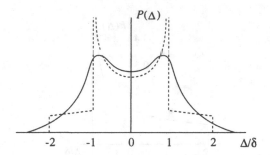

Figure 6.3 Calculated absorption line for powdered gypsum.

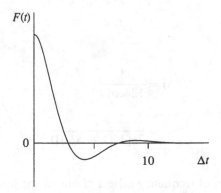

Figure 6.4 Transverse proton relaxation of powdered gypsum.

result in a broadening of the resonances (as discussed above). Nevertheless in Pake's original measurements, the proton absorption spectrum of single crystals of gypsum clearly displayed four peaks for appropriate orientations as shown in Figure 6.5.

6.1.5 Three and more spins

It is possible to perform calculations of the lineshapes of three and also of four isolated spins (Andrew and Bersohn, 1950). For increasing numbers, however, the labour involved prohibits such an enterprise. In fact, it should be clear that these calculations are of limited value because the intermolecular interactions result in a smearing-out of whatever detail might be calculated. This is demonstrated dramatically for the case of three spins. Even for this relatively simple system the unbroadened spectrum may be quite complex. The solid line in Figure 6.6 shows the calculated spectrum for three spins $\frac{1}{2}$ averaged over all orientations. This should be compared with the two-spin spectrum shown in Figure 6.3. Note how the addition of just one spin makes the profile so much more complicated. The important point, however, is the effect of broadening caused by

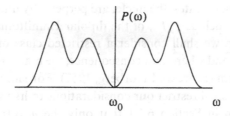

Figure 6.5 Observed absorption line for a gypsum crystal.

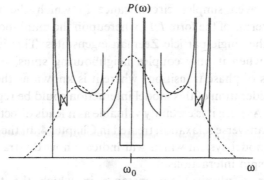

Figure 6.6 Absorption spectrum for three spins.

the intermolecular interactions. Once this is taken into account, very little detail is seen to remain. The dotted line of Figure 6.6 is typical of what is observed.

At this stage a new approach seems desirable, particularly for solid assemblies of larger numbers of spins. Fine details are not so important. We shall consider, in the next sections, how this may be done.

6.2 A class of solvable systems

6.2.1 Local field models

In the previous sections we saw how assemblies of very small numbers of spins could be treated in an exact manner. It proved possible, through solution of the equations of motion, to obtain the transverse relaxation/absorption line shape for these systems. However, one serious lesson learned from that exercise was that increasing numbers of spins became prohibitively difficult to treat in that way. The difficulty stemmed from the many-body nature of the energy eigenstates of the system; the total z component of spin was a good quantum number, but no so the z components of spin of the constituent dipoles. In other words, the many-body eigenstates are not simply products of the single-particle Zeeman eigenstates; the nuclei are perpetually undergoing mutual spin flips by virtue of terms, such as I_+I_-, of the dipolar Hamiltonian.

In the present section we shall consider a restricted class of systems where this difficulty does not occur and where, as a consequence, the many-body eigenstates have a particularly simple form (Lowe and Norberg, 1957). For simplicity, as we shall do often in this chapter, we shall restrict our considerations to like spins of magnitude $\frac{1}{2}$. We have already shown in Section 6.1.1 that only the $m = 0$ term of the dipolar Hamiltonian, D_0 need be considered when treating transverse relaxation in solids (although a fuller justification will emerge in the next chapter). Here we shall be concerned with the even simpler circumstance in which the relevant interspin interaction has only terms of the form I_zI_z, whereupon the many-body eigenstates *are* simply products of the single-particle Zeeman eigenstates. This interaction Hamiltonian, particularly when it only couples neighbouring spins, has received much attention by students of phase transitions where it is known as the Ising model.

To start with, consideration of this model interaction could be regarded simply as a pedagogical exercise. As such it is certainly valuable as it leads directly to the (classical) local field model of transverse relaxation treated in Chapter 4. In the pedagogical spirit we shall consider a model system which will indicate how the transverse relaxation depends on the interspin interactions.

There is, however, a physical circumstance in which the truncation of the dipole–dipole Hamiltonian to the I_zI_z part is justified. In a system where the nuclear spins are diluted in a non-magnetic matrix, mutual spin flips become very much more infrequent because of the greater separation of the spins. Furthermore since, in such a dilute system, the spins may be assumed to be distributed randomly, a given spin will interact mainly with its nearest neighbour rather than with many other particles. In such cases this simpler representation of the interspin interaction is acceptable. Using this, it will prove possible to calculate the profile of the transverse relaxation or, equivalently, the shape of the CW absorption line.

6.2.2 *The relaxation function*

Within the generality which the discussion of the previous section permits we shall write the Hamiltonian representing the internuclear interaction as

$$\mathcal{H} = \frac{1}{2}\sum_{k\neq l}I_z^kI_z^lB_{kl},\tag{6.16}$$

where, in the dipole–dipole case, from Equation (6.12)

$$B_{kl} = \frac{\mu_0}{4\pi}\frac{3}{2}\frac{\gamma^2\hbar^4}{r_{kl}^3}(1 - 3\cos^2\theta_{kl}).\tag{6.17}$$

The factor $\frac{1}{2}$ in the sum ensures that each pair of spins is counted only once.

We shall take the transverse relaxation function $F(t)$ written in terms of I_x:

$$F(t) = \text{Tr}\{I_x(t)I_x(0)\}/\text{Tr}\{I_x^2\} \tag{6.18}$$

and to start with we concentrate our attention on the behaviour of a single spin, say the jth spin. Its (thermally averaged) evolution may be described by a relaxation function $F_j(t)$, where

$$F_j(t) = \text{Tr}\{I_x^j(t)I_x^j(0)\}/\text{Tr}\{I_x^{j2}\} \tag{6.19}$$

and the time dependence of $I_x^j(t)$ is given by

$$I_x^j(t) = \exp\left(\frac{i\mathscr{H}t}{\hbar}\right)I_x^j\exp\left(\frac{-i\mathscr{H}t}{\hbar}\right),$$

with the Hamiltonian of Equation (6.16).
Writing the trace explicitly as the sum over states s:

$$F_j(t) = \sum_s \langle s|\exp\left(\frac{i\mathscr{H}t}{\hbar}\right)I_x^j\exp\left(\frac{-i\mathscr{H}t}{\hbar}\right)I_x^j|s\rangle$$

and since $\mathscr{H}$ is diagonal, this may be expressed as

$$F_j(t) = \sum_{r,s} \langle s|\exp\left(\frac{i\mathscr{H}t}{\hbar}\right)|s\rangle\langle s|I_x^j|r\rangle\langle r|\exp\left(\frac{-i\mathscr{H}t}{\hbar}\right)|r\rangle\langle r|I_x^j|s\rangle$$

$$= \sum_{r,s} |\langle s|I_x^j|r\rangle|^2\exp\left(\frac{it}{\hbar}(\mathscr{H}_{ss} - \mathscr{H}_{rr})\right). \tag{6.20}$$

The states r and s are products of the individual particle I_z eigenstates and so they may be labelled with their eigenvalues (Cohen-Tannoudji *et al.*, 1977):

$$|r\rangle = |r_1, r_2, \ldots, r_j \ldots, r_N\rangle,$$
$$|s\rangle = |s_1, s_2, \ldots, s_j \ldots, s_N\rangle,$$

where

$$I_z^j|s\rangle = s_j|s\rangle$$

and

$$s_j = \pm\tfrac{1}{2},$$

since we are restricting consideration to spin $\tfrac{1}{2}$.
The diagonal elements of our simple Hamiltonian are easily found as

$$\left.\begin{array}{l} \mathcal{H}_{rr} = \dfrac{1}{2}\displaystyle\sum_{k \neq l} r_k r_l B_{kl}, \\[4mm] \mathcal{H}_{ss} = \dfrac{1}{2}\displaystyle\sum_{k \neq l} s_k s_l B_{kl}. \end{array}\right\} \qquad (6.21)$$

The matrix representation of the operator I_x^j (in the I_z basis) is given by

$$\frac{1}{2}\begin{pmatrix} 0 & 1 \\ 1 & 0 \end{pmatrix}.$$

If this operates on an up spin it flips it down and if it operates on a down spin it flips it up. In other words it inverts the sign of the jth eigenvalue. We may succinctly express this as

$$\langle s_j | I_x^j | r_j \rangle = \frac{1}{2}\delta_{s_j,\,-r_j} \qquad (6.22)$$

and its square:

$$|\langle s_j | I_x^j | r_j \rangle|^2 = \frac{1}{4}\delta_{s_j,\,-r_j}.$$

This is telling us that in the expression for $F_j(t)$, Equation (6.20), the states r and s which contribute to the sum are those for which only the jth index differs

$$\begin{array}{l} r_j = -s_j \\ r_i = s_i \text{ for } i \neq j. \end{array}$$

Now both the k and the l labels will have an opportunity of becoming equal to the label of interest j. Thus *either*

$$k = j \text{ in which case } \quad \begin{array}{l} r_k = -s_j, \\ r_l = s_l, \end{array} \quad \text{to be summed over;}$$

or

$$l = j \text{ in which case } \quad \begin{array}{l} r_l = -s_j, \\ r_k = s_k, \end{array} \quad \text{to be summed over.}$$

Thus the relevant contributions give

$$\mathcal{H}_{ss} - \mathcal{H}_{rr} = \frac{1}{2}\sum_l B_{jl} s_j (s_l + r_l) + \frac{1}{2}\sum_k B_{kj} s_j (s_k + r_K)$$

$$= 2s_j \sum_k B_{jk} s_k,$$

since $B_{jk} = B_{kj}$ and it is understood that the sum excludes the $k = j$ term.

We obtain, then, the expression for $F_j(t)$:

$$F_j(t) \propto \sum_s \left[\exp\left(\frac{2its_j}{\hbar}\right) \Sigma_k B_{jk} s_k \right], \tag{6.23}$$

a result which admits of a straightforward classical interpretation.

To see the meaning of Equation (6.23) let us assume that we actually known the quantum state our spin system is in. That is imagine that we know all the single-particle eigenvalues s_k which specify the state s. The energy of the jth particle, i.e. its energy of interaction with the other particles, is given from Equation (6.21) by

$$E_j = s_j \sum_k s_k B_{jk}.$$

Now the magnetic moment (z component) of the jth spin is given by

$$\mu_j = \gamma \hbar s_j,$$

so that we may write the energy as

$$E_j = \mu_j \frac{\Sigma_k s_k B_{jk}}{\gamma \hbar}.$$

But this bears a striking similarity to the expression for the energy of a magnetic moment in a magnetic field. Thus we may identify the magnetic field at the site j as

$$b_j = \frac{\Sigma_k s_k B_{jk}}{\gamma \hbar}.$$

In the case of the dipole–dipole interaction the local field indeed corresponds to the z component of the sum of the dipole fields of the other spins at the site j.

From this perspective we see that the quantum expression we have derived for $F_j(t)$ is the expected classical relaxation averaged over all possible states of the system, each state being weighted equally. This is as it should be since Equation (6.18) for $F(t)$ was obtained, in the last chapter, as a thermal average: appropriate when we don't know the actual state of the system. The high temperature approximation used in its derivation is reflected here by all possible states having equal probability.

This interpretation of the expression for the evolution of the jth spin facilitates a further step forward. Equation (6.23) is understood as an average over all possible spin orientations, attention being focused on the jth spin. However, in performing such an average it is clearly immaterial which particular spin is chosen as the jth; an identical result would be obtained regardless of the chosen reference spin. Thus in reality the derived result, Equation (6.23), gives the relaxation for the entire assembly of spins. Similarly, we may arbitrarily fix the state of the jth spin, setting $s_j = +\frac{1}{2}$, say, since the interaction energy of a given state is unchanged if all spins are inverted. Thus we may write the relaxation function for this model as

$$F(t) \propto \sum_s \left[\exp\left(\frac{it}{\hbar} \Sigma_k B_{jk} s_k \right) \right].$$

(6.24)

We shall now proceed to an evaluation of the sum over states in our expression for $F(t)$. Each spin may be in one of two states, described by $s_k = +\frac{1}{2}$ or $s_k = -\frac{1}{2}$. For N particles there are 2^N different many-body states to be considered (really $2^N/2$ since the sum excludes the $k = j$ term and we already fixed $s_j = +\frac{1}{2}$).

$$\sum_s \equiv \sum_{s_1 = \pm\frac{1}{2}, s_2 = \pm\frac{1}{2}, \ldots, s_N = \pm\frac{1}{2}} .$$

Thus

$$F(t) \propto \sum_{s_1 = \pm\frac{1}{2}, s_2 = \pm\frac{1}{2}, \ldots, s_N = \pm\frac{1}{2}} \exp\left(\frac{it}{\hbar} \sum_k B_{jk} s_k \right)$$

$$\propto \sum_{s_1 = \pm\frac{1}{2}, s_2 = \pm\frac{1}{2}, \ldots, s_N = \pm\frac{1}{2}} \prod_k \exp\left(\frac{it}{\hbar} B_{jk} s_k \right)$$

$$\propto \left[\sum_{s_1 = \pm\frac{1}{2}} \exp\left(\frac{it}{\hbar} B_{j1} s_1 \right) \right] \left[\sum_{s_2 = \pm\frac{1}{2}} \exp\left(\frac{it}{\hbar} B_{j2} s_2 \right) \right] \ldots,$$

but each of these terms gives a cosine:

$$F(t) \propto \left[\cos\left(\frac{B_{j1} t}{2\hbar} \right) \right] \left[\cos\left(\frac{B_{j2} t}{2\hbar} \right) \right] \ldots.$$

The final expression for $F(t)$ is thus found as

$$F(t) = \prod_k \cos\left(\frac{B_{jk} t}{2\hbar} \right),$$

(6.25)

where we note that the $t = 0$ normalisation is correct here.

Box 6.3 Generalisation from spin $\frac{1}{2}$

The result of this section may be summarised by saying that the quantum calculation of transverse relaxation gives exactly the same behaviour as expected from classical arguments (accepting discrete spin orientations). However, all quantum calculations have been done for the special case of spin $\frac{1}{2}$. How important is this and what happens for larger spins? It will be recalled that the restriction to spin $\frac{1}{2}$ was quite crucial in the evaluation of the matrix elements of I_x and the consequence that only states of opposite s_j were coupled. Clearly that would no longer be the case for larger spin values. The calculation would be much more complicated but the above-stated conclusion would be the same. It is just that its mathematical expression would be more complex and we would not have the very elegant expression, Equation (6.25), for $F(t)$.

6.2.3 A model interaction

We shall now examine a particular model (Heims, 1965) for the interactions between spins: a particular form for the dependence of the interaction parameter B_{jk} on the indices j and k. We consider the one-dimensional case of a chain of spins where the interaction halves for each successive 'neighbour', i.e. $B_{jk} \propto 1/2^{|j-k|}$, and we shall write

$$B_{jk} = 2B\hbar/2^{|j-k|}, \tag{6.26}$$

so that the expression for $F(t)$ is then

$$F(t) = \prod_k \cos(Bt/2^{|k|}), \tag{6.27}$$

where we have labelled our representative jth spin to be the zeroth.

This particular form for the interspin interaction does not have any physical justification; it is not claimed to be representative, quantitatively, of a real system. Its principal merit is that it provides a soluble expression for the relaxation function. The product in Equation (6.27) can be evaluated analytically, as we shall see below. Nevertheless, it does have the expected qualitative behaviour of decreasing with increasing interspin separation. (Of course for the dipolar interaction B_{jk} would vary as $|j-k|^{-3}$.)

To evaluate Equation (6.27) we shall consider a circle of N spins (so that all spins may be regarded as equivalent). The spin index will run from $-N/2$ to $N/2$. Then since the cosine is an even function of its argument and recalling that the $k = 0$ term is excluded from the product, we obtain

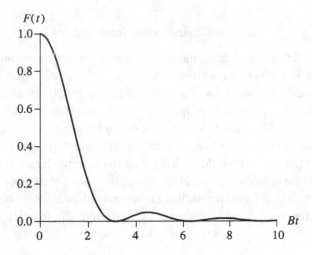

Figure 6.7 Transverse relaxation for model interspin interaction.

$$F(t) = \left[\prod_{k=1}^{N/2} \cos(Bt/2^k) \right]^2. \tag{6.28}$$

Evaluation of this product is based on the identity

$$\prod_{k=1}^{N} \cos(x/2^k) = \frac{\sin(x)/x}{\sin(x/2^N)/(x/2^N)}. \tag{6.29}$$

Proof of this is established in Problem 6.1. In the limit $N \to \infty$ this product reduces to $\sin(x)/x$ so that the expression for $F(t)$ becomes

$$F(t) = \left[\frac{\sin(Bt)}{Bt} \right]^2, \tag{6.30}$$

which is plotted in Figure 6.7.

Observe that the relaxation function, while decaying to zero, does not do so in a monotonic fashion; there is structure in the relaxation. This sort of behaviour accords, certainty qualitatively, with the transverse relaxation observed in solids; recall the fluorine relaxation in CaF_2 crystals, shown in Figure (4.2) and the proton relaxation in gypsum, shown in Figure (6.4).

Box 6.4 Poincaré recurrence

The relaxation function $F(t)$ is seen to decay to zero; this is very important and it is intimately connected with the Second Law of Thermodynamics. However, a crucial step in ensuring this is taking the limit of $N \to \infty$ in Equation (6.29). For finite sized systems it will be seen that while the transverse magnetisation appears to decay to zero, much as described by Equation (6.30), after a very long time it will grow again and regain its original magnitude. The time for this to happen, known as the Poincaré recurrence time, is essentially the inverse of the lowest common multiple of the different precession frequencies of the different spins. This is the time for all the sinusoidal oscillations to come, once again, into phase. True irreversibility occurs only for infinite systems: the 'thermodynamic limit'. However, for systems of a few tens of particles the recurrence time is already many centuries so that while finite systems do not rigorously exhibit irreversibility, the problem is usually only academic.

6.2.4 Magnetically dilute solids

The treatment of the previous section gave an expression for the transverse relaxation function for a static assembly of spins whose interactions were described by a Hamiltonian involving products of only the z components of the magnetic dipole moments. This section considers a realistic application of such an interaction. We explained in Section 6.2.1 that the truncation of the dipolar interaction to the $I_z I_z$ part was permissible where the nuclear dipoles were diluted in a non-magnetic lattice. Thus we shall examine the form of the transverse relaxation in such magnetically dilute solids and since we will be building on the results of the last section: in particular Equation (6.25), the dipoles are considered to have spin $\frac{1}{2}$. The distribution of spins in the solid is assumed to be random so that we approach the product in Equation (6.25) from a statistical point of view (Anderson, 1965).

In considering the relaxation of

$$F(t) = \prod_{k=1}^{N} \cos\left(\frac{B_{jk}t}{2\hbar}\right)$$

it is important to appreciate that this is a product of a very larger number of terms. This means that $F(t)$ will have decayed away when the individual cosines have dropped only very slightly from unity. Let us denote the individual contributions to the precession frequency of our spin by ω_k, i.e.

$$\omega_k = B_{jk}2/\hbar, \tag{6.31}$$

whereupon we will write $F(t)$ as

$$F(t) = \prod_{k=1}^{N} \{1 + [\cos(\omega_k t) - 1]\}, \tag{6.32}$$

where the term $[\cos(\omega_k t) - 1]$ is very small indeed. By expanding the product and resumming the terms we may express this as

$$F(t) = \exp\left\{\sum_{k=1}^{N} [\cos(\omega_k t) - 1] + \ldots\right\},\tag{6.33}$$

where the higher order terms may be neglected compared with the first in the limit of large N. Regarding the sum over N particles as N times the mean value, the expression for $F(t)$ takes the form

$$F(t) = \exp[N\langle\cos(\omega t) - 1\rangle],\tag{6.34}$$

where here ω is the 'local precession frequency' caused by one spin at the site of another. From this we see that to find the transverse relaxation function $F(t)$ we must evaluate the average $\langle\cos(\omega t) - 1\rangle$ which necessitates knowledge of the distribution function for ω. Now ω is given, from Equation (6.17), by

$$\omega = \frac{\mu_0}{4\pi}\frac{3}{4}\frac{\gamma^2\hbar^3}{r^3}(1 - 3\cos^2\theta)$$

$$= \alpha\frac{(1 - 3\cos^2\theta)}{r^3},\tag{6.35}$$

where we have introduced the constant α for simplicity. In terms of this expression for α, the average becomes an average over position; the proportion of spins producing a local frequency between ω and $\omega + d\omega$ is given by the fraction of space having radius between r and $r + dr$ and azimuthal angle between θ and $\theta + d\theta$. That is the average may be evaluated as

$$\langle\ldots\rangle = \int_{\text{all space}} \ldots \frac{2\pi\sin\theta d\theta r^2 dr}{V},\tag{6.36}$$

where V is the volume of the system. Thus we have

$$\langle\cos(\omega t) - 1\rangle = \frac{2\pi}{V}\int_0^\pi d\theta \int_0^\infty dr \left\{\cos\left[\frac{\alpha(1 - 3\cos^2\theta)t}{r^3}\right] - 1\right\}r^2\sin\theta.\tag{6.37}$$

The integral over r may be evaluated through the substitution

$$x = \frac{\alpha(1 - 3\cos^2\theta)}{r^3}t,\tag{6.38}$$

giving

$$\int\limits_0^\infty dr \left\{ \cos\left[\frac{\alpha(1 - 3\cos^2\theta)t}{r^3}\right] - 1 \right\} r^2 dr = \left|\frac{1}{3}\alpha(1 - 3\cos^2\theta)t\right| \int\limits_0^\infty \frac{\cos x - 1}{x^2} dx, \quad (6.39)$$

where the absolute value of $\alpha(1 - 3\cos 2\theta)t/3$ must be used since the cosine (on the left hand side) is an even function of its argument. The integral on the right hand side is a pure number: $-\pi/2$, giving (for positive t)

$$\langle \cos(\omega t) - 1 \rangle = -\frac{\pi^2\alpha}{3V}t \int\limits_0^\pi \left|1 - 3\cos^2\theta\right| \sin\theta d\theta. \quad (6.40)$$

The θ integral evaluates to $8/(3\sqrt{3})$, so that

$$\langle \cos(\omega t) - 1 \rangle = -\frac{8\pi^2\alpha}{9\sqrt{3}V}t, \quad (6.41)$$

giving an exponential transverse relaxation:

$$F(t) = \exp(-t/T_2) \quad (6.42)$$

where

$$\left.\begin{aligned}\frac{1}{T_2} &= \frac{8\pi^2\alpha}{9\sqrt{3}V} \\ &= \frac{\mu_0}{4\pi}\frac{N}{V}\frac{2\pi^2\gamma^2\hbar^3}{3\sqrt{3}}.\end{aligned}\right\} \quad (6.43)$$

This is a remarkable result for a number of reasons. Within the validity of the model, we have derived an exact expression for the relaxation. Furthermore we have found that the relaxation profile is exponential. We shall comment in the next section and later in this chapter on the observation that the relaxation here is exponential. For the present we note that the relaxation is monotonic, with no zeros. This is unusual for solids; both the proton decay in gypsum and the fluorine decay in calcium fluoride were seen to relax in an oscillatory manner.

This section claims to be modelling magnetically dilute solid. It might be worthwhile at this point to recall exactly where the assumption of dilution is fed into the formalism. It happens in two places. Firstly recall that the interspin interaction is taken to be of the form I_zI_z, and in Section 6.2.1 we explained that it was only for a diluted assembly of spins that the dipole interaction could be approximated by just that part. Then we replaced the sum over spins in Equation (6.33) by a statistical average, Equation (6.34), where the treatment which followed assumed a random distribution. It is only with random dilution that the regular distribution of spins (particularly for the important close neighbours) ceases to be important. Thus the effect of dilution differs from that obtained by uniformly expanding a filled lattice. For these reasons the derivation is appropriate for a dilute solid.

6.2.5 Dilute solids in n dimensions

The main purpose of this section is to set the scene and establish some results which will be of use in future discussion. However, there are example of two- and one-dimensional systems which can be, and often are, studied using NMR. Spins absorbed on solid surfaces and within intercalated compounds are essentially two-dimensional, while spins trapped along grain boundaries and dislocation lines in crystals may well form one-dimensional arrays. Some long chain compounds may also approximate a one-dimensional system.

The previous section showed that dilution of the magnetic species in a solid led to a transverse relaxation which was exponential. In this section we shall see that this conclusion only applies in three dimensions. One-dimensional and two-dimensional systems have their own characteristic form for the relaxation of transverse magnetisation. What we will do, then, in this section is to generalise the results of the previous section to n dimensions.

Equation (6.34) for $F(t)$ is unchanged, but here we shall neglect the orientation dependence of the dipole interaction by simplifying Equation (6.35) to

$$\omega = \alpha/r^3.$$

This will ease the discussion, although, in fact, it is a straightforward matter to include the dependence on θ, as considered in Problem 6.3. With the above simplifying assumption the positional average, given in Equation (6.36), becomes

$$\langle\ldots\rangle = \int_{\text{all space}} \ldots \frac{4\pi r^2 \mathrm{d}r}{V}$$

since $4\pi r^2 \mathrm{d}r$ is the volume of the shell of radius r and thickness $\mathrm{d}r$. In general the probability of finding a spin between distance $r + \mathrm{d}r$ is given by the volume of the shell divided by the volume of the whole system:

$$P\mathrm{d}r = \frac{1}{V}\frac{\mathrm{d}V}{\mathrm{d}r}\mathrm{d}r.$$

Now in n dimensions the 'volume' of a hypersphere of radius r is given by

$$V_n = \frac{2\pi^{n/2}r^n}{n\Gamma(n/2)} \tag{6.44}$$

(Huang, 1987), and on differentiation we find

$$\frac{dV_n}{dr} = \frac{2\pi^{n/2}r^{n-1}}{\Gamma(n/2)},$$

so that the procedure for finding the spatial average generalises to

$$\langle\ldots\rangle = \frac{2\pi^{n/2}}{\Gamma(n/2)V}\int_0^\infty \ldots r^{n-1}dr, \tag{6.45}$$

where Γ is the gamma function. The generalisation of Equation (6.37) is now

$$\langle\cos(\omega t) - 1\rangle = \frac{2\pi^{n/2}}{\Gamma(n/2)V}\int_0^\infty \left[\cos\left(\frac{\alpha t}{r^3}\right) - 1\right]r^{n-1}dr. \tag{6.46}$$

In the three-dimensional case one notes that the time dependence for the relaxation becomes explicit immediately upon substituting for r as we saw in Equation (6.39); performing the integral over x gives only a numerical coefficient. Thus we expect to find the variation of $F(t)$ directly on substituting $r = (\alpha t/x)^{1/3}$. This gives

$$\langle\cos(\omega t) - 1\rangle = \frac{2\pi^{n/2}|\alpha t|^{n/3}}{3\Gamma(n/2)V}\int_0^\infty \frac{\cos(x) - 1}{x^{(n/3)+1}}dx. \tag{6.47}$$

From this we see that, since the integral is clearly negative, the transverse magnetisation decays as $\exp(-t^{n/3})$: pure exponential relaxation in three dimensions but not in one and two dimensions.

The integrals over x for $n = 1$, 2 and 3 are given by

$$-3\Gamma(2/3)\sin(\pi/3)/4, \quad -3\Gamma(1/3)\sin(\pi/6)/5 \quad \text{and} \quad -\pi/2 \text{ respectively.}$$

On evaluating the gamma functions and the sines we obtain finally the relaxation functions

$$F(t) = \begin{cases} \exp\left(-0.586\dfrac{N}{l}|\alpha t|^{1/3}\right) & \text{1 dimension} \\[2mm] \exp\left(-1.695\dfrac{N}{A}|\alpha t|^{2/3}\right) & \text{2 dimensions} \\[2mm] \exp\left(-6.580\dfrac{N}{V}|\alpha t|\right) & \text{3 dimensions} \end{cases} \tag{6.48}$$

where l is the length of the one-dimensional system, A is the area of the two-dimensional

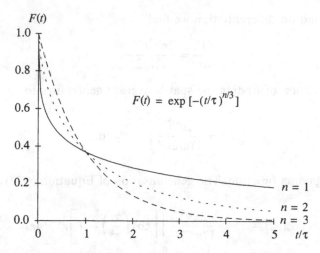

Figure 6.8 Transverse relaxation for magnetically dilute solids in n dimensions.

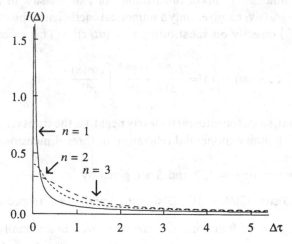

Figure 6.9 NMR absorption lineshapes for magnetically dilute solids in n dimensions.

system and V the volume of the three-dimensional system. The shape of these relaxation profiles is shown in Figure 6.8.

In one dimension there is a very rapid initial decay, but this slows down so that at long times the relaxation becomes rather slow. The corresponding behaviour in the frequency domain is shown in Figure 6.9; the absorption line has wide tails together with a narrow central peak. This is related to the fewer neighbours that a spin sees in one dimension. In higher dimensions a spin sees increasing numbers of distant neighbours, contributing smaller fields and so broadening out the absorption line in the locality of the peak.

6.3 The method of moments

6.3.1 Rationale for the moment method

The original calculation and use of moments was in relation to the absorption spectra of CW NMR and it is to that field that the true meaning of the word moment relates. The problem was to calculate from first principles the NMR absorption lineshape of macroscopic assemblies of nuclear spins. Although it was not possible, in the general case, to obtain a closed expression for the precise form of the line, it was realised quite early on by Waller in 1932 and Van Vleck in 1948 that formal expressions for the moments of the line could be obtained.

The moments of a function $f(x)$ are defined by the expression

$$m_n = \int_{-\infty}^{\infty} x^n f(x)\mathrm{d}x \tag{6.49}$$

for integral n. We see immediately that the zeroth moment m_0 is simply the area of the function f. Also, the first moment m_1 is the sort of thing one encounters when calculating the turning moment or torque of a distributed force $f(x)$ and the second moment m_2 is closely related to the moment of inertia of a body.

In probability theory the function $f(x)$ would be the probability density of the variable x. In this case m_0 will be unity assuming the probability function to be normalised. The first moment is the mean value of x, the second moment the mean square value $\langle x^2 \rangle$ etc.

Alternatively, and of more importance to us, are what are known as *central moments*. If the function $f(x)$ is peaked at some value x_0 then it is the distance of x from x_0 that is considered. The central moments are defined as

$$M_n = \int_{-\infty}^{\infty} (x - x_0)^n f(x)\mathrm{d}x. \tag{6.50}$$

In probability theory usage then M_0 would be unity as before. The first central moment M_1 will be zero if the distribution is symmetric about its peak x, and the second central moment M_2 is the mean square deviation or variance of the distribution.

In an analogous fashion the central moments of a CW NMR line may be considered. In this case it is natural to evaluate the moments about the Larmor frequency ω_0. In the NMR context since (unless explicitly stated to the contrary) only central moments will be considered, we shall leave the adjective central implied, and henceforth the central moments will be referred to simply as moments.

It is instructive at this stage to make the connection with pulsed NMR and the transverse relaxation function, the FID shape $F(t)$. This will enable us to appreciate both the power and the limitations of the moment method. It will also facilitate the derivation of quantum mechanical expressions for the moments.

The absorption of power in a CW NMR experiment is related to the imaginary part of the magnetic susceptibility $\chi''(\omega)$. This was established in Problem 1.2. Also we have seen in Section 2.6.3 that for linear response χ'' is related to the relaxation of the transverse magnetisation through

$$m(t) \propto \int_{-\infty}^{\infty} \chi''(\omega)\exp(-i\omega t)d\omega. \tag{6.51}$$

If we now express the frequency in terms of the deviation from the Larmor frequency

$$\omega = \omega_0 + \Delta, \tag{6.52}$$

then we may change variables in the integral to obtain

$$m(t) \propto \exp(-i\omega_0 t) \int_{-\infty}^{\infty} \chi''(\omega_0 + \Delta)\exp(-i\Delta t)d\Delta, \tag{6.53}$$

where Δ is the distance from the centre of the line. The effect of the change of variables has been to factorise out the precession at ω_0. The integral remaining thus gives the envelope of the relaxation: the FID shape, $F(t)$. Since $F(t)$ is normalised to be unity at zero time we may write it as

$$F(t) = \frac{\displaystyle\int_{-\infty}^{\infty} \chi''(\omega_0 + \Delta)\exp(-i\Delta t)d\Delta}{\displaystyle\int_{-\infty}^{\infty} \chi''(\omega_0 + \Delta)d\Delta}.$$

This simplifies by introducing a normalised lineshape function $I(\Delta)$ as

$$I(\Delta) = \frac{\chi''(\omega_0 + \Delta)}{\displaystyle\int_{-\infty}^{\infty} \chi''(\omega_0 + \Delta')d\Delta'} \tag{6.54}$$

in terms of which the FID is given by

$$F(t) = \int_{-\infty}^{\infty} I(\Delta)\exp(-i\Delta t)d\Delta. \tag{6.55}$$

In this way we have obtained the result that the FID is the Fourier transform of the absorption lineshape function – apart from a factor of 2π which is conspicuous by its absence (see the conventions adopted for Fourier transforms as outlined in Appendix A).

Box 6.5 Comment

The observant reader may wonder at the choice of $\chi''(\omega)$ for the lineshape function considering that the power absorbed in a CW experiment is proportional to $\omega\chi''(\omega)$ as shown in Problem 1.2. From the practical point of view this is no problem because $\chi''(\omega)$ is a function sharply peaked at ω_0 and thus $\omega_0\chi''(\omega)$ and $\omega\chi''(\omega)$ have very similar shapes. From the theoretical point of view (and the real motivation for the particular choice of shape function), it is when using precisely this expression for $I(\Delta)$ that one has the Fourier transform relation between $F(t)$ and $I(\Delta)$.

Also, it should be noted that the $F(t)$ we are using here is not the causal relaxation function, which is zero for negative times, but the time-symmetric function: $F(t) = F^*(-t)$. The quantum mechanical expression for $F(t)$ provides this automatically for us.

It is the Fourier transform relation between $F(t)$ and $I(\Delta)$ which is central to our perspective of the moment method. It enables us to relate the moments of line to the derivatives of $F(t)$. These derivatives may be obtained from our established expression for $F(t)$, such as Equation (5.17).

Let us take Equation (6.55) and on the right hand side expand the exponential as a power series

$$\exp(-i\Delta t) = 1 - i\Delta t + i^2\Delta^2 t^2/2 - \ldots$$

$$= \sum_{n=0}^{\infty} \frac{-i^n\Delta^n t^n}{n!}$$

to give

$$F(t) = \int_{-\infty}^{\infty} d\Delta I(\Delta) \sum_{n=0}^{\infty} \frac{-i^n\Delta^n t^n}{n!}.$$

Upon rearrangement we then have

$$F(t) = \sum_{n=0}^{\infty} \frac{i^n t^n}{n!} \int_{-\infty}^{\infty} I(\Delta)\Delta^n d\Delta \tag{6.56}$$

and the integral here we immediately recognise as the nth moment. Thus we obtain the important result

$$F(t) = \sum_{n=0}^{\infty} \frac{-i^n M_n t^n}{n!} ; \tag{6.57}$$

the moments are essentially the coefficients of the power series expansion of the transverse relaxation function. Knowledge of *all* the moments would thus enable one to predict the form of $F(t)$, assuming it is a sufficiently 'well-behaved' function.

The above result may be expressed more clearly by writing $F(t)$ also as its power series expansion

$$F(t) = F(0) + tF'(0) + t^2 F''(0)/2! + \ldots$$

$$= \sum_{n=0}^{\infty} \frac{t^n F^{(n)}(0)}{n!},$$

whereupon we find

$$\sum_{n=0}^{\infty} \frac{F^{(n)}(0)t^n}{n!} = \sum_{n=0}^{\infty} \frac{-i^n M_n t^n}{n!}.$$

Here we may equate powers of t which gives

$$F^{(n)}(0) = -i^n M_n$$

or

$$M_n = i^n F^{(n)}(0). \tag{6.58}$$

The nth moment is thus essentially the nth derivative of $F(t)$ at the origin.

Since the definition of $F(t)$ in Section (2.6.3) implies the symmetry property

$$F(t) = F^*(-t), \tag{6.59}$$

it follows that at $t = 0$ the even derivatives are real while the odd derivatives are imaginary. Thus, as must be the case, all moments are real. Of course, when the lineshape $I(\Delta)$ is symmetric, when $F(t)$ is entirely real, then all odd moments are zero.

The main traditional use of moments was in the calculation of transverse relaxation in solids: i.e. the shape of $I(\Delta)$ or $F(t)$. Some of these applications will be covered later in this chapter. In fluids, on the other hand, the shape of $F(t)$ usually resembles an exponential and one may use moments to approximate the relaxation time T_2. However, in this case the cumulant method, treated in Chapter 7, may then be used, which has the advantage of (*a*) giving a more physical understanding of the shape of the relaxation functions, and (*b*) permitting a coherent treatment of both transverse and longitudinal relaxation.

When we get down to practicalities in the following sections we shall obtain formal expression for the moments, mainly for systems with symmetrical lineshapes. In principle such moments may be evaluated for a given system. However, while M_2 is relatively easy to calculate, M_4 is somewhat more difficult and M_6 is so difficult that there are very few recorded results of such calculations. The author is aware of only one published calculation of M_8, which will be discussed in Section 8.1.3. Because of the difficulties involved, it is unlikely that many of the higher moments will be evaluated even using computers.

This appears remarkably depressing! A Taylor series expansion about $t = 0$ with so few terms will have dubious value outside the neighbourhood of the origin, while for long times any finite power series will clearly diverge. This is to be contrasted with the reasonable success of some moment calculations in approximating lineshapes, as demonstrated in Section 8.1. The explanation of this paradox: why moment expansions may be valid away from the neighbourhood of the origin, is discussed in Section 6.3.7. But at the simplest level, if the low order moments, say M_2 and M_4, are different for competing 'trial' shapes then a reasonable choice might be made on this basis.

6.3.2 Formal expressions for the moments

It is a straightforward matter to obtain expressions for the moments. In essence, we simply require the derivatives of $F(t)$. The expression for $F(t)$ is given by Equation (6.1):

$$F(t) = \frac{\text{Tr}\{I_+(t)I_-(0)\}}{\text{Tr}\{I_+I_-\}}. \tag{6.60}$$

This is real: an even function of time. This is a consequence of the high temperature approximation used in its derivation. In this case odd moments vanish. Upon successive differentiation the moments are obtained as

$$M_n = i^n \frac{\text{Tr}\left\{\frac{d^n}{dt^n}I_+(t)\Big|_{t=0} I_-(0)\right\}}{\text{Tr}\{I_+I_-\}}.$$

The derivative of a quantum mechanical operator is found by taking he commutator with the Hamiltonian operator; this is the Heisenberg equation:

$$\frac{dI}{dt} = \frac{i}{\hbar}[\mathscr{H},I],$$

but at this stage we must remember that the operator $I_+(t)$ of Equation (6.1) is in an interaction picture. Recall that the precession has been factorised out and, as quoted in Equation (6.4), the relevant Hamiltonian to use in the commutator is the truncated dipolar Hamiltonian, D_0:

$$\frac{dI_+}{dt} = \frac{i}{\hbar}[D_0,I_+].$$

Upon successive differentiation, through application of this formula, we obtain the general expression for the moments:

$$M_n = \left(\frac{-1}{\hbar}\right)^n \mathrm{Tr}\{\underbrace{[D_0,\ldots[D_0,I_+]\ldots]}_{n \text{ commutators}}I_-\}/\mathrm{Tr}\{I_+I_-\}. \tag{6.61}$$

A simplification may be obtained through a rearrangement of the commutator of the multiple trace. If the first commutator in the trace is expanded one obtains

$$\mathrm{Tr}\{D_0\underbrace{[D_0\ldots[D_0,I_+]\ldots]}_{n-1 \text{ commutators}}I_-\} - \mathrm{Tr}\{\underbrace{[D_0,\ldots[D_0,I_+]\ldots]}_{n-1 \text{ commutators}}D_0I_-\}.$$

An important property of the trace is that we can cyclically permute the operators within it. Thus in the first term we can move the first D_0 around to the end giving

$$\mathrm{Tr}\{[D\ldots[D_0,I_+]\ldots]I_-D_0\} - \mathrm{Tr}\{[D_0,\ldots[D_0,I_+]\ldots]D_0I_-\}$$

or

$$\mathrm{Tr}\{[D_0\ldots[D_0,I_+]\ldots](I_-D_0 - D_0I_-)\}.$$

But on the right here we see another commutator. Thus we have

$$\mathrm{Tr}\{\underbrace{[D_0\ldots[D_0,I_+]\ldots]}_{n-1 \text{ commutators}}\underbrace{[I_-,D_0]}_{1 \text{ commutator}}\}$$

What we have done is to move one commutator from the I_+ to the I_-. This may be done repeatedly.

For the case of even moments, we may have $n/2$ commutators on the I_+ and $n/2$ on the I_-. It therefore follows in this case that we may write M_n as

$$M_n = \left(\frac{-1}{\hbar}\right)^n \mathrm{Tr}\{\underbrace{[D_0,\ldots[D_0,I_+]\ldots]}_{n/2 \text{ commutators}}\underbrace{[\ldots[I_-,D_0],\ldots D_0]}_{n/2 \text{ commutators}}\}/\mathrm{Tr}\{I_+I_-\}. \tag{6.62}$$

Recall that we had an expression, Equation (5.11), for the FID in terms of the real I_x before the introduction of the complex I_+ and I_-. Using that as the starting point for differentiation we obtain expressions similar to that above but with I_x instead of I_+ and

I_-. In this case if the second commutator is reversed (introducing a factor of $-1^{n/2}$) the two multiple commutators are equivalent and we then have

$$M_n = \frac{(-1)^{n/2}}{\hbar^n} \operatorname{Tr}\{[\underbrace{D_0, \ldots [D_0, I_x] \ldots]}_{n/2 \text{ commutators}}{}^2\}/\operatorname{Tr}\{I_x^2\}. \tag{6.63}$$

We shall use these two forms interchangeably. The former has the advantage of being slightly easier to calculate with, while the latter has the benefit of being a more compact expression. In particular, we have for the second and the fourth moments:

$$\left.\begin{aligned} M_2 &= \frac{-1}{\hbar^2} \operatorname{Tr}\{[D_0, I_x]^2\}/\operatorname{Tr}\{I_x^2\} \\[2mm] M_4 &= \frac{1}{\hbar^4} \operatorname{Tr}\{[D_0, [D_0, I_x]]^2\}/\operatorname{Tr}\{I_x^2\}. \end{aligned}\right\} \tag{6.64}$$

When expressions such as these were originally derived in the early days of magnetic resonance, it created quite a stir. The normal way to solve a quantum mechanical problem is to find the stationary states: the eigenstates of the Hamiltonian. Recall that even for the simple case of two spins, considered at the start of this chapter, the use of a diagonal representation was vital in performing the calculation. Clearly for a many particle system this is impossible. However, expressions in terms of traces obviate the need to know the eigenstates. An important property of the trace is that it is independent of representation. In other words *any* suitable basis set may be used. The spin states are naturally taken as the eigenstates of $\mathscr{H}_z$ (or equivalently I_z) and in this way realistic calculations on many-body systems were made for the first time.

From our point of view, clearly the same considerations apply to the calculation of the relaxation functions $F(t)$ and $L(t)$ as these are expressed in terms of traces: Equations (5.17) and (5.18). However, in these cases even with the freedom to choose a basis set the difficulties are generally insuperable. Calculation of $F(t)$ would be equivalent to the evaluation of *all* the moments.

6.3.3 Calculation of dipolar moments

We have stated above that the calculation of moments of increasing order becomes rapidly more difficult. In this section we shall give a brief outline of how the moments are calculated. From Equations (6.62)–(6.64) we see that the key steps in the calculation of moments are the evaluation of commutators of D_0 with the spin operators. For simplicity we will restrict consideration to an assembly of identical nuclei of spin $\frac{1}{2}$. In this case D_0 is given, from Equation (5.26), by

$$D_0 = -\frac{\mu_0}{4\pi} \frac{\hbar^2 \gamma^2}{2} \sum_{i \neq j} \frac{1 - 3\cos^2\theta_{ij}}{r_{ij}^3} T_{ij}^0,$$

where the spin flip operator T_{ij}^0 is defined in Equation (5.24). Commutators of D_0 with the spin operators are most easily evaluated through use of the properties of the T_{ij}^m operators given in Equation (5.25). It is convenient, and customary, to denote by b_{ij} the spatial part of the dipolar interaction:

$$b_{ij} = \frac{1 - 3\cos^2\theta_{ij}}{r_{ij}^3}.$$ (6.65)

A straightforward, but tedious, calculation of the single commutators required for M_2 then gives the result

$$M_2 = \left(\frac{\mu_0}{4\pi}\right)^2 \frac{\hbar^2\gamma^4}{4} \sum_k b_{jk}^2.$$ (6.66)

Double commutators are required for M_4. A considerably more complicated calculation then gives the result

$$M_4 = \left(\frac{\mu_0}{4\pi}\right)^4 \frac{\hbar^4\gamma^8}{16}\left[3\left(\sum_k b_{jk}^2\right)^2 - \frac{1}{3N}\sum_{jkl \neq} b_{jk}^2(b_{jl} - b_{kl})^2 - 2\sum_k b_{jk}^4\right],$$ (6.67)

where $jkl \neq$ indicates that the three indices must refer to different sites; no pairs are allowed to be equal. For further details of these calculations the reader is referred to Van Vleck's original 1948 paper and Abragam's 1961 book.

From Equations (6.66) and (6.67) and the corresponding expressions for higher order moments, we see that the evaluation of the moments has now reduced to the calculation of lattice sums for the b_{ij} combinations. Specific examples will be treated in later sections. For the present we restrict discussion to the following points.

If the second and third terms on the right hand side of Equation (6.67) were ignored then we would have $M_4 = 3M_2^2$, which would be consistent with a Gaussian lineshape (see the following section). Numerical evaluation of the moment sums for a solid lattice of nuclear spins typically gives a ratio M_4/M_2^2 of between 2 and 3, indicating that the Gaussian profile might be a reasonable approximation.

6.3.4 Special shapes and their moments

At this stage it is beneficial to pause and to consider some typical model FID shapes (or absorption lineshapes) and to examine the moments of such functions. This will give us a little insight when actual values of moments are calculated, and we should then be in a position to suggest what sort of functions could possibly correspond to such cases.

It is helpful to keep in mind the Fourier transform relation between the FID and the absorption lineshape:

$$F(t) = \int_{-\infty}^{\infty} I(\Delta)\exp(-i\Delta t)d\Delta. \tag{6.68}$$

Let us first consider the Gaussian function, so prevalent in probability theory and encountered in Section 4.4.3 as a transverse relaxation function. In this case we have the Fourier transform pair

$$F(t) = \exp\left(\frac{-t^2}{2T^2}\right) \qquad I(\Delta) = \frac{T}{(2\pi)^{\frac{1}{2}}}\exp\left(\frac{-\Delta^2 T^2}{2}\right). \tag{6.69}$$

Both $F(t)$ and $I(\Delta)$ are Gaussian functions; the width of one is the inverse of the width of the other (the 'uncertainty relation', Appendix A). From the expansion of $F(t)$ (or by direct integration of $I(\Delta)$) the moments are found to be

$$M_2 = 1/T^2 \qquad M_4 = 3/T^4$$

and in general

$$M_{2n} = 1 \times 3 \times 5 \times \ldots \times (2n-1)/T^{2n}$$

or

$$M_{2n} = 1 \times 3 \times 5 \times \ldots \times (2n-1)M_2^n. \tag{6.70}$$

From the experimental point of view we know that exponential relaxation is a common occurrence. An exponential $F(t)$ corresponds to a Lorentzian $I(\Delta)$. These are the functions characteristic of a simply damped harmonic oscillator:

$$F(t) = \exp\left(\frac{-|t|}{T}\right) \qquad I(\Delta) = \frac{T}{\pi(1 + \Delta^2 T^2)}. \tag{6.71}$$

Note that it is the modulus of t that appears in the expression. $F(t)$ must be an even function. However, there is now a problem in finding the moments. $F(t)$ is discontinuous at the origin so that its derivatives there are undefined and, correspondingly, in the frequency domain if the moments are calculated directly from $I(\Delta)$ then except for M_0 the integrals diverge.

The explanation for this is that quantum mechanics forbids exponential decay at short times: the relaxation functions must be smooth and continuous everywhere. Exponential relaxation functions, so frequently observed in many areas of nature, are hydrodynamic phenomena. In other words there is a microscopic coherence time such that for times much less than this the exponential law breaks down. Such behaviour will, in fact, be derived in the following chapter and, of course, it has already been seen in the local field/semiclassical treatment of $F(t)$ in Section (4.3.2).

For the present we rescue the moment method from this disaster in the traditional NMR manner, by truncating the wings of the Lorentzian.

Box 6.6 Moments – why bother?

You may wonder why one should bother to rescue the moment method. The first part of the answer is that moments remain the only rigorous quantities which can be calculated for such systems. The fact is that the use of moments is rather more successful than might at first be expected. Fundamentally we are trying to use short time information to make inferences about long time behaviour. The justification for this will be considered in Section 6.3.7.

Let us look directly at the shape function $I(\Delta)$ and the expression for the moments

$$M_n = \int_{-\infty}^{\infty} \Delta^n I(\Delta) d\Delta.$$

Evaluating the integral directly, all even moments higher than the zeroth diverge; there is too much power in the tails of the Lorentzian.

The fact that for short times $F(t)$ must flatten off translates to the frequency domain as the requirement that at very large frequencies $I(\Delta)$ must decay faster than the Δ^{-2} of the Lorentzian. In fact it must decay faster than any inverse power of frequency for all moments to converge. Thus $I(\Delta)$ must at large frequencies decay exponentially or faster. A straightforward, if somewhat unphysical, approach is simply to truncate $I(\Delta)$ to zero at some sufficiently high frequency. We thus consider the shape function

$$I(\Delta) = \begin{cases} \dfrac{T}{\pi(1 + \Delta^2 T^2)} & |\Delta| < \Delta_c, \\ 0 & |\Delta| > \Delta_c, \end{cases} \tag{6.72}$$

where $\Delta_c \gg 1/T$ is the cut-off frequency. The relaxation function corresponding to this is cumbersome to evaluate, but the moments may be found to leading order of $1/T\Delta_c$ by direct integration:

$$M_2 = \frac{2\Delta_c}{\pi T}, \quad M_4 = \frac{2\Delta_c^3}{3\pi T},$$

$$M_{2n} = \left(\frac{\pi \Delta_c T}{2}\right)^{n-1} \frac{M_2^n}{2n - 1}. \tag{6.73}$$

We see that whereas the Gaussian decay function is characterised by one parameter:

the relaxation time T, the relaxation corresponding to the truncated Lorentzian $I(\Delta)$ involves an extra parameter: the cut off frequency Δ_c. Eliminating this between the expressions for M_2 and M_4 gives an expression for the relaxation time as

$$T = \frac{2\sqrt{3}}{\pi} \frac{M_4^{1/2}}{M_2^{3/2}}. \qquad (6.74)$$

Accepting that only in very rare circumstances can moments higher than M_4 be evaluated, the question naturally arises as to what one can infer on the basis of knowledge of M_2 and M_4 only. At a very elementary level, considering the dimensionless quantity M_4/M_2^2, we have found for the two decay functions considered:

Gaussian decay $\qquad M_4/M_2^2 = 3$;
modified exponential $M_4/M_2^2 = \pi\Delta_c T/3 \gg 1$.

This leads to the conventional wisdom whereby one tries a Gaussian when M_4/M_2^2 is of the order of 3 and a modified exponential decay (truncated Lorentzian lineshape) when M_4/M_2^2 is much greater than 1. Knowledge of higher moments may support or disprove such assertions but it cannot prove that the choice is correct.

The extent to which these choices are justified will become clear in the following sections and chapters.

6.3.5 A second look at magnetically dilute solids

It is possible to approach the problem of magnetically dilute solids, treated in Section 6.2.4, in terms of moments. This will give an insight into the utility and the possible pitfalls of the moment method. Recall that in a magnetically dilute solid we have a lattice, a small fraction f of whose sites are occupied by a nuclear spin. It was that condition which, in Section 6.2.1, permitted the approximation of the dipolar interaction by its $I_z I_z$ part. This way of studying the dilute solid was first described by Kittel and Abrahams in 1953.

The moments for the dilute system are expressed in terms of the fractional concentration f. In the expression for M_2 (Equation (6.66)) there is a single sum over lattice sites k. If only a fraction f of these sites, at random, actually have a nuclear spin then the sum will be that fraction f of the sum with all sites occupied:

$$M_2(f) = f M_2(1).$$

Turning now to Equation (6.67) for M_4 we see that the first term on the right hand side will be proportional to f^2. The sum in the second term is over three indices, but one also divides by the number of spins. So this term is also proportional to f^2. However, the sum in the third term is over a single index and so it is proportional to f. Thus when the concentration of spins f is small then it is the third term which dominates: M_4 will be proportional to f. Similarly with higher order moments, the leading term will be in f. One has

$$M_n(f) \propto f. \tag{6.75}$$

If we use this result in the application of the 'conventional wisdom' discussed in the previous section we find the ratio M_4/M_2^2 as

$$M_4/M_2^2 \propto 1/f.$$

So when f is small this ratio will be large. Under these circumstances we are advised to adopt a Lorentzian lineshape. In Section 6.2.4 we saw that the lineshape of a three-dimensional dilute solid did indeed have a Lorentzian form.

This result has, in the past, been cited as a confirmation of the moment method. But as we learned in Section 6.2.5, the result is peculiar to three dimensions. In one and two dimensions – in chains and films – the situation is markedly different. Fundamentally, the reason for the failure is that knowledge of the second and fourth moments alone is *not* sufficient to make any meaningful inference about the shape of the resonance line. The fact that M_4/M_2^2 is very large does not necessarily mean that all the moments satisfy the Lorentzian criterion, Equation (6.73). The case of the three-dimensional dilute solid must be regarded as no more than a fortunate coincidence.

6.3.6 Real systems

We have already encountered some examples of transverse relaxation in real solid systems. In Figure 4.2 we saw the fluorine relaxation in calcium fluoride crystals and in Figure 6.4 we saw the proton relaxation in gypsum. The simple (classical) model of transverse relaxation presented in Chapter 4 led to a monotonic decay which tended to a Gaussian form in the rigid lattice limit. At the time we indicated some of the weaknesses of the arguments which led to this result, and certainly the model calculations of this chapter are in accord with the phenomenon of oscillations in the relaxation function, observed in both calcium fluoride and gypsum. We examined the gypsum relaxation in some detail at the start of this chapter; it provided a good example of the Pake doublet. Calcium fluoride has been much studied over the years. The reasons for this and the results of the investigations will be considered in depth in Chapter 8.

In gypsum the dominant interaction was between *pairs* of protons. This led to the splitting of the resonance into two lines which were then smeared out by the distribution of orientations of the pairs. Then a further broadening of the resonance resulted from the *inter*molecular interactions. In contrast to this, in calcium fluoride we have a true many-body system with each nuclear spin interacting with every other. We saw that fine detail was smoothed out of the resonance line and it was really for that

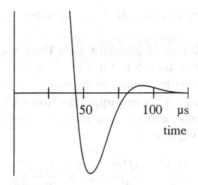

Figure 6.10 Proton relaxation in n-H_2.

reason that the moment method was introduced. We shall give one further example in this section for completeness, which will be useful for making comparisons.

Figure 6.10 shows the transverse proton relaxation in n-H_2. These measurements, at 4.2 K, were made by Metzger and Gaines in 1966. Once again one observes the characteristic damped oscillatory shape of the relaxation. However, while the zeros of the calcium fluoride relaxation appear evenly spaced, those of the hydrogen relaxation are not.

6.3.7 Widom's Theorem

In this section (which may be omitted on a first reading) we consider the question of what *exact* results may be obtained from moment calculations, in particular, from a limited number of moments. The theorem we shall derive is based on unpublished work of Alan Widom of Northeastern University.

By way of introduction let us consider the FID of calcium fluoride. The shape of the FID shown in Figure 4.2. While a Gaussian function gives a fair approximation of the resonance absorption $I(\Delta)$, it does not work so well for the FID. The Gaussian $F(t)$ cannot reflect the oscillations in the relaxation function. So what does one do?

Abragam made the remarkable observation that the fluorine relaxation in calcium fluoride was described extremely well by the function

$$F(t) = \exp\left(-\frac{a^2 t^2}{2}\right)\frac{\sin(bt)}{bt}$$

when the parameters a and b are fitted to conform to the calculated M_2 and M_4. Is this a coincidence? Although the *specific* case of calcium fluoride will be considered in detail in Section 8.1, in this section we will discuss the generalities. In particular, to what extent is $F(t)$ determined by M_2 and M_4? This question may be answered in the following way.

If one knows a finite number of moments then these may be assembled into a polynomial in t for an approximation to the FID. The true FID is an infinite series requiring knowledge of an infinite number of moments. To what extent can the finite series be said to approximate $F(t)$? We may appeal to Taylor's Theorem, which proves a power series expansion together with information on the error introduced in terminating after a finite number of terms.

$$F(t) = 1 - \frac{M_2 t^2}{2!} + \frac{M^4 t^4}{4!} - \ldots + \frac{M_{n-2} t^{n-2}}{(n-2)!} + R_n,$$

where R_n is the remainder term due to truncating the series at the nth term.

Now the remainder is given in the Lagrange form (Kreyszig 1993) by

$$R_n = \frac{F^{(n)}(\tau) t^n}{n!},$$

where τ is some time between 0 and t. The nth derivative of $F(t)$ may be related to $I(\Delta)$ through

$$F^{(n)}(\tau) = \int_{-\infty}^{\infty} I(\Delta)(-i\Delta)^n \exp(-i\Delta\tau) d\Delta$$

or, since only even moments need be considered,

$$F^{(n)}(\tau) = \int_{-\infty}^{\infty} I(\Delta)(-1)^{n/2} \Delta^n \cos(\Delta\tau) d\Delta.$$

We are able to derive a bound on $F^{(n)}(\tau)$ based on the important property that $I(\Delta)$ must always be positive; passive systems dissipate energy. Taking the modulus of this expression, we have:

$$|F^{(n)}(\tau)| = \int_{-\infty}^{\infty} I(\Delta)\Delta^n |(-1)^{n/2} \cos(\Delta\tau)| d\Delta,$$

but since the cosine goes only between $+1$ and -1 it follows that

$$|(-1)^{n/2} \cos\Delta\tau| \leq 1$$

and since

$$\int\limits_{-\infty}^{\infty} I(\Delta)\Delta^n d\Delta = M_n,$$

we have the result

$$|F_n(\tau)| \le M_n$$

and thus for the remainder R_n we conclude

$$|R_n| \le \frac{t^n M_n}{n!}.$$

This result has an illuminating graphical interpretation. Let us start with the zeroth approximation. Here we have

$$F(t) = R_0,$$

where

$$R_0 \le M_0 = 1.$$

The condition on $F(t)$ then is that

$$|F(t)| \le 1,$$

so on a graph we may shade the forbidden area; the allowed area is between $+1$ and -1 as shown in Figure 6.11.

Now let us consider the next approximation: the second. Here we have

$$F(t) = 1 + R_2,$$

where

$$|R_2| \le M_2 t^2/2.$$

In this case the area under $1 - M_2 t^2/2$ is now also forbidden, as shown in Figure 6.12.

The next approximation is the fourth. In this case:

$$F(t) = 1 - M_2 t^2/2 + R_4$$

where

$$|R_4| \le M_4 t^4/4!,$$

and now the area above $1 - M_2 t^2/2 + M_4 t^4/4!$ is also forbidden (see Figure 6.13). In this way we see how successive approximations, further moments, eliminate more areas of the plane. We shall refer to this result as Widom's Theorem.

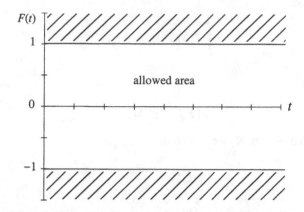

Figure 6.11 Allowed area resulting from knowledge of *no* moments.

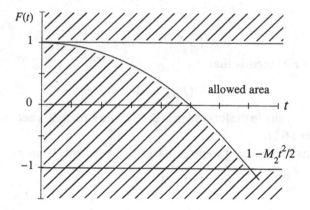

Figure 6.12 Allowed area resulting from knowledge of the second moment.

This result gives rigorous bounds on what the form of the FID may be on knowledge of a given number of moments. In particular it demonstrates the extent to which finite moment expansions are valid away from the neighbourhood of the origin. It is possible that on the basis of knowing only M_2 and M_4 one has a sufficient condition for oscillations in $F(t)$. At least one zero must occur in $F(t)$ if $M_4/M_2^2 < 3/2$ (Problem 6.4).

As an example of the utility of the Widom Theorem we show in Figure 6.14 the various approximants to the calcium fluoride FID using up to the eighth moment – the highest one calculated. Observe how successive approximants 'peel off' from the true behaviour. Although the first zero occurs at about 21 μs, we see that knowledge of the

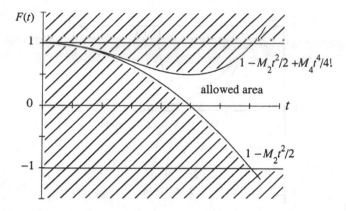

Figure 6.13 Allowed area resulting from knowledge of the second and fourth moments.

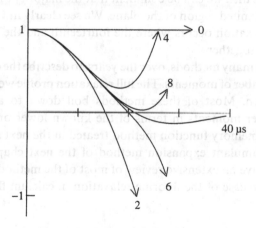

Figure 6.14 Approximants to the calcium fluoride FID based on second, fourth, sixth and eighth moments.

eighth moment is still not quite sufficient to guarantee its existence. As a final example we show a Gaussian function and its calculated approximants using up to the sixteenth moment (Figure 6.15).

This, then, is the 'rehabilitation' of the moment method, promised in Box 6.6. It is clear that knowledge of the first few moments really does confine the magnetisation to a limited part of the relaxation plane and when the excluded part of the plane approaches

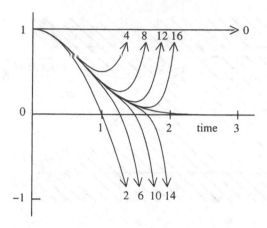

Figure 6.15 Approximants to a Gaussian relaxation.

closely to the zero line then we can be confident that the majority of the relaxation will
have occurred in the limited region of the plane. We see clearly in Figure 6.15 that the
major part of the relaxation occurs where the fourteenth and the sixteenth approximants are very close together.

There have evolved many methods, over the years, to describe the entire relaxation in
terms of a limited number of moments. The full relaxation profile would be known if *all*
moments were known. Most of these methods boil down to approximating the
unknown higher order moments in terms of the known lower order moments in a
systematic way. The memory function method treated in the next section is one such
technique and the cumulant expansion method of the next chapter is another. In
Section 8.1 we shall give an extensive review of most of the methods which have been
applied to the specific case of the fluorine relaxation in calcium fluoride.

6.4 Memory functions and related methods

6.4.1 The memory equation

There have been a number of techniques developed for the calculation (really
estimation) of relaxation functions from the limited information available in the form of
a finite number of moments. As we have discussed before, spin systems are of interest in
this respect because they are both sufficiently complex to yield interesting irreversible
behaviour and at the same time sufficiently simple to permit some analytic calculation.
Of the different techniques developed, a number have evolved which can be expressed
in terms of the so-called memory equation.

The memory equation for the relaxation function $F(t)$ is an equation of motion which expresses the derivative of $F(t)$ as a linear function of $F(t)$ for all earlier times. The equation of motion is written in terms of a *memory function* $\Phi(t)$ as:

$$\frac{\mathrm{d}}{\mathrm{d}t}F(t) = -\int_0^t \Phi(\tau)F(t-\tau)\mathrm{d}\tau. \qquad (6.76)$$

In some approaches this equation is *derived* from a quantum mechanical expression for $F(t)$. Such calculations, of course, give an expression for $\Phi(t)$ in terms of the system parameters. However, derivation of Equation (6.76) is not actually necessary; the memory equation can simply be regarded as the *definition* of the memory function $\Phi(t)$ (Zwanzig, 1961).

One merit of the memory function formalism is that it gives a clear indication of the emergence of exponential relaxation as a consequence of the separation of the time scales of the relaxation function and the memory function. If $\Phi(t)$ decays to zero much faster than does $F(t)$ then Equation (6.76) may be approximated by

$$\frac{\mathrm{d}}{\mathrm{d}t}F(t) = -F(t)\int_0^\infty \Phi(\tau)\mathrm{d}\tau, \qquad (6.77)$$

which indicates exponential relaxation with a decay rate T^{-1} of

$$\frac{1}{T} = \int_0^\infty \Phi(\tau)\mathrm{d}\tau, \qquad (6.78)$$

the area under the memory function.

Essentially, interest has been transferred from the relaxation function to the memory function. In general this is no help; the details of $\Phi(t)$ must be determined in order to find the behaviour of $F(t)$. However, particularly in fluid systems, there *is* a separation of the time scales of $F(t)$ and $\Phi(t)$ and then only the gross features of $\Phi(t)$ are required to give the fine details of $F(t)$. In general, if one starts with some limited knowledge of $F(t)$, in the form of a finite number of moments, then limited features of $\Phi(t)$ may be determined. One can then use *plausible* assumptions to complete the picture of $\Phi(t)$, to work back to an expression for $F(t)$. When the separation of time scales is applicable, the required assumptions are minimal.

6.4.2 Laplace transformation

The memory function equation of motion has the form of a (causal) convolution. As such, it is conveniently manipulated after Laplace transformation. We define the transforms of $F(t)$ and $\Phi(t)$ to be $f(s)$ and $\phi(s)$:

$$f(s) = \int_0^\infty F(t)\exp(-st)dt,$$

$$\left.\right\}\qquad(6.79)$$

$$\phi(s) = \int_0^\infty \Phi(t)\exp(-st)dt.$$

In terms of these the memory of equation of motion, Equation (6.76), can be expressed as

$$f(s) = \frac{F(0)}{s + \phi(s)}$$

or, since the system is almost always normalised so that $F(0)$ is unity,

$$f(s) = \frac{1}{s + \phi(s)}.\qquad(6.80)$$

From this equation it can be seen how simple the relation between the relaxation function and the memory function is in the Laplace domain and exponential relaxation of $F(t)$ is seen to arise when $\phi(s)$ is a constant.

6.4.3 Moment expansions

The definition of moments arises from the absorption function $I(\Delta)$, which is the Fourier transform of the relaxation function $F(t)$:

$$I(\Delta) = \frac{1}{2\pi} \int_{-\infty}^\infty F(t)\exp(i\Delta t)dt\qquad(6.81)$$

and the inverse of this transform is:

$$F(t) = \int_{-\infty}^\infty I(\Delta)\exp(-i\Delta t)d\Delta.\qquad(6.82)$$

The nth moment M_n is defined by

$$M_n = \int_{-\infty}^\infty \Delta^n I(\Delta)d\Delta,\qquad(6.83)$$

where the normalisation $F(0) = 0$ is assumed, which now ensures that the area under $I(\Delta)$, the zeroth moment, is unity. By the expansion of the exponential function in Equation (6.82) we see, as we have already established, that

$$F(t) = \sum_{n=0}^{\infty} \frac{(-i)^n}{n!} M_n t^n \qquad (6.84)$$

and we may write a similar time expansion of the memory function $\Phi(t)$ in terms of 'moments' K_n as

$$\Phi(t) = \sum_{n=0}^{\infty} \frac{(-i)^n}{n!} K_n t^n. \qquad (6.85)$$

It should be emphasised that the K_n are not true moments as K_0 is not equal to unity: $\Phi(t)$ has not been normalised to unity at $t = 0$.

Since the Laplace transform of t^n is $n!/s^{n+1}$ we may evaluate $f(s)$ and $\phi(s)$ directly from Equations (6.84) and (6.85) to give the asymptotic expansions

$$\left. \begin{aligned} f(s) &\sim \sum_{n=0}^{\infty} \frac{(-i)^n M_n}{s^{n+1}} \\ \phi(s) &\sim \sum_{n=0}^{\infty} \frac{(-i)^n K_n}{s^{n+1}}. \end{aligned} \right\} \qquad (6.86)$$

This shows that the short time behaviour of $F(t)$ is related to the large-s behaviour of $f(s)$, and similarly for $\Phi(t)$ and $\phi(s)$.

Comparing Equation (6.79) and $f(s)$ and Equation (6.81) for the lineshape function $I(\Delta)$ we see that $I(\Delta)$ may be found from

$$I(\Delta) = \frac{1}{\pi} \Re_e f(s = -i\Delta),$$

which can be expressed as

$$I(\Delta) = \frac{\varphi'(\Delta)}{[\Delta - \pi\varphi''(\Delta)]^2 + [\pi\varphi'(\Delta)]^2}, \qquad (6.87)$$

where $\varphi'(\Delta)$ and $\varphi''(\Delta)$ are the cosine and sine transforms of the memory kernel:

$$\varphi'(\Delta) = \frac{1}{\pi} \int_0^{\infty} \Phi(t)\cos(\Delta t)dt, \qquad (6.88)$$

$$\varphi''(\Delta) = \frac{1}{\pi} \int_0^{\infty} \Phi(t)\sin(\Delta t)dt. \qquad (6.89)$$

6.4.4 Relations between the moments of F(t) and Φ(t)

The relaxation function and the memory function are related: by Equation (6.76) in the time domain, or by Equation (6.80) in the s domain. Using these relations to connect the time expansions of $F(t)$ and $\Phi(t)$ or the s asymptotic expansions of $f(s)$ and $\phi(s)$, relations between the moments M and K may be established. Thus expressing the Ms in terms of the Ks we have:

$$
\left.
\begin{aligned}
M_0 &= 1, \\
M_1 &= 0, \\
M_2 &= K_0, \\
M_3 &= K_1, \\
M_4 &= K_2 + K_0^2, \\
M_5 &= K_3 + 2K_0K_1, \\
M_6 &= K_4 + K_1^2 + 2K_0K_2 + K_0^3, \\
M_7 &= K_5 + 2K_1K_2 + 2K_0K_3 + 3K_0^2K_1, \\
M_8 &= K_6 + K_2^2 + 2K_1K_3 + 2K_0K_4 + 3K_0K_1^2 + 3K_0^2K_2 + K_0^4,
\end{aligned}
\right\}
\tag{6.90}
$$

where we note that $M_0 = 1$, as it was specified above.

6.4.5 Gaussian memory function

We have seen that a reasonable zeroth approximation to the rigid lattice lineshape is a Gaussian. This relies on knowledge of M_2 only. A Gaussian approximation for the memory function would be expected to be an improvement (Mehring, 1976). Thus we are led to consider an expression for $\Phi(t)$ of the form

$$
\Phi(t) = K_0 \exp\left(-\frac{K_2 t^2}{2K_0}\right)
\tag{6.91}
$$

or, using Equations (6.90) to express this in terms of the moments M_n,

$$
\Phi(t) = M_2 \exp\left[-\frac{(M_4 - M_2^2)t^2}{2M_2^{\,5}}\right].
\tag{6.92}
$$

The cosine and sine transforms of $\Phi(t)$ are found to be

$$
\left.
\begin{aligned}
\varphi'(\Delta) &= \frac{1}{(2\pi)^{\frac{1}{2}}} \frac{M_2^{3/2}}{(M_4 - M_2^2)^{\frac{1}{2}}} \exp\left(-\frac{M_2\Delta^2}{2(M_4 - M_2^2)}\right), \\[2ex]
\varphi''(\Delta) &= \frac{M_2^2\Delta}{\pi(M_4 - M_2^2)} \exp\left(-\frac{M_2\Delta^2}{2(M_4 - M_2^2)}\right) {}_1F_1\left[\frac{1}{2};\frac{3}{2};\frac{M_2\Delta^2}{2(M_4 - M_2^2)}\right],
\end{aligned}
\right\}
\tag{6.93}
$$

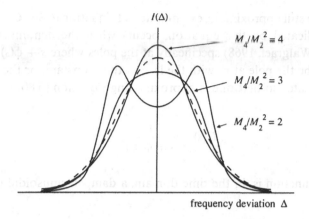

Figure 6.16 Lineshapes for various ratios of M_4/M_2^2. Dashed line is a Gaussian.

where $_1F_1$ is a confluent hypergeometric function (Abramowitz and Stegun, 1965). These expressions may then be substituted into Equation (6.87) to obtain the lineshape function. Unfortunately the hypergeometric function is sufficiently intractable to preclude an analytic expression for the relaxation in the time domain.

Figure 6.16 shows predicted absorption lineshapes for various ratios of the parameter M_4/M_2^2. The dashed curve shows the profile of a Gaussian line. When the ratio M_4/M_2^2 is 4 we observe a slight narrowing as compared with the Gaussian. For $M_4/M_2^2 = 3$, the peak of the line exhibits appreciable flattening, while for $M_4/M_2^2 = 2$ there are two peaks to the curve. This is rather suggestive of the gypsum powder lineshape; however, that is probably just fortuitous. Comparison with the lineshape of calcium fluoride will be made in Section 8.1.

The direct use of this technique, where one makes a guess for the memory function instead of making a guess for the relaxation function, is not a great improvement in treating relaxation in the general case, although it is successful when the memory function decays sufficiently rapidly (separation of time scales). In that case it leads, of course, to exponential relaxation. This is particularly so in the presence of motion, as we shall see in the following chapter.

The structure of the memory equation is such that if there is a time beyond which the memory kernel has decayed to zero while the relaxation function has not, then for longer times the relaxation function will decay exponentially. This is inferred from a crude examination of the Laplace transformed memory equation, Equation (6.80), where one approximates $\phi(s)$ by $\phi(0)$. However, if this separation of time scales is not valid then it is still possible to discuss the long time behaviour of the relaxation through a deeper, although still approximate, examination of Equation (6.80). Clearly the main behaviour, as indicated by this equation, occurs when the denominator is small (Borckmans and Walgraef, 1968) specifically at the poles where $s + \phi(s)$ vanishes. At long times it will be the pole at s_0 with the largest time constant, or the smallest real part, which dominates so that one is approximating Equation (6.80) by

$$F(s) = \frac{1}{s + \phi(s_0)}.$$

Now a one-pole function is, in the time domain, a damped sinusoidal oscillation

$$F(t) = [a\cos(\Omega t) + b\sin(\Omega t)]\exp\left(-\frac{t}{\tau}\right), \tag{6.94}$$

where $1/\tau$ and Ω are the real and imaginary parts of s_0, which must be found from the particular memory function. In the Gaussian case one must resort to numerical evaluation.

So while to zeroth order one concludes that the memory function formalism implies an exponential decay at sufficiently long times, to first order we conclude that there may be sinusoidal oscillations in the exponential decay.

In the following subsection we shall give a more physical explanation of this behaviour.

6.4.6 Exponential memory function

Since the memory function is itself a form of relaxation function, one should be able to evaluate it with the aid of a further memory function. That procedure could be continued recursively and it leads to an expression for relaxation, in the s plane, as a continued fraction (Mori, 1965). We will not consider this here, but we shall take the first step by considering an exponential memory function which is known to be appropriate asymptotically to zeroth order. We shall see that such a memory function will actually lead to Equation (6.94) as the exact solution.

We thus consider a memory function of the form

$$\Phi(t) = M_2 \exp(-t/\tau),$$

where τ is a characteristic time. We shall take this as a parameter for the system rather than relating it to moments of the resonance, since we do not expect this behaviour to continue right down to $t = 0$. In the s domain this $\Phi(t)$ corresponds to

$$\phi(s) = \frac{M_2}{s + 1/\tau},$$

so, from Equation (6.80), we see that $f(s)$ is given by

$$f(s) = \frac{1}{s + \dfrac{M_2}{s + 1/\tau}}$$

$$= \frac{1 + s\tau}{M_2\tau + s + s^2\tau}.$$

This may be split into partial fractions:

$$f(s) = \frac{\frac{1}{2}[1 + (1 - 4M_2\tau^2)^{-\frac{1}{2}}]}{s + [1 - (1 - 4M_2\tau^2)^{\frac{1}{2}}]/2\tau} + \frac{\frac{1}{2}[1 - (1 - 4M_2\tau^2)^{-\frac{1}{2}}]}{s + [1 + (1 - 4M_2\tau^2)^{\frac{1}{2}}]/2\tau},$$

from which the behaviour of the relaxation function is immediately apparent; it is the sum of two (possibly complex) exponentials. The numerators give the amplitude of each component, while in the denominator the Laplace variable s is added to the respective relaxation rate. Thus according to the sign of the $1 - 4M_2\tau^2$, the FID will either comprise the sum of two decaying exponentials or an exponentially damped sinusoid.

If we denote the two (possibly complex) relaxation rates by γ_+ and γ_-

$$\gamma_\pm = -\frac{1 \pm (1 - 4M_2\tau^2)^{\frac{1}{2}}}{2\tau} \tag{6.95}$$

then the FID may be written as

$$F(t) = \frac{\gamma_- \exp(\gamma_+ t) - \gamma_+ \exp(\gamma_- t)}{\gamma_- - \gamma_+}. \tag{6.96}$$

When τ is large $1 - 4M_2 t^2$ will be negative, leading to oscillations at a frequency Ω given by

$$\Omega = \frac{(4M_2\tau^2 \quad 1)^{\frac{1}{2}}}{2\tau}$$

with $F(t)$ given by

$$F(t) = \left[\frac{\sin(\Omega t)}{2\Omega\tau} + \cos(\Omega t)\right]\exp\left(-\frac{t}{2\tau}\right).$$

We see that this corresponds to the general asymptotic result, Equation (6.94), but now we have values for the various parameters, which follow from this particular model. We also have a criterion for the existence of oscillations, when $4M_2\tau^a$ is greater than unity, but this result is model dependent.

7

Relaxation in liquids

7.1 Transverse relaxation: moments

7.1.1 Relaxation in liquids

The previous chapter was devoted to a consideration of solid systems, where the spin magnetic moments were immobile, located at fixed positions. We examined the transverse relaxation which occurred in such cases and we saw that the relaxation profile could become quite complex. In the present chapter we turn our attention to fluid systems. Here the spins are moving and as a general rule the relaxation profile becomes much simpler. Our fundamental task will be to examine the effect of the particle motion upon the relaxation.

There is an important distinction, from the NMR point of view, between liquids and gases. In a gas the atoms or molecules spend the majority of their time moving freely, only relatively occasionally colliding with other particles. The atoms of a liquid, however, are constantly being buffeted by their neighbours. This distinction is relevant for relaxation mediated by interparticle interactions; clearly their effects will be much attenuated in gases. On the other hand, when considering relaxation resulting from interactions with an inhomogeneous external magnetic field, interparticle collisions are unimportant except insofar as they influence the diffusion coefficient of the fluid. However, since diffusion can be quite rapid in gases, some motional averaging of the inhomogeneity of the NMR magnet's B_0 field can then occur. As a consequence, the usual spin-echo technique will no longer recover the transverse magnetisation lost due to the imperfect magnet.

In molecules there are two types of interparticle interaction which must be considered. Interactions between nuclear spins in the *same* molecule are averaged away relatively inefficiently by molecular motion. The movement of the particles is highly correlated, with the spins tumbling about each other, while remaining in close proximity; only the θ and ϕ parts of the interparticle coordinates are modulated, the r is constant. This is different from the weaker effects of interactions between nuclear spins in *different* molecules where all r, θ and ϕ coordinates are modulated by the motion. Since the intramolecular interactions are so much stronger, these determine the relaxation of a first approximation. The effects of the intermolecular interactions are then to add some further smoothing to the absorption lineshape, just as we saw in the case of solids.

We start by examining transverse relaxation in mobile systems, with a brief consideration of the method of moments. This will be followed by the introduction of the cumulant expansion technique. Here parallels with the classical treatment of Chapter 4 will be seen most clearly; one of our aims will be to place the arguments of that chapter on a quantum mechanical footing. However, we will also treat longitudinal relaxation in a similar way. The result will be a unified view of spin–spin and spin–lattice relaxation. This may be compared with the classical treatment of Appendix E. Later in the chapter, with the benefit of hindsight, we shall return to the moment method and examine some general features of relaxation in fluid systems from this point of view.

7.1.2 Interaction picture

In Chapter 5 the expression for the transverse relaxation function was derived:

$$F(t) = \text{Tr}\{I_+(t)I_-(0)\}/\text{Tr}\{I_+I_-\}. \tag{7.1}$$

The time evolution of the spin operator $I_+(t)$ here, which we wrote as

$$I_+(t) = S(t)I_+S^+(t), \tag{7.2}$$

had the Larmor precession factorised out from it. In other words, $F(t)$ describes the *envelope* of the oscillating transverse magnetisation.

In the absence of motion, where the Hamiltonian was written as the sum of the Zeeman and the dipole terms

$$\mathcal{H} = \mathcal{H}_z + \mathcal{H}_d, \tag{7.3}$$

the full quantum mechanical evolution is generated by the operator

$$U(t) = \exp[i(\mathcal{H}_z + \mathcal{H}_d)t/\hbar].$$

The precession is factorised out by going to an interaction picture (transforming to the rotating frame) and the interaction picture evolution operator $S(t)$ was defined by

$$\exp[i(\mathcal{H}_z + \mathcal{H}_d)t/\hbar] = S(t)\exp(i\mathcal{H}_z t/\hbar),$$

where from Appendix C we saw that $S(t)$ could be expressed as

$$S(t) = \exp\left[\frac{i}{\hbar}\int_0^t \mathcal{H}_d(\tau)d\tau\right], \tag{7.4}$$

the time dependence of $\mathcal{H}_d(\tau)$ here being given by

$$\mathcal{H}_d(\tau) = \exp(i\mathcal{H}_z\tau/\hbar)\mathcal{H}_d\exp(-i\mathcal{H}_z\tau/\hbar). \tag{7.5}$$

We examined the nature of the dipolar interaction in Section 5.5 where we saw that for like spins the interaction could be decomposed into its spin-flip components as

$$\mathcal{H}_d = \sum_{m=-2}^{2} D_m \tag{7.6}$$

where, from Equations (5.24), (5.25) and (5.26), the D_m have the following commutation relation with the Zeeman Hamiltonian:

$$[D_m, \mathcal{H}_z] = m\omega_0 D_m. \tag{7.7}$$

From this it follows that the equation of motion for $D_m(\tau)$ may be integrated to give

$$D_m(\tau) = \exp(im\omega\tau)D_m \tag{7.8}$$

and thus

$$\mathcal{H}_d(\tau) = \sum_{m=-2}^{2} D_m\exp(im\omega_0\tau). \tag{7.9}$$

It is this expression for $\mathcal{H}_d(\tau)$ which must be integrated in Equation (7.4) for the evolution operator $S(t)$.

When the integration is performed the rapid oscillations of the $m \neq 0$ terms cause their contributions to become vanishingly small; the $m = 0$ term thus makes the overwhelming contribution. This was the justification in keeping only the D_0 part of the dipolar Hamiltonian in Equation (6.4) and the ensuing treatment of Chapter 6.

The above discussion summarised the situation when the motion of the particles could be ignored. Now we turn to the general case, where $\mathcal{H}_m$ cannot be neglected. The full evolution operator is now

$$U(t) = \exp[i(\mathcal{H}_d + \mathcal{H}_m + \mathcal{H}_z)t/\hbar]$$

and we must, as before, factorise out the Larmor precession to give an interaction picture $S(t)$ for use in our expression for the transverse relaxation function $F(t)$. The most straightforward procedure is simply to add $\mathcal{H}_m$ to $\mathcal{H}_d$. Thus we are factorising the precession term and defining $S(t)$ by

$$\exp[i(\mathcal{H}_d + \mathcal{H}_z)t/\hbar] = S(t)\exp(i\mathcal{H}_z t/\hbar). \tag{7.10}$$

In this case the previously established results for the solid case (particularly the calculation of moments) can be carried over to the present case simply by the substitution

$$\mathscr{H}_d \to \mathscr{H}_d + \mathscr{H}_m. \tag{7.11}$$

We shall adopt this procedure in the next section on moments, but then, in later sections, we will find a different partition to be of use.

7.1.3 Moments and motion

An extensive discussion of moments was presented in the last chapter in relation to transverse relaxation in solids. General expressions were derived for the moments of a dipolar system. We saw that (at high temperatures) the odd moments vanished while the even moments could be expressed as traces of commutators of various operators. It is worth emphasising again here the important point about using moments. Whereas the full evolution of a system cannot be calculated, moments can be evaluated without knowledge of the system's exact eigenstates; the trace is invariant under a change of representation. By this means some of the gross features of the absorption line or the relaxation profile may be calculated from first principles.

In order to evaluate the moments for a system of mobile nuclear dipoles we shall adopt the partitioning of the Hamiltonian contributions according to Equation (7.10). We can then carry over the moment results of the previous chapter, simply by adding $\mathscr{H}_m$ whenever the dipolar Hamiltonian occurs. Thus Equation (6.63) becomes

$$M_n = \frac{(-1)^{n/2}}{\hbar^n} \frac{\mathrm{Tr}\{[D_0 + \mathscr{H}_m, \ldots [D_0 + \mathscr{H}_m, I_x] \ldots]^2\}}{\mathrm{Tr}\{I_x^2\}} \tag{7.12}$$

for even n. But an important simplification is possible here. Since $\mathscr{H}_m$ commutes with I_x, it follows that in evaluating the innermost commutator the term in $\mathscr{H}_m$ will vanish. Thus only the D_0 need be retained in the innermost commutator. No such cancellations will occur in the other commutators, but there, if the motion is rapid, the dipolar Hamiltonian may be neglected in comparison with that for the motion. The second and the fourth moments are given by

$$M_2 = \frac{1}{\hbar^2} \frac{\mathrm{Tr}\{[D_0, I_x]^2\}}{\mathrm{Tr}\{I_x^2\}}, \tag{7.13}$$

$$M_4 = \frac{1}{\hbar^4} \frac{\mathrm{Tr}\{[D_0 + \mathscr{H}_m, [D_0, I_x]]^2\}}{\mathrm{Tr}\{I_x^2\}} \tag{7.14}$$

or, for rapid motion i.e. $\mathscr{H}_m \gg D_0$,

$$M_4 = \frac{1}{\hbar^4} \frac{\mathrm{Tr}\{[\mathscr{H}_m, [D_0, I_x]]^2\}}{\mathrm{Tr}\{I_x^2\}}. \tag{7.15}$$

We see that the second moment is independent of the motion of the spins. This is an important result. As we saw, mathematically this result follows because the motion

Hamiltonian commutes with the Zeeman Hamiltonian. But physically what does it mean? Motion is something which happens in the ordinary x–y–z space whereas the Zeeman interaction operates in spin space. These are entirely separate spaces and thus their operators commute. It is only the existence of the 'interaction' Hamiltonian, here dipolar, with both spin and space variables, which permits the motion to have an effect on the spin variables. This couples the otherwise non-interacting subsystems. Section 7.1.5 will support this conclusion from a slightly different point of view.

The effect of motion is seen to be to increase M_4 (and higher order moments), while leaving M_2 unchanged. Since M_4 will then become large compared with its rigid lattice value of ≈ 3, we see that this is *consistent* with having a Lorentzian line profile or an exponential free induction decay.

7.1.4 Descriptions of motion

It is usually the case that a quantum description of the motion of atoms or molecules in a fluid is inappropriate. The motion is very complex and ultimately some sort of statistical approach must be used. Clearly the precise details of the motion Hamiltonian are unimportant. Moreover, it is likely that the motion is actually visualised classically rather than as a quantum process.

A convergence between the classical and the quantum views may be seen through the consideration of a different interaction picture. We shall 'factorise out' the particle motion as well as the Larmor precession. Thus the definition of $S(t)$ is now taken as

$$\exp[i(\mathscr{H}_d + \mathscr{H}_m + \mathscr{H}_z)t/\hbar] = S(t)\exp(i\mathscr{H}_m t/\hbar)\exp(i\mathscr{H}_z t/\hbar), \qquad (7.16)$$

where we recall that $\mathscr{H}_m$ and $\mathscr{H}_z$ commute. Using this form for generating the evolution of $I_+(t)$ in the expression for $F(t)$ we see that since the $\exp(i\mathscr{H}_m t/\hbar)$ commutes with the spin operator I_- the motion evolution operator cancels from the expression when the trace is taken; $\mathscr{H}_m$ remains only in its effect on $S(t)$.

In the present case $S(t)$ is given, as before, by

$$S(t) = \exp\left[\frac{i}{\hbar}\int_0^t \mathscr{H}_d(\tau)\mathrm{d}\tau\right], \qquad (7.17)$$

but now the time dependence of $\mathscr{H}_d$ is more complex. The results of Appendix C give, in this case,

$$\mathscr{H}_d(\tau) = \exp(i\mathscr{H}_m\tau/\hbar)\exp(i\mathscr{H}_z\tau/\hbar)\mathscr{H}_d\exp(-i\mathscr{H}_z\tau/\hbar)\exp(-i\mathscr{H}_m\tau/\hbar). \qquad (7.18)$$

Now we know the effect of $\mathscr{H}_z$ on $\mathscr{H}_d$. This was shown in Equations (7.5) – (7.9):

$$\exp(i\mathscr{H}_z t/\hbar)\mathscr{H}_d\exp(-i\mathscr{H}_z t/\hbar) = \sum_{m=-2}^{2} D_m\exp(im\omega_0\tau).$$

So in this case we have

$$\mathscr{H}_d(\tau) = \sum_{m=-2}^{2} D_m(\tau)\exp(im\omega_0\tau), \tag{7.19}$$

where the time evolution of the dipolar operators D_m is generated only by the motion Hamiltonian

$$D_m(\tau) = \exp(i\mathscr{H}_m\tau/\hbar)D_m\exp(-i\mathscr{H}_m\tau/\hbar). \tag{7.20}$$

But what is the effect of the motion Hamiltonian on the dipole operators? The motion Hamiltonian causes the coordinates of the particles to vary with time: it induces the particles' motion. Writing

$$D_m = D_m[\mathbf{r}]$$

where $\mathbf{r}$ represents the operators for the coordinates of the particles, we can make this explicit since

$$\exp(i\mathscr{H}_m\ \tau/\hbar)D_m[\mathbf{r}]\exp(-i\mathscr{H}_m\tau/\hbar) = D_m[\mathbf{r}(\tau)],$$

where

$$\mathbf{r}(\tau) = \exp(i\mathscr{H}_m\tau/\hbar)\mathbf{r}\exp(-i\mathscr{H}_m\tau/\hbar).$$

So the time dependence of $D_m(\tau)$ is simply that due to the motion of the particles: the variation of the r, θ and φ in the expression for the D. The origin of the motion is thus seen to be unimportant; it can be quantum mechanical or it can be treated as a classical variation of the coordinates, the effect is the same.

7.1.5 Semiclassical expressions for moments

The effect of motion on the moments can be treated in this semiclassical way. In this case there is no motion Hamiltonian, as in Equation (7.12). Rather, in evaluating the derivatives of the relaxation function the time dependence of $\mathscr{H}_d$, through the variation of the coordinates, must be taken into account explicitly. In discussing moments recall that it is more convenient to express $F(t)$ in terms of I_x rather than I_+ and I_-. We are thus working with

$$F(t) = \mathrm{Tr}\{I_x(t)I_x(0)\}/\mathrm{Tr}\{I_x^2\}.$$

Since the moments are given by the derivatives of $F(t)$, we see that the operator $I_x(t)$ must be differentiated. The derivative is given, from Heisenberg's equation, as

$$I'_x(t) = \frac{i}{\hbar}[D_0(t),I_x(t)],$$

where in accordance with the earlier discussion we use only the adiabatic part D_0 of the internuclear interaction. For clarity we have indicated explicitly the time variation of the terms. Because of the time dependence of $D_0(t)$ the next derivative has an extra term:

$$I_x''(t) = \left(\frac{i}{\hbar}\right)^2 [D_0(t),[D_0(t),I_x(t)]] + \frac{i}{\hbar}[D_0'(t),I_x(t)].$$

From this we see that in this case the second moment is given by

$$M_2 = -F''(0)$$

$$= \frac{1}{\hbar^2}\frac{\text{Tr}\{[D_0,[D_0,I_x]]I_x\}}{\text{Tr}\{I_x^2\}} - \frac{i}{\hbar}\frac{\text{Tr}\{[D_0',I_x]I_x\}}{\text{Tr}\{I_x^2\}}.$$

Note that it is the second term which contains a reference to the time variation of D_0, in the time derivative. However, this trace is seen to vanish on expanding the commutator and cyclically permuting the operators. The remaining first term is precisely that calculated in the immobile case: there is no reference to motion in the expression for the second moment. We thus again find that the second moment is independent of any motion of the nuclear spins:

$$M_2 = \frac{1}{\hbar^2}\text{Tr}\{[D_0,I_x]^2\}/\text{Tr}\{I_x^2\}$$

Turning now to the fourth moment, we proceed in a similar way. The calculation is a little more complicated, but having found the fourth derivative of I_x, this leads to the following expression for M_4.

$$M_4 = \frac{1}{\hbar^4}\text{Tr}\{[D_0,[D_0,I_x]]^2\}/\text{Tr}\{I_x^2\} +$$

$$\frac{3}{\hbar^2}\text{Tr}\{[D_0',I_x]^2\}/\text{Tr}\{I_x^2\} + \frac{4}{\hbar^2}\text{Tr}\{[D_0,I_x][D_0'',I_x]\}/\text{Tr}\{I_x^2\}. \tag{7.21}$$

The first term here is the value of M_4 in the absence of motion. However, the second and third terms involve the derivatives of D_0 so they will depend on particle motion. The effect of differentiating D_0 with respect to time may be approximated, in an average way, by multiplying by $-1/\tau$ where τ is a measure of the correlation time for the atomic motion. The remaining commutators in the two terms then reduce to M_2. It follows that M_4 is given, at least qualitatively, by

$$M_4 = M_4(\text{rigid lattice}) + \text{const} \times \frac{M_2}{\tau^2}. \tag{7.22}$$

The effect of motion is again seen to increase the magnitude of M_4 over its rigid lattice value.

Similar arguments may be applied to higher order moments where, for rapid motion, the leading term is found to vary as

$$M_{2(n+1)} \sim \frac{M_2}{\tau^{2n}}, \tag{7.23}$$

but the numerical coefficient is hard to evaluate. On very general grounds, then, we see that the moments greater than M_2 are substantially increased by the effects of motion although precise numerical calculations may be difficult.

Since the effect of motion is seen to be to increase M_4 from its rigid lattice value of ≈ 3, this is *consistent* with having a Lorentzian line profile or an exponential FID. Accordingly, we can estimate the resultant relaxation time for the truncated Lorentzian model from Equation (6.74)

$$\frac{1}{T_2} = \frac{\pi}{2\sqrt{3}} \frac{M_2^{3/2}}{M_4^{1/2}},$$

which then gives

$$\frac{1}{T_2} = \frac{\pi}{2\sqrt{3}} M_2 \tau.$$

The numerical constant is unimportant, but the dependence of T_2 on the second moment and the correlation time is the familiar form, characteristic of motional averaging, which we have seen many times.

7.1.6 Conclusions

Whichever way the moments are treated, quantum mechanically or semiclassically, we conclude that M_2 is independent of motion while M_4 and higher moments increase with the speed of the motion. M_4 thus increases from its rigid lattice value and then a (truncated) Lorentzian is a possible shape for the resonance line profile – an exponential FID. Since this is precisely what is observed in practice, things look promising. However, we have not *proved* that the Lorentzian is the actual shape, and in this respect things are unsatisfactory. Recall the dilute solid model of the previous chapter. There we saw it was possible to be misled by moment calculations in the one- and two-dimensional cases. So as yet we must regard the line-shape/FID problem in fluids as not fully settled. The following section will address this problem from a slightly different perspective, drawing on many ideas from Chapter 4.

7.2 Cumulant expansion treatment of relaxation

7.2.1 Introduction to the method

We know that it is necessary to evaluate a large number of moments in order to describe the relaxation profile away from $t = 0$. We also know that the evaluation of higher order moments becomes prohibitively complex. We may regard the motion as having a randomising effect on the internuclear interaction. Thus one way ahead might be to adopt a probabilistic approach to the calculation of the moments. This certainly should be easier than direct evaluation, and then conclusions about the relaxation could be drawn. We shall, however, consider a more direct approach to the relaxation profile. Furthermore the method to be developed will prove to be equally applicable to the study of longitudinal relaxation and in this way both the transverse and the longitudinal cases may be treated in a uniform manner.

A quantitative description of the effects of motion on transverse relaxation was provided by the 'local field' models of Chapter 4. In particular, by the adoption of the Gaussian phase assumption the relaxation profile could be followed as the speed of the atomic motion varied. It was an important result, from that treatment, that in the case of rapid motion the transverse relaxation became exponential in form. In this section we start to build on the achievements of Chapter 4 by recasting those arguments in the formalism of quantum mechanics.

Real systems are usually too complex for their behaviour to be found from an exact solution of their equations of motion. Frequently some approximation is used. Of course the trick is to make realistic approximations: to simplify the problem without 'throwing out the baby with the bath water'. Sometimes an expansion is made in terms of a small parameter. The moment method is of this type. Here the small parameter is time and one can really only infer information about our system in the neighbourhood of the origin $t = 0$. Precisely how large this neighbourhood is was shown in Section 6.3.6. It is not immediately apparent that the Gaussian phase model of Chapter 4 is an approximation of this type; no expansion was made and no small parameter was identified. However, there was a hint in Box 4.1 where we showed that the method was equivalent to performing an expansion, essentially in powers of the local fields, followed by a resummation of (what we considered to be) the dominant terms.

From the quantum mechanical point of view our prototypical system is described by the Hamiltonian

$$\mathcal{H} = \mathcal{H}_z + \mathcal{H}_m + \mathcal{H}_d.$$

In the absence of the dipolar interaction the Zeeman Hamiltonian is responsible for an infinitely sharp resonance at the Larmor frequency – and the motion Hamiltonian has no influence. The effect of $\mathcal{H}_d$ is both to broaden resonance and to mediate the influence of the motion. In practice the resonance is still fairly pronounced; in other words the dipolar Hamiltonian is, relatively speaking, a small quantity.

The immediate implication of this discussion is that we should consider an expansion in powers of $\mathcal{H}_d$. However, it is clear that an expansion to any finite order in $\mathcal{H}_d$ will diverge at long times and in reality it is precisely this long time or hydrodynamic behaviour that we are interested in; this accounts for the majority of the relaxation in mobile systems, as was shown in Section 4.3.4.

A new approach to this problem was pioneered by Ryogo Kubo in 1953 (see also Kubo and Tomita (1954)) whereby convergence was guaranteed for expansions of finite order. The essence of Kubo's method was to write the relaxation function as the exponential of another function:

$$F(t) = \exp[\Psi(t)].$$

This equation may be regarded as the definition of the function $\Psi(t)$. Then the perturbation expansion is performed on $\Psi(t)$; this function is approximated by an expansion in powers of the coupling Hamiltonian $\mathcal{H}_d$.

From the *mathematical* point of view this is neither worse nor better than the direct expansion of $F(t)$, but by this means (so long as Ψ is found to be negative) we have fed in some extra knowledge about the physical behaviour of $F(t)$ and thereby ensured its convergence. Thus from the *physical* point of view this is an eminently sensible procedure.

Problem 4.3 implies that this procedure is equivalent to the Gaussian phase averaging method of Chapter 4. Thus expanding $\Psi(t)$ to second order in $\mathcal{H}_d$ is equivalent to the assumption that the phase angle which a spin processes through is a random function with a Gaussian distribution. This is certainly reasonable in fluids, considering the large number of particles contributing to the local field of a given spin.

The term *cumulant* is used in the classical theory of probability (Cramer, 1946). Given a distribution function $I(\omega)$ and its Fourier transform $F(t)$, we saw that the moments of $I(\omega)$ were related to derivatives of $F(t)$. In a similar way the cumulants are defined in terms of the logarithm of F, and *generalised* moment and cumulant expansions are performed in powers of a quantity other than our t. One use of cumulant expansions is in the demonstration that distributions may often be approximated by a Gaussian shape. This follows when cumulants higher than the second may be neglected.

Our generalised cumulant expansion procedure may be specified in the following way. We introduce an expansion parameter ε so that the Hamiltonian is

$$\mathcal{H} = \mathcal{H}_z + \mathcal{H}_m + \varepsilon\mathcal{H}_d, \tag{7.24}$$

whereupon the longitudinal and the transverse relaxation functions $L(t)$ and $F(t)$ are now functions of ε also:

$$L = L(t, \varepsilon), \quad F = F(t, \varepsilon).$$

The following arguments are the same for both $L(t,\varepsilon)$, and $F(t,\varepsilon)$. We shall define $\Psi(t,\varepsilon)$ by

$$F(t,\ \varepsilon) = \exp[\Psi(t,\ \varepsilon)],$$

and we expand Ψ in powers of ε. Thus

$$\Psi(t,\varepsilon) = \Psi(t,0) + \varepsilon\Psi'(t,0) + \varepsilon^2\Psi''(t,0)/2 + \dots,$$

where the prime indicates differentiation with respect to ε. Let us examine the terms in the expansion. The first is

$$\Psi(t,0) = \ln F(t,0),$$

which is zero since $F(t,0) = 1$; in the absence of $\mathcal{H}_d$ there is no relaxation. The second term contains $\Psi'(t,0)$. This may be expressed as

$$\frac{\partial\Psi(t,0)}{\partial\varepsilon} = \frac{\partial}{\partial\varepsilon}\ln F(t,0)$$

$$= \frac{1}{F(t,0)}\frac{\partial}{\partial\varepsilon}F(t,0)$$

$$= \frac{\partial}{\partial\varepsilon}F(t,0)$$

since $F(t,0) = 1$. It is shown in the following section that this derivative of $F(t,\varepsilon)$ is zero. The next term contains the second derivative, $\Psi''(t,0)$, which may be expressed as

$$\frac{\partial^2\Psi(t,0)}{\partial\varepsilon^2} = \frac{\partial^2}{\partial\varepsilon^2}\ln F(t,0)$$

$$= \frac{\partial}{\partial\varepsilon}\frac{1}{F(t,0)}\frac{\partial}{\partial\varepsilon}F(t,0)$$

which evaluates to

$$\frac{\partial^2\Psi(t,0)}{\partial\varepsilon^2} = \frac{\partial^2}{\partial\varepsilon^2}F(t,0).$$

Thus to lowest non-vanishing order of expansion we may write

$$F(t) = \exp\left[\frac{1}{2}\frac{\partial^2 F(t,0)}{\partial\varepsilon^2}\right] \tag{7.25}$$

with a similar expression for the longitudinal relaxation

$$L(t) = \exp\left[\frac{1}{2}\frac{\partial^2 L(t,0)}{\partial \varepsilon^2}\right].$$ (7.26)

7.2.2 Evaluation of the terms

In this section we shall evaluate in a consistent way the various derivatives of the relaxation functions with respect to ε, the 'strength' of the dipolar coupling. We will work with the transverse relaxation function; the results for the longitudinal case follow simply by the replacement of I_+ and I_- in the expressions relating to $F(t)$ by I_z.

We start, then, with the general expression for transverse relaxation:

$$F(t) = \mathrm{Tr}\{I_+(t)I_-(0)\}/\mathrm{Tr}\{I_+I_-\},$$

where we recall that the time evolution of $I_+(t)$ is given by

$$I_+(t) = \exp\left[\frac{i\varepsilon}{\hbar}\int_0^t \mathcal{H}_d(\tau)d\tau\right]I_+\exp\left[-\frac{i\varepsilon}{\hbar}\int_0^t \mathcal{H}_d(\tau)d\tau\right]$$

and at this stage we do not need to worry about the nature of the time development of $\mathcal{H}_d$.

The derivatives of $F(t)$ with respect to ε are best found, not by direct differentiation, but by expanding the exponentials in the evolution operators and then collecting the various powers of ε. Thus we expand

$$\exp\left[\frac{i\varepsilon}{\hbar}\int_0^t \mathcal{H}_d(\tau)d\tau\right] = 1 + \frac{i\varepsilon}{\hbar}\int_0^t \mathcal{H}_d(\tau)d\tau - \frac{\varepsilon^2}{2\hbar^2}\int_0^t\int_0^t \mathcal{H}_d(\tau_1)\mathcal{H}_d(\tau_2)d\tau_1 d\tau_2 + \ldots$$

so that the relaxation function becomes

$$F(t) = \mathrm{Tr}\left\{I_+\left[1 + \frac{i\varepsilon}{\hbar}\int_0^t \mathcal{H}_d(\tau)d\tau - \frac{\varepsilon^2}{2\hbar^2}\int_0^t\int_0^t \mathcal{H}_d(\tau_1)\mathcal{H}_d(\tau_2)d\tau_1 d\tau_2 + \ldots\right]I_-\right.$$

$$\left. \times \left[1 + \frac{i\varepsilon}{\hbar}\int_0^t \mathcal{H}_d(\tau)d\tau - \frac{\varepsilon^2}{2\hbar^2}\int_0^t\int_0^t \mathcal{H}_d(\tau_1)\mathcal{H}_d(\tau_2)d\tau_1 d\tau_2 + \ldots\right]^+\right\}.$$

Upon collecting terms and using the cyclic invariance of the trace we find

Coefficient of $\varepsilon^0 = 1$ (as expected)

Coefficient of $\varepsilon^1 = \dfrac{i}{\hbar}\displaystyle\int_0^t \mathrm{Tr}\{[\mathscr{H}_d(\tau),I_+]I_-\}\,d\tau/\mathrm{Tr}\{I_+I\}$

Coefficient of $\varepsilon^2 = -\dfrac{1}{2\hbar^2}\displaystyle\int_0^t d\tau_1 \int_0^t d\tau_2 \mathrm{Tr}\{[I_+,\mathscr{H}_d(\tau_2)][\mathscr{H}_d(\tau_1),I_-]\}/\mathrm{Tr}\{I_+I_-\}.$

Let us examine the coefficient of ε. This is the first derivative of $F(t)$, which we expect to evaluate to zero. The trace may be written as

$$\mathrm{Tr}\{[I_+,I_-],\mathscr{H}_d\} = 2\mathrm{Tr}\{I_z\mathscr{H}_d\}.$$

By symmetry this must vanish, as may be seen by inverting the z axis. The trace must remain unchanged, $\mathscr{H}_d$ is unchanged while I_z has transformed to $-I_z$. Thus:

$$\mathrm{Tr}\{I_z\mathscr{H}_d\} = -\mathrm{Tr}\{I_z\mathscr{H}_d\} = 0.$$

Turning now to the term in ε^2, this gives us the second derivative

$$\frac{\partial^2 F(t,0)}{\partial\varepsilon^2} = -\frac{1}{\hbar^2}\int_0^t d\tau_1 \int_0^t d\tau_2 \mathrm{Tr}\{[I_+,\mathscr{H}_d(\tau_2)][\mathscr{H}_d(\tau_1),I_-]\}/\mathrm{Tr}\{I_+I_-\}$$

with the corresponding expression for $L(t)$. However, the trace depends on the time variables only as $\tau_2 - \tau_1$.

Box 7.1 Stationarity

The above trace may be regarded as an autocorrelation function, as discussed in Appendix B. If the time variation of the function is purely quantum mechanical then the fact that it depends only on $\tau_2 - \tau_1$ follows from cyclically permuting the time evolution operators within the trace. On the other hand if the motion is treated as a classical modulation of the coordinates of $\mathscr{H}_d$ then the $\tau_2 - \tau_1$ variation follows because we are dealing with the correlation function of a stationary random function. Either way, the time dependence turns out to have the same characteristic form.

Since the trace varies as $\tau_2 - \tau_1$ it follows that we can change variables and perform one of the integrals, just as was done in Section 4.2.4 to obtain

$$\frac{\partial^2 F(t,0)}{\partial\varepsilon^2} = -\frac{2}{\hbar^2}\int_0^t (t-\tau)\frac{\mathrm{Tr}\{[I_+,\mathscr{H}_d(\tau)][\mathscr{H}_d(0),I_-]\}}{\mathrm{Tr}\{I_+I_-\}}\,d\tau$$

and similarly

$$\frac{\partial^2 L(t,0)}{\partial \varepsilon^2} = -\frac{2}{\hbar^2}\int_0^t (t-\tau)\frac{\mathrm{Tr}\{[I_z,\mathscr{H}_d(\tau)][\mathscr{H}_d(0),I_z]\}}{\mathrm{Tr}\{I_z^2\}}\,d\tau.$$

So from Equations (7.25) and (7.26) we obtain the expressions for the relaxation functions:

$$
\left.
\begin{aligned}
F(t) &= \exp\left(-\int_0^t (t-\tau)\frac{\mathrm{Tr}\{[I_+,\mathscr{H}_d(\tau)][\mathscr{H}_d(0),I_-]\}}{\hbar^2\mathrm{Tr}\{I_+I_-\}}\,d\tau\right) \\[2mm]
L(t) &= \exp\left(-\int_0^t (t-\tau)\frac{\mathrm{Tr}\{[I_z,\mathscr{H}_d(\tau)][\mathscr{H}_d(0),I_z]\}}{\hbar^2\mathrm{Tr}\{I_z^2\}}\,d\tau\right).
\end{aligned}
\right\}
\tag{7.27}
$$

It is instructive, at this point, to make connection with the memory function method treated in Section 6.4.1. If we, temporarily, denote the correlation function appearing in Equations (7.27) by $G(t)$, then we have the familiar

$$F(t) = \exp\left[-\int_0^t (t-\tau)G(\tau)d\tau\right].$$

This may be differentiated with respect to time, giving an 'equation of motion' for $F(t)$ which may be written

$$\frac{\mathrm{d}}{\mathrm{d}t}F(t) = -\int^t G(\tau)F(t)d\tau.$$

Now this is not identical to the memory equation, Equation (6.76), but it is similar. Moreover, it does become identical once the separation of time scales has been invoked; the upper limit of the integral goes to infinity and we make the comparison with Equation (6.77). This indicates to us that there are indeed similarities in the methods even though the approaches are so different. Recall the memory equation was suggested on the general grounds of linearity, causality etc. However, the present result, as did the parallel one from Chapter 4, followed from arguments about specifics of the system – the random fields etc.

7.2.3 Transverse correlation functions

Transverse relaxation is described in Equation (7.27) in terms of the quantum mechanical trace

$$\frac{\text{Tr}\{[I_+,\mathscr{H}_d(t)][\mathscr{H}_d(0),I_-]\}}{\hbar^2\text{Tr}\{I_+I_-\}}.$$

(7.28)

If we expand $\mathscr{H}_d$ into its various spin-flip components D_m then the time evolution separates as expressed in Equation (7.19):

$$\mathscr{H}_d(t) = \sum_{m=-2}^{2} D_m(t)\exp(im\omega_0 t):$$

(7.29)

precession, the effect of the Zeeman interaction, is contained in the exponentials while the $D_m(t)$ varies through the bodily motion of the particles. The trace thus becomes a sum of oscillating components so that Expression (7.28) becomes

$$\sum_{m,n=-2}^{2} \exp(im\omega_0 t)\frac{\text{Tr}\{[I_+,D_m(t)][D_n(0),I_-]\}}{\hbar^2\text{Tr}\{I_+I_-\}}.$$

(7.30)

We saw the expression for the dipolar components D_m in Equation (5.26):

$$D_M = \frac{\mu_0}{4\pi}\left(\frac{4\pi}{5}\right)^{1/2}\hbar^2\gamma^2\sum_{i\neq j}(-1)^m\frac{Y_2^{-m}(\Omega_{ij})}{r_{ij}^3}T_{ij}^m,$$

so that we must evaluate the commutators of the operators I_+ and I_- with the spin-flip tensor operators T_{ij}^m. However, this is provided for us in the identities of Equations (5.25). Thus we obtain

$$\left.\begin{aligned}[I_+,D_m(t)] &= [6-m(m+1)]^{1/2}\frac{\mu_0}{4\pi}\left(\frac{4\pi}{5}\right)^{1/2}\hbar^2\gamma^2\sum_{i\neq j}(-1)^m\frac{Y_2^{-m}[\Omega_{ij}(t)]}{r_{ij}^3(t)}T_{ij}^{m+1},\\[I_-,D_n] &= [6-n(n-1)]^{1/2}\frac{\mu_0}{4\pi}\left(\frac{4\pi}{5}\right)^{1/2}\hbar^2\gamma^2\sum_{k\neq l}(-1)^n\frac{Y_2^{-n}[\Omega_{kl}(0)]}{r_{kl}^3(0)}T_{kl}^{n-1}.\end{aligned}\right\}$$

(7.31)

Assuming motion to be described by classical means, the T are the only quantum mechanical operators, and the trace depends on their product.

In evaluating $\text{Tr}\{T_{ij}^{m+1}T_{kl}^{n-1}\}$ the trace will be zero unless spin is conserved in the argument. Thus only the cases where $n = -m$ will contribute. Then Expression (7.30) becomes

$$\sum_{m=-2}^{2}\frac{[6-m(m+1)]}{2}\exp(im\omega_0 t)\left(\frac{\mu_0}{4\pi}\right)^2\frac{4\pi\hbar^2\gamma^4}{5N}\sum_{i\neq j,k\neq l}\frac{Y_2^{m*}[\Omega_{ij}(t)]Y_2^m[\Omega_{kl}(0)]}{r_{ij}^3(t)r_{kl}^3(0)}\Gamma_{ijkl},$$

(7.32)

where

$$\Gamma_{ijkl} - N\frac{\text{Tr}\{T_{ij}^{-m}T_{kl}^{m}\}}{\text{Tr}\{I_z^2\}} = \frac{3}{2}(\delta_{ik}\delta_{jl} + \delta_{il}\delta_{jk}),$$

(7.33)

independent of the spin-flip index m. The delta functions permit a simplification, giving

$$\sum_{m=-2}^{2}\frac{[6 - m(m + 1)]}{2}\exp(im\omega_0 t)\left(\frac{\mu_0}{4\pi}\right)^2\frac{12\pi\hbar^2\gamma^4}{5N}\sum_{i\neq j}\frac{Y_2^{m*}[\Omega_{ij}(t)]\,Y_2^{m}[\Omega_{ij}(0)]}{r_{ij}^3(t)r_{ij}^3(0)}.$$

(7.34)

We observe that the second summation is precisely of the form of an autocorrelation function. This leads us to define a set of dipolar correlation functions

$$G_m(t) = \left(\frac{\mu_0}{4\pi}\right)^2\frac{12\pi\hbar^2\gamma^4}{5N}\sum_{i\neq j}\frac{Y_2^{m*}[\Omega_{ij}(t)]\,Y_2^{m}[\Omega_{ij}(0)]}{r_{ij}^3(t)r_{ij}^3(0)},$$

(7.35)

so that Expression (7.30) may be written

$$\sum_{m=-2}^{2}\frac{[6 - m(m + 1)]}{2}\exp(im\omega_0 t)G_m(t),$$

giving the transverse relaxation function as

$$F(t) = \exp\left\{-\int_0^t (t - \tau)\sum_{m=-2}^{2}[3 - m(m + 1)/2]\exp(im\omega_0\tau)G_m(\tau)d\tau\right\}.$$

(7.36)

This is an important result which contains much information. Later sections will be devoted to the interpretation of this expression and an exploration of the parallels with the classical treatment of Chapter 4. However, we must first pause to consider, in the next section, the longitudinal case in a similar manner. We shall thus see that $L(t)$ may also be expressed in terms of the same correlation functions $G_m(t)$. The succeeding sections on the nature of the correlation functions will then be seen to be of relevance to both longitudinal and transverse relaxation.

7.2.4 Longitudinal correlation functions

This section parallels the previous treatment of transverse relaxation. However, the longitudinal calculation is, in fact, rather simpler. Longitudinal relaxation is described in Equation (7.27) in terms of the quantum mechanical trace

$$\frac{\text{Tr}\{[I_z,\mathcal{H}_d(t)][\mathcal{H}_d(0),I_z]\}}{\hbar^2\text{Tr}\{I_z^2\}}.$$

(7.37)

As in the previous section we expand $\mathcal{H}_d$ into its various spin-flip components,

Equation (7.29). The result is that Expression (7.37) becomes

$$\sum_{m=-2}^{2} \exp(im\omega_0 t) \frac{\text{Tr}\{[I_z, D_m(t)][D_{-m}(0), I_z]\}}{\hbar^2 \text{Tr}\{I_z^2\}},\qquad(7.38)$$

where we have anticipated a little, by acknowledging the requirement of spin conservation within the trace. Again the commutators are provided by Equations (5.25) which may be written in the form

$$[I_z, D_m] = mD_m.$$

Thus Expression (7.37) simplifies to

$$\sum_{m=-2}^{2} m^2 \exp(im\omega_0 t) \frac{\text{Tr}\{D_m(t)D_{-m}(0)\}}{\hbar^2 \text{Tr}\{I_z^2\}},$$

and it is a straightforward matter to show that this may be expressed in terms of the $G_m(t)$ of the previous section,

$$\sum_{m=-2}^{2} m^2 \exp(im\omega_0 t) G_m(t),$$

giving the longitudinal relaxation function as

$$L(t) = \exp\left[-\int_0^t (t-\tau) \sum_{m=-2}^{2} m^2 \exp(im\omega_0 \tau) G_m(\tau) d\tau\right].\qquad(7.39)$$

As a bonus we have a more direct interpretation of the correlation functions $G_m(t)$. They are simply the autocorrelation functions of the various dipolar components (appropriately normalised):

$$G_m(t) = \frac{\text{Tr}\{D_m(t)D_{-m}(0)\}}{\hbar^2 \text{Tr}\{I_z^2\}}.\qquad(7.40)$$

7.2.5 *Properties of $G_m(t)$*

The result of the cumulant expansion method is that immediate attention is shifted from the relaxation functions $F(t)$ and $L(t)$ to the autocorrelation functions of the coupling interaction $G_m(t)$. This is similar to the result of Chapter 4 where attention was shifted to the local field autocorrelation function $G(t)$ introduced there. However, here we have a complication in that there is now more than one autocorrelation function to be considered. In this section, and further in Appendix F, we shall consider some general features of the $G_m(t)$ correlation functions.

From the definition of $G_m(t)$ in Equation (7.35) we immediately observe some important properties of the dipolar autocorrelation functions. Taking the complex conjugate and using the properties of the spherical harmonics we find

(i) $G_m^*(t) = G_{-m}(t)$.

The stationarity property of the correlation functions gives

(ii) $G_{-m}(t) = G_m(-t)$.

However, the correlation functions satisfy time reversal invariance:

(iii) $G_m(-t) = G_m(t)$,

which, taken together with the above two properties, implies that the correlation functions are real:

(iv) $G_m(t)$ are real.

This means that $G_m(t) = G_{-m}(t)$ so that there are only three rather than five functions to be considered.

The next property to be examined is a little more complicated to derive; the details are relegated to Appendix F. The expression of the dipolar Hamiltonian components in terms of spherical harmonics, Equation (5.26), was done partly to make explicit the symmetry properties of the interaction. One consequence of this symmetry is that for rotationally invariant systems the correlation functions become independent of the index m. In other words, in that case there is only one function, which we can denote simply by $G(t)$, to be considered. We state this fifth property:

(v) Rotational invariance $\Rightarrow G_m(t)$ independent of m.

Thus where appropriate we shall drop the index m.

Since the $G_m(t)$ are autocorrelation functions, the discussion of Section 4.3.1 are relevant here. Such functions are usually 'bell-shaped', decaying monotonically to zero as shown in Figure 4.6 and, as explained in Section 4.3.1, there will be a characteristic time τ_c such that for times much shorter than this there will be very little decay: $G_m(t) \sim G_m(0)$ for $t \ll \tau_c$, while for times much longer than this the function will have decayed appreciably: $G_m(t) \sim 0$ for $t \gg \tau_c$. This correlation time is the time scale for the decay of the autocorrelation function; it is a measure of the 'speed' of the atomic motion, insofar as it effects the NMR properties.

It must be emphasised that for a given atomic motion the correlation time will be dependent on the actual mechanism responsible for the relaxation – on the source of the local magnetic fields. The correlation time is the time scale over which the fields are observed to vary. From the discussions of Chapter 4 we know that when the relaxation is caused by inhomogeneities in an externally applied magnetic field the relevant correlation time is approximately the time it takes for the particles to move through a 'significant' field variation. This is typically the time it takes for a particle to diffuse a macroscopic distance. However, for the internuclear dipolar interaction, the correlation time is very roughly the time it takes for two particles to move past each other, as we shall see more clearly later in this chapter. In other words, in this case, it is the time for a particle to move a microscopic distance.

7.3 The relaxation times

7.3.1 Adiabatic T_2

We now return to the main business of studying the relaxation functions. Here we start the examination of transverse relaxation, as obtained in Section 7.2.3, namely Equation (7.36).

There are a number of different time and frequency quantities relevant to our discussion. There is the Larmor frequency ω_0, there is the measure of the local fields $\omega_{loc} \sim M_2^{\frac{1}{2}}$ and there is the correlation time τ_c. We must consider a number of different cases according to the relative magnitudes of these quantities. In this section we shall take the case where the motion is slower than the Larmor frequency, i.e. where $\omega_0 \tau_c$ is large.

Inspection of Equation (7.36) shows that when $\omega_0 \tau_c \gg 1$ there will be many oscillations of the exponential while the $G_m(\tau)$ has hardly changed. Thus the oscillating contributions to the integral will cancel. Only the non-oscillatory $m = 0$ term will be significant; the other terms may be neglected. This is the *adiabatic* approximation, so-called because the effect of the retained D_0 term in $\mathcal{H}_d$ is to leave the system's eigenstates unchanged, only shifting the energy levels. This is a convenient simplification; the full treatment, including the effects of the non-adiabatic terms, those for which $m \neq 0$, will be given in Section 7.3.3. The transverse relaxation function in the adiabatic case simplifies to

$$F(t) = \exp\left[-3 \int_0^t (t - \tau)G_0(\tau)\mathrm{d}\tau\right]. \tag{7.41}$$

For further reference, note that since the second moment is given by minus the second derivative of $F(t)$ at $t - 0$, we have

$$M_2 = 3G_0(0). \tag{7.42}$$

Observe that apart from the factor of 3, Equation (7.41) is identical to the classical result derived in Chapter 4. We can then carry over the bulk of the discussions of Sections 4.3.2–4.3.4. These may be summarised as follows:

(a) For 'slow' motion, i.e. when $M_2\tau_c^2$ is much greater than unity, $G_0(\tau)$ hardly changes from its zero-time value throughout the relaxation. We may then take the $G_0(0)$ out of the integral. The remaining integration is trivial to perform, giving

$$F(t) = \exp(-3G_0(0)t^2/2)$$

or

$$F(t) = \exp(-M_2t^2/2),$$

a Gaussian relaxation, independent of the correlation time τ_c. Although the relaxation is not exponential here, the time scale for the decay is clearly $M_2^{-1/2}$ so that for this case we may generalise the idea of a relaxation time, writing the T_2 as

$$\frac{1}{T_2} \sim M_2^{1/2}.$$

(b) For faster motion, i.e. when $M_2\tau_c^2$ is much less than unity, $G_0(\tau)$ will have decayed to zero so that no error is made in extending the upper limit of the integral to infinity, giving exponential relaxation

$$F(t) = \exp\left(-\frac{t}{T_2}\right),$$

where

$$\frac{1}{T_2} = 3\int_0^\infty G_0(\tau)d\tau. \qquad (7.43)$$

We must extend the discussion to see how this result is modified when the motion becomes even more rapid so that then $\omega_0\tau_c$ becomes less than unity. Those results will, however, make more sense with a knowledge of the behaviour of the longitudinal relaxation under similar circumstances. We thus pause to consider this first.

At this point we see that the aims set out in Section 7.1.6 have been achieved. There, the conclusion was that an exponential relaxation profile was simply consistent with the calculated information. Here we have shown that the relaxation profile is indeed exponential, and as a bonus we have a statement of the conditions under which this will be so.

7.3.2 Longitudinal relaxation

The expression for the relaxation of longitudinal magnetisation was given in Equation (7.39). Note that there is no contribution from the $m = 0$ term of the autocorrelation function; one must have net spin flips for longitudinal relaxation to occur. The condition that the upper limit of the integral may be extended to infinity is that $G(t)$ should decay faster than $L(t)$, in other words, $\tau_c \ll T_1$. This is invariably the case, as we shall see below. Thus the relaxation will be exponential, with time constant T_1 given by

$$1/T_1 = \sum_{m=-2}^{2} m^2 \int_0^\infty G_m(t)\exp(im\omega_0 t)\mathrm{d}t.$$

The m^2 in this equation implies equal contributions from the $+m$ and the $-m$ terms. As a result we see (as was probably expected) that $1/T_1$ is a real quantity, and we can combine the positive and negative m terms by extending the lower limit of the integrals to minus infinity:

$$1/T_1 = \sum_{m=-1}^{2} m^2 \int_{-\infty}^\infty G_m(t)\exp(im\omega_0 t)\mathrm{d}t.$$

Here we observe the Fourier transform of the autocorrelation functions. In Appendix A we see that in general the Fourier transform of an autocorrelation function is known as a spectral density function, and we examine some of its characteristic properties. We are thus led to define the dipolar spectral density functions as

$$J_m(\omega) = \int_{-\infty}^\infty G_m(t)\exp(i\omega t)\mathrm{d}t, \tag{7.44}$$

in terms of which the expression for T_1 becomes:

$$1/T_1 = J_1(\omega_0) + 4J_2(2\omega_0). \tag{7.45}$$

We shall defer discussion of this result until we have obtained the corresponding expression for T_2.

7.3.3 Non-adiabatic T₂

Returning to the consideration of transverse relaxation we see that when the motion is more rapid that the Larmor period, i.e. $\omega_0 \tau_c \ll 1$, the correlation functions in Equation (7.36) will have decayed significantly before a full cycle of the exponential's oscillations, so that there is no longer a cancellation of the $m \neq 0$ terms. Of course, in this case the integral can be extended to infinity so that the expression for $F(t)$ becomes:

$$F(t) = \exp(-t) \sum_{m=-2}^{2} [3 - m(m+1)/2] \int_0^\infty G_m(\tau)\exp(im\omega_0\tau)\mathrm{d}\tau,$$

which, again, represents an exponential relaxation. Observe, however, that in this case there is not an equal contribution from the $+m$ and $-m$ terms in the sum. As a consequence it is not a purely real function. Separating the real and imaginary parts as.

$$\frac{1}{T_2} - i\Delta = \sum_{m=-2}^{2} [3 - m(m+1)/2] \int_0^\infty G_m(\tau)\exp(im\omega_0\tau)d\tau \qquad (7.46)$$

the relaxation function is

$$F(t) = \exp(-t/T_2)\exp(i\Delta t). \qquad (7.47)$$

We have the decay time for the relaxation, T_2, while Δ corresponds to a shift of the frequency of precession from the Larmor frequency ω_0. This shift is always a very small quantity and it can usually be ignored. It is, however, measurable and we shall examine this further in Section 7.3.5.

Using the various symmetry properties of the $G_m(t)$ we can write the real part of Equation (7.46), the relaxation time T_2, as

$$\frac{1}{T_2} = \sum_{m=-2}^{2} [3 - m^2/2] \int_0^\infty G_m(t)\exp(im\omega_0 t)dt,$$

which is observed to be symmetric in m.

Using the property $G_{-m}(t) = G_m(-t)$ enables us to incorporate the $-m$ terms in the sum by extending the integrals to $-\infty$, as was done for T_1, giving

$$\frac{1}{T_2} = \frac{3}{2} \int_{-\infty}^\infty G_0(t)dt + \frac{5}{2} \int_{-\infty}^\infty G_1(t)\exp(i\omega_0 t)dt + \int_{-\infty}^\infty G_2(t)\exp(2i\omega_0 t)dt.$$

As with the T_1 calculation we can express these integrals in terms of the dipolar spectral density functions:

$$1/T_2 = \frac{3}{2}J_0(0) + \frac{5}{2}J_1(\omega_0) + J_2(2\omega_0). \qquad (7.48)$$

The first term is that derived in the adiabatic case, Equation (7.43), although here we have interpreted the integral as the zero-frequency Fourier transform. The J_1 and the J_2 terms become important when $\omega_0\tau_c \ll 1$; they are negligible in the opposite limit.

Had we used the expression for $F(t)$ in terms of I_x, Equation (6.18), instead of that in terms of I_+ and I_-, then no imaginary part would have appeared. That expression yields the correct value for the relaxation time without giving the frequency shift. It is only when using the complex representation of transverse magnetisation that a shift in the phase of the magnetisation vector can be accounted for by a multiplicative factor $\exp(i\varphi)$.

7.3.4 Behaviour of the relaxation times

We are now in a position to summarise and discuss the behaviour of the relaxation times, based upon the expressions derived in the previous sections. In particular we shall see how both relaxation times vary with the motion of the spin-carrying particles. In order, initially, to draw some qualitative conclusions it is occasionally expedient in this section to neglect any possible rotational anisotropy, whereupon the spectral density functions J_0, J_1 and J_2 are identical. In such cases the functions will be written without their distinguishing subscript.

Regardless of their precise details, the autocorrelation functions and their spectral densities are essentially bell shaped functions and, as they are Fourier transforms of each other, they will satisfy an 'uncertainty principle' relating their widths (Appendix A). The correlation time τ_c was defined as the width of $G(t)$: its area divided by its height

$$\tau_c = \frac{1}{G(0)} \int_0^\infty G(t)\mathrm{d}t.$$

Thus τ_c^{-1} is a measure of the width of $J(\omega)$. In other words,

$$\left.\begin{array}{l} J(\omega) \sim J(0) \text{ for } \omega \ll \tau_c^{-1} \\ J(\omega) \sim 0 \quad \text{ for } \omega \gg \tau_c^{-1} \end{array}\right\} \tag{7.49}$$

Now $J(0)$ is twice the area under $G(t)$ (the integral goes from $-\infty$ to $+\infty$). So since

$$M_2 = G(0)/3,$$

we have

$$J(0) = \frac{2}{3}M_2\tau_c. \tag{7.50}$$

This looks suggestive of the expression for T_2, as it should.

Away from the rigid lattice region, i.e. when $\tau_c \ll M_2^{-\frac{1}{2}}$, T_2 is given by

$$1/T_2 = \frac{3}{2}J_0(0) + \frac{5}{2}J_1(\omega_0) + J_2(2\omega_0). \tag{7.51}$$

If we first consider the case where $\tau_c \gg \omega_0^{-1}$ or $\omega_0 \gg \tau_c^{-1}$ then, by Equation (7.49), $J(\omega)$ will be vanishingly small so that

$$1/T_2 = \frac{3}{2}J_0(0),$$

which, from Equation (7.50), becomes

$$1/T_2 = M_2 \tau_c, \tag{7.52}$$

the usual adiabatic motional narrowing expression, Equation (4.41).

Now let us consider the case in which the motion is faster than the Larmor period, i.e. when $\tau_c \ll \omega_0^{-1}$ or $\omega_0 \ll \tau_c^{-1}$. In this case, from Equation (7.49)

$$J(\omega) \sim J(0).$$

Then Equation (7.51) gives

$$\frac{1}{T_2} = \left[\frac{3}{2} + \frac{5}{2} + 1\right] \frac{2}{3} M_2 \tau_c$$

$$= \frac{10}{3} M_2 \tau_c. \tag{7.53}$$

The functional form of T_2 is thus similar whether the motion is faster or slower than the Larmor period, but there is a factor of 10/3 between the two régimes. When the distinction must be made between the $J_m(\omega)$, in the absence of rotational invariance, the numerical factor will no longer be precisely 10/3.

Turning now to T_1, consider first the fast motion situation, $\tau_c \ll \omega_0^{-1}$ or $\omega_0 \ll \tau_c^{-1}$. As we just saw for T_2, then

$$J(\omega) \sim J(0),$$

so that

$$1/T_1 = [1 + 4]J(0)$$

$$= \frac{10}{3} M_2 \tau_c, \tag{7.54}$$

which we see is identical to the expression for T_2 in the same régime. Clearly the equality of T_1 and T_2 in the fast motion case holds only for rotationally invariant systems, where the three spectral density functions are similar.

Box 7.2 The equality of T_2 and T_2

It would appear, from the above treatment, that the equality of T_1 and T_2 which happens for fast motion, $\omega_0 \tau_c \ll 1$, is accidental. We shall now show that this result follows in a general way. As a restatement of the results of the previous sections we may say that away from the rigid lattice region, in the frame rotating at the Larmor frequency, a non-equilibrium magnetisation in the direction α will relax to equilibrium with a time constant T_α:

$$\frac{1}{T_\alpha} = \frac{\int_0^\infty \mathrm{Tr}\{[I_\alpha,\mathcal{H}_d(t)][\mathcal{H}_d(0),I_\alpha]\}\mathrm{d}t}{\hbar^2 \mathrm{Tr}\{I_\alpha^2\}}$$

where I_α is the spin operator along the direction α. Decomposition of $\mathcal{H}_d$ into its spin-flip components D_m then gives the oscillating exponentials $\exp(im\omega_0 t)$ within the integral, in accordance with Equation (7.29).

For fast motion, when τ_c is smaller than the Larmor period ω_0^{-1}, the correlations in $\mathcal{H}_d$ will have decayed away and the trace will have vanished before a cycle of precession. In that case the $\exp(im\omega_0 t)$ may be ignored in the integral so that the D_m may be recombined to give $\mathcal{H}_d$. We then have the freedom to choose a new decomposition of $\mathcal{H}_d$ corresponding to a different direction of quantisation. Let us take the quantisation axis to be in the direction α. In this case the expression for the correlation functions are given just as in Equation (7.40), although the G_m for different m values will differ according to the direction α. However, we saw that rotational invariance makes the functions independent of m. In that case the correlation functions will all be the same. Their integrals will thus be the same and so the relaxation time for magnetisation in any direction α will be the same.

We thus see that in an isotropic system, when the motion is fast so that $\omega_0\tau_c \ll 1$, magnetisation in any direction (in the rotating frame) will decay at the same rate and, in particular, we then have $T_1 = T_2$.

$J(\omega)$ is very small, and reducing for motion slower than ω_0. Then $1/T_1$ will be decreasing, although its precise variation with τ_c will depend on the functional form of $J(\omega)$. However, since T_1 has changed direction, it must have a minimum when τ_c is in the vicinity of ω_0. Section 7.5.2 discusses this further.

The behaviour of the relaxation times is summarised in Figure 7.1 and in Table 7.1. Observe that the minimum of T_1 appears in the same region as that where the '10/3' effect occurs, when the speed of motion matches the Larmor frequency. The value of T_1 at the minimum may be estimated. Here T_1^{-1} and T_2^{-1} are both approximately equal to $M_2\tau_c$ and since $\tau_c \sim \omega_0^{-1}$ at the minimum, we find

$$T_1^{(min)} \sim \omega_0/M_2, \tag{7.55}$$

which is only an approximate result. There will be a numerical factor, of the order of unity, which depends on the precise shape of the $J(\omega)$.

The observation of minima in T_1 can form the basis of a useful experimental tool. By

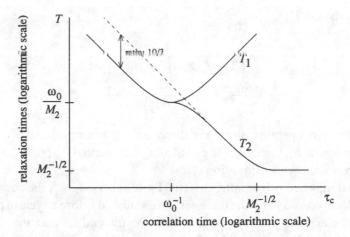

Figure 7.1 Variation of T_1 and T_2 with motion.

Table 7.1. *Variation of T_1 and T_2 with motion.*

		$\tau_c \rightarrow$	
		ω_0^{-1}	$M_2^{-1/2}$
$\dfrac{1}{T_1}$	$\dfrac{10}{3} M_2 \tau_c$	$\sim \dfrac{M_2}{\omega_0}$ $T_1 \min$	T_1 increasing, depending on the shape of $J(\omega)$
$\dfrac{1}{T_2}$	$\dfrac{10}{3} M_2 \tau_c$	$\sim \dfrac{M_2}{\omega_0}$	$M_2 \omega_c$ $\sim M_2^{1/2}$ non-exponential

varying, say, the temperature of a system the correlation time is altered. At the point where a minimum in T_1 is observed, that is where the correlation time has become equal to the Larmor period. In this way observations of T_1 minima form the basis of a simple technique for the spectroscopy of atomic dynamics. Furthermore, from the value of T_1 at the minimum, the value of M_2 can be estimated. From this it can be checked that it is the expected relaxation mechanism which is responsible for the relaxation. This is important since for a given 'speed' of motion the correlation time will be different for different relaxation mechanisms.

7.3.5 The frequency shift

The relaxation function $F(t)$ describes the decay of the magnitude of the rotating magnetisation, the Larmor precession having been factored out. The evolution of the (complex) transverse magnetisation may be expressed, from Equation (5.19), as

$$M(t) \propto \exp(i\omega_0 t)F(t).$$

In Equations (7.46) and (7.47) we saw that there could be a small imaginary contribution to the expression for the relaxation:

$$F(t) = \exp(-t/T_2)\exp(i\Delta t),$$

from which we see that $M(t)$ evolves according to

$$M(t) \propto \exp[i(\omega_0 + \Delta)t]\exp(-t/T_2).$$

The quantity Δ thus corresponds to an increase in the precession frequency. From Equation (7.46) we see that the shift is given by

$$\Delta = - \sum_{m=-2}^{2} [3 - m(m + 1)/2] \int_0^\infty G_m(t)\sin(m\omega_0 t)dt$$

and since $G_{-m}(t) = G_m(t)$ and $\sin(-m\omega t) = -\sin(m\omega t)$, this gives

$$\Delta = \int_0^\infty G_1(t)\sin(\omega_0 t)dt + 2 \int_0^\infty G_2(t)\sin(2\omega_0 t)dt. \qquad (7.56)$$

We see that the frequency shift is given by the sine transform of the correlation functions. This is in contrast to the relaxation times which are given by the cosine transforms.

Let us now consider, in a general way, how the frequency shift depends on the speed of the atomic motion. In the process we shall also obtain an expression for the magnitude of Δ so that we can, as promised, demonstrate that the shift is almost always sufficiently small so that it may be ignored.

Consider first the case where the motion is much faster than the Larmor period. Then in Equation (7.56) the sines will have hardly grown from zero value before the $G(t)$ have decayed. Thus there will be no contribution to the integral: the shift is zero in the fast motion limit. In the opposite case the $G(t)$ will have changed little while the sines undergo many oscillations. Then we may take $G(0)$ out of the integral giving, since $G(0) = M_2/3$,

$$\Delta = \frac{M_2}{3} \left[\int_0^\infty \sin(\omega_0 t)dt + 2 \int_0^\infty \sin(2\omega_0 t)dt \right].$$

Evaluating the integrals this gives

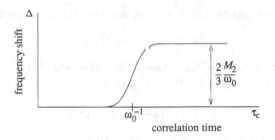

Figure 7.2 Variation of frequency shift with motion.

$$\Delta = \frac{2}{3}\frac{M_2}{\omega_0}, \tag{7.57}$$

which is the value of the shift in the slow motion limit.

The general behaviour of the shift is shown in Figure 7.2. We see that the frequency shift occurs in the region of the T_1 minimum. Furthermore the maximum value of the shift is of the same order of magnitude as the inverse of T_1 or T_2 at the T_1 minimum.

For such a shift to be observable it should be greater than, or at least comparable with, the linewidth. The product $T_2\Delta$ may thus be regarded as a measure of the 'observability' of the shift. This will be peaked in the region of the T_1 minimum. On the fast motion side where $\tau_c \ll \omega_0^{-1}$, there Δ goes to zero, while on the slow motion side where $\tau_c \gg \omega_0^{-1}$, the value of T_2 will become very small. In the region of the maximum we have $\Delta \sim M_2/\omega_0$ and $T_2 \sim \omega_0/M_2$ which gives $T_2\Delta \sim 1$. Thus we conclude that the shift is just observable in the vicinity of the T_1 minimum; in general it may be ignored.

7.4 Dipolar $J(\omega)$ and $G(t)$

7.4.1 Rotation of a diatomic molecule

In the next few sections we shall consider the form of the autocorrelation function and its spectral density for a number of model descriptions of the dynamics of physical systems. This will enable us to understand the relaxation times T_1 and T_2 for those systems in terms of the parameters of those descriptions. Some of the models we treat will have solutions for $J(\omega)$ and $G(t)$ which can be expressed analytically. Other will not be fully soluble in this way and we shall then consider some of the general conclusions which can be made, even in such cases.

Although we are interested particularly in many-particle systems, we shall start with a simple example of a diatomic molecule undergoing rotational diffusion, tumbling about amongst its neighbours. This would be an appropriate model for explaining the proton relaxation in water, for example.

The expression for the dipolar autocorrelation function is given by Equation (7.35). In the present case we are dealing with only two nuclear spins and their separation r remains constant. In the sum over i and j either particle can be labelled by either index; there is a double counting of the interaction. Allowing for this, and since we have rotational invariance, the expression for $G(t)$ may be written

$$G(t) = \frac{12\pi}{20} \left(\frac{\mu_0}{4\pi}\right)^2 \frac{\hbar^2 \gamma^4}{r^6} \langle Y_2^{m*}[\Omega(t)] Y_2^m[\Omega(0)]\rangle, \tag{7.58}$$

where the angle brackets indicate an average over a probabilistic description of the motion and $\Omega(t)$ is the orientation of the vector joining the spins at time t. Here r is constant, the hydrogen–hydrogen separation in the water molecule. The average may be expressed as

$$\langle \ \ \rangle = \frac{1}{4\pi} \int d\Omega \int d\Omega_0 P(\Omega,\Omega_0,t) Y_2^{m*}(\Omega) Y_2^m(\Omega_0), \tag{7.59}$$

where the $1/4\pi$ is the probability of a given initial orientation.

The characteristics of the motion are contained in the probability distribution function $P(\Omega,\Omega_0,t)$. We consider the simplest model of rotational diffusion. In Appendix G it is shown that such a model leads to the following expression for $P(\Omega,\Omega_0,t)$:

$$P(\Omega,\Omega_0,t) = \sum_{l,n} Y_l^{n*}(\Omega) Y_l^n(\Omega_0)\exp(-t/\tau_l), \tag{7.60}$$

where

$$\tau_l = \frac{a^2}{l(l+1)D}. \tag{7.61}$$

Here a is the radius of the molecule and D is the rotational diffusion coefficient, which from Stoke's Law is given in terms of the viscosity η by

$$D = kT/8\pi a\eta. \tag{7.62}$$

The average is given by

$$\langle \ \ \rangle = \frac{1}{4\pi} \int d\Omega \int d\Omega_0 \sum_{l,n} Y_l^{n*}(\Omega_0) Y_l^n(\Omega) Y_2^{m*}(\Omega) Y_2^m(\Omega_0)\exp(-t/\tau_l).$$

The integrals may now be performed separately:

$$\int Y_l^n(\Omega) Y_2^{m*}(\Omega)d\Omega = \delta_{l,2}\delta_{n,m},$$

thus only the $l = 2$, $n = -m$ term of the sum remains:

$$\langle\ \rangle = \frac{1}{4\pi}\exp(-t/\tau_2),\tag{7.63}$$

independent of m as expected.

We have thus found a simple exponential form for the autocorrelation function in this case; $G(t)$ is given by

$$G(t) = \frac{3}{20}\left(\frac{\mu_0}{4\pi}\right)^2\frac{\hbar^2\gamma^4}{r^6}\exp(-t/\tau_c)\tag{7.64}$$

or

$$G(t) = \frac{M_2}{3}\exp(-t/\tau_c),\tag{7.65}$$

with correlation time τ_c

$$\tau_c = a^2/6D\tag{7.66}$$

and second moment

$$M_2 = \frac{9}{20}\left(\frac{\mu_0}{4\pi}\right)^2\frac{\hbar^2\gamma^4}{r^6}.\tag{7.67}$$

The correlation time may be evaluated from a knowledge of the physical constant and the various properties of water:

$kT = 4.04 \times 10^{-21}\,\text{J}$ for $T = 20\,^\circ\text{C}$,
$\eta = 1.04 \times 10^{-3}\,\text{N}\,\text{s}\,\text{m}^{-2}$ for water at $T = 20\,^\circ\text{C}$,
$a = 1.40 \times 10^{-10}\,\text{m}$,

from which we find $\tau_c = 2.96 \times 10^{-12}\,\text{s}$. This is a very short correlation time, and for all reasonable resonance frequencies the fast motion limit would apply, where $T_1 = T_2$ and the precise form of the correlation functions are not required for finding the relaxation times. In such a case we have

$$\frac{1}{T_1} = \frac{1}{T_2} = \frac{10}{3}M_2\tau_c.\tag{7.68}$$

The expression for M_2 is given in Equation (7.67), and since

$\gamma = 2.675 \times 10^8\,\text{s}^{-1}\,\text{T}^{-1}$,
$\hbar = 1.055 \times 10^{-34}\,\text{J}\,\text{s}$,
$r = 1.514 \times 10^{-10}\,\text{m}$,

this gives $M_2 = 2.12 \times 10^{10}\,\text{s}^{-2}$, from which we obtain a relaxation time of $4.78\,\text{s}$.

We have evaluated the contribution to the relaxation from the intramolecular interaction in water. While this is the dominant effect, there will be a smaller contribution from intermolecular interactions, best understood after the discussion of later sections in this chapter. Thus we must regard 4.78 s as an upper limit on the relaxation time. This is consistent with an observed T_1 of 3.6 s for pure water.

7.4.2 Rotation of polyatomic molecules

The relaxation of three- and four-spin assemblies undergoing rotational diffusion can have a non-exponential decay. Unfortunately it is not possible to demonstrate this using our currently established techniques since there are quantities of importance (quasi-constants of the motion) other than the total magnetisation; there are correlations between the various particles. The full machinery of the density operator, as treated in Chapter 9, must then be used.

In such cases both the transverse and the longitudinal relaxation can be shown to comprise the sum of a number of exponentials (Hubbard 1948). However, with larger numbers of spins, as was seen with relaxation in the solid, the various features become 'washed out' and the relaxation, once again, takes on a simple exponential form.

If we consider a molecule with n resonant nuclear spins then the spin relaxation is influenced by $n \times (n - 1)$ interspin interactions. On making the assumption that the correlations between the motions of the different pairs have a minimal effect, the results of the previous section may be carried over so that

$$G(t) = \frac{M_2}{3} \exp(-t/\tau_c),$$ (7.69)

where here M_2 has the value corresponding to the many-spin molecule. The correlation time τ_c will be as given by Equations (7.66) and (7.62) in terms of the parameters a and η, the 'radius' of the molecule and the viscosity of the liquid.

In the case of water, treated in the last section, the motion was always 'fast' so that $\omega_0 \tau_c$ is vanishingly small. It then followed that the form of the spectral density junction $J(\omega)$ was unimportant in finding the relaxation times; only its zero-frequency value was needed. With a more viscous system however, where the correlation time can be much longer, knowledge of the mathematical form for the spectral density function is required to obtain values for the relaxation times. Upon Fourier transformation of Equation (7.69) we have

$$J(\omega) = \frac{2}{3} \frac{M_2 \tau_c}{1 + \omega^2 \tau_c^2},$$ (7.70)

so that T_1 and T_2 are given by

$$\frac{1}{T_1} = M_2 \tau_c \left(\frac{2/3}{1 + \omega^2 \tau_c^2} + \frac{8/3}{1 + 4\omega^2 \tau_c^2} \right),$$

$$\frac{1}{T_2} = M_2 \tau_c \left(1 + \frac{5/3}{1 + \omega^2 \tau_c^2} + \frac{2/3}{1 + 4\omega^2 \tau_c^2} \right). \tag{7.71}$$

At a given frequency the variation of the relaxation times with τ_c, as given by these equations, follows the general behaviour of the curves in Figure 7.1. On a log–log plot the curve of T_2, and that of T_1 to the left of the minimum have a slope of -1, which is independent of the particular model and the consequent shape of $J(\omega)$. To the right of the minimum the slope of the T_1 line is $+1$ for this specific model having a Lorentzian $J(\omega)$.

The exponential correlation function – Equation (7.69) – the resultant Lorentzian spectral density function – Equation (7.70) – and the expressions for T_1 and T_2 which follow – Equations (7.71) – are a consequence of the rotational diffusion model treated in the last section. In the following sections we shall consider other dynamical models, mainly relating to translational motion. We will see that while the calculated autocorrelation functions generally display a smooth monotonic decay, their mathematical description can be very complex indeed. Sometimes, in such cases, the exponential model is still adopted as a reasonable, mathematically tractable, approximation to the system. Thus Equations (7.71) are very often used outside their domain of strict validity.

At sufficiently high frequencies or small correlation times the Lorentzian form for $J(\omega)$ must break down. Diffusion, whether rotational or translational, is a hydrodynamic description of motion. It is the result of a large number of collisions between particles and for times shorter than the collision time the diffusive description is entirely inappropriate. For such short times the correlation function cannot vary exponentially; and, in fact, the first time derivative must vanish at $t = 0$.

7.4.3 Relaxation by translational diffusion

The random thermal motion of molecules in a liquid may be understood, on the macroscopic scale, as a self-diffusion process. This provides a mechanism for spin relaxation through the resultant modulation of the internuclear dipolar interactions between molecules. While always present, for molecules containing only one resonant nucleus this will be the dominant relaxation process. Thus translational diffusion will have an influence on relaxation and the measurement of relaxation times can therefore be used to investigate the diffusion process.

In Chapter 4 we saw that translational diffusion could be studied by the technique of spin echoes in a magnetic field gradient. It was demonstrated that from measurements of the echo relaxation the diffusion coefficient could be obtained. This section will

show how diffusion motion affects the dipolar relaxation times, providing a complementary technique for the study of diffusion; see also Torrey (1953, 1954).

We have the dipolar autocorrelation function for N interacting nuclear dipoles given
by Equation (7.35):

$$G_m(t) = \left(\frac{\mu_0}{4\pi}\right)^2 \frac{6\pi\hbar^2\gamma^4}{5N} \sum_{i<j} \frac{Y_2^{m*}[\Omega_{ij}(t)]\, Y_2^m[\Omega_{ij}(0)]}{r_{ij}^3(t)r_{ij}^3(0)},$$

where, in contrast to the treatment of the previous section, here the r_{ij} can vary. This is
truly a many-body problem and there will be many terms contributing to the sum. We
are thus led immediately to the adoption of a probabilistic approach whereby the sum
over particles is replaced by an integral over an appropriate distribution function

$$\sum_{i,j} \rightarrow \alpha \int d\mathbf{r} \int d\mathbf{r}_0 P(\mathbf{r},\mathbf{r}_0,t)g(\mathbf{r}_0).$$

Here $P(\mathbf{r},\mathbf{r}_0,t)$ is the probability that a pair of particles, initially of separation $\mathbf{r}_0$, will
after a time t be separated by $\mathbf{r}$ and $\alpha g(\mathbf{r}_0)$ gives the probability that at time $t = 0$ a pair
of particles will be found with separation $\mathbf{r}_0$: α is the spin density and for a rotationally
invariant system $g(r)$ is the radial distribution function.

A number of approximations will expedite our evaluation of $G(t)$ in closed form. We
shall defer discussion and the possible relaxation of some of the approximations until a
later section. For diffusive motion $P(\mathbf{r},\mathbf{r}_0,t)$ may be found by solution of a diffusion
equation subject to the appropriate boundary conditions. For

$$P(\mathbf{r},\mathbf{r}_0,t) = \delta(\mathbf{r} - \mathbf{r}_0)$$

the solution is

$$P(\mathbf{r},\mathbf{r}_0,t) = (8\pi Dt)^{-3/2}\exp(-|\mathbf{r} - \mathbf{r}_0|^2/8Dt), \qquad (7.72)$$

an extra factor of 2 multiplying the expected $4Dt$ because $\mathbf{r}$ is the separation of two
particles *both* moving with diffusion coefficient D.

On the assumption that the spin distribution is uniform, $g(r)$ may be approximated
by a step function

$$g(r) = \begin{cases} 0, & r \leqslant a \\ 1, & r > a \end{cases} \qquad (7.73)$$

where a is the 'hard core' diameter of the spin-carrying molecule or atom.

Unfortunately the diffusion propagator, Equation (7.72), does not preserve the form
of the radial distribution function $g(r)$. For long times it tends to a uniform distribution,
allowing particles to approach closer than their hard core dimensions. This difficulty

can be circumvented, within the approximation of Equation (7.73), by restricting the range of $|\mathbf{r}|$ in the integral for $G(t)$ from becoming less than a. We then obtain

$$G_m(t) = \left(\frac{\mu_0}{4\pi}\right)^2 \frac{3\pi a\hbar^2\gamma^4}{5} \int_{|\mathbf{r}|>a} d\mathbf{r} \int_{|\mathbf{r}_0|>a} d\mathbf{r}_0 P(\mathbf{r},\mathbf{r}_0,t) \frac{Y_2^{m*}(\Omega) Y_2^m(\Omega_0)}{r^3 r_0^3}. \tag{7.74}$$

To evaluate the integrals, $P(\mathbf{r},\mathbf{r}_0,t)$ from Equation (7.72) is first expressed as a Fourier integral using

$$\exp\left(-\frac{|\mathbf{r}-\mathbf{r}_0|^2}{8Dt}\right) = \left(\frac{2Dt}{\pi}\right)^{3/2}\int \exp(-2k^2Dt)\exp[i\mathbf{k}\cdot(\mathbf{r}-\mathbf{r}_0)]d^3k.$$

Then the exponentials are expressed as Bessel function expansions:

$$\exp(i\mathbf{k}\cdot\mathbf{r}) = 4\pi\left(\frac{\pi}{2kr}\right)^{1/2}\sum_{l,m} i^l Y_2^{m*}(\Omega) Y_2^m(\Omega') J_{l+\frac{1}{2}}(kr),$$

where Ω' is the orientation of the vector $\mathbf{k}$ and the J is a Bessel function. Substituting the resultant $P(\mathbf{r},\mathbf{r}_0,t)$ into Equation (7.74) and using the orthogonality properties of the spherical harmonics gives

$$G(t) = \left(\frac{\mu_0}{4\pi}\right)^2 \frac{3\pi a\hbar^2\gamma^4}{5} \int_0^\infty k\exp(-2k^2Dt)\left[\int_a^\infty \frac{J_{5/2}(kr)}{k^{3/2}}\right]^2 dk.$$

Observe that the index m has vanished, since the diffusion was taken to be isotropic.

It is convenient, at this stage, to introduce a measure of the correlation time which we shall denote by τ. This is defined as the time it would take for a particle to 'diffuse' the hard core distance:

$$\tau = a^2/2D. \tag{7.75}$$

The integral in the square bracket above may be performed and we can then write $G(t)$ as

$$G(t) = 3G(0)\int_0^\infty \exp\left(-\frac{t}{\tau}x^2\right) J_{3/2}^2(x)\frac{dx}{x}, \tag{7.76}$$

where, of course, $3G(0)$ is the second moment M_2:

$$3G(0) = M_2 = \left(\frac{\mu_0}{4\pi}\right)^2 \frac{3\pi a\hbar^2\gamma^4}{5}. \tag{7.77}$$

Now the integral in Equation (7.76) cannot be evaluated in terms of the usual range of elementary functions. It can, however, be expressed in terms of the $_3F_3$ generalised hypergeometric function (Abramowitz and Stegun 1965) as

$$\frac{G(t)}{G(0)} = \frac{1}{6\pi^{1/2}}\left(\frac{\tau}{t}\right)^{3/2}\ _3F_3\left[2,\frac{5}{2},\frac{3}{2},\frac{5}{2},\frac{5}{2},4;\ -\frac{\tau}{t}\right]. \tag{7.78}$$

Fundamentally, of course, there is no difference between writing the expression for $G(t)$ in terms of the integral in Equation (7.76) or as the hypergeometric function of Equation (7.78). The merit of the latter form is, however, that the various tabulated properties of the function can then be used when required, as is the case with sines, cosines, Bessel functions, spherical harmonics etc. Recall that we encountered the $_1F_1$ hypergeometric function in the previous chapter.

To find the relaxation times we need the spectral density functions: the Fourier transform of $G(t)$. This may be found from Equation (7.78) using tables or it may be evaluated by performing the Fourier integral on Equation (7.76). The integral over x is done after that over t, the result being

$$J(\omega) = 2M_2\tau u^{-5}\{u^2 - 2 + [(u^2 + 4u + 2)\cos(u) + (u^2 - 2)\sin(u)]\exp(-u)\}, \tag{7.79}$$

where

$$u = (2\omega\tau)^{\frac{1}{2}}. \tag{7.80}$$

It is of interest to compare this result with the exponential/Lorentzian form which, as we stated previously, was an often-used approximation.

Box 7.3 Definition of the correlation time

In discussing the general principles of the relaxation process and in the various dynamical models there is a microscopic time which is used to characterise the system. For the rotational and translational diffusion models this time was taken as that necessary for a particle to move a given angle and distance respectively.

The correlation time was given a precise definition in terms of the area under the autocorrelation function:

$$\tau_c = \frac{1}{G(0)}\int_0^\infty G(t)dt, \tag{7.81}$$

(the width of the function being given by the area divided by the height). In terms of this definition we saw that the adiabatic T_2 took on the particularly simple form

$$\frac{1}{T_2} = \frac{3}{2}J(0) = M_2\tau_c. \tag{7.82}$$

In comparing the consequences of different relaxation models it is important to adopt a consistent definition for the microscopic characteristic time. We shall take Equation (7.81) for this definition.

The τ_c defined for rotational relaxation, treated in the previous two sections, does actually also satisfy the condition of Equation (7.81). However, the characteristic time for translational diffusion, i.e. the τ defined in Equation (7.75), does not. There is a numerical coefficient which may be determined from consideration of the zero-frequency value of $J(\omega)$. In the limit $\omega \to 0$, Equations (7.79) and (7.80) give

$$J(0) = \frac{4}{5}G(0)\tau, \tag{7.83}$$

while from Equation (7.81), in terms of τ_c, we have

$$J(0) = 2G(0)\tau_c.$$

Thus the translational diffusion τ is related to the correlation time τ_c by

$$\tau = \frac{5}{2}\tau_c. \tag{7.84}$$

Then the expression for $J(\omega)$ is still given by Equation (7.79), while the parameter u is now expressed as

$$u = (5\omega\tau_c)^{\frac{1}{2}}. \tag{7.85}$$

In Figure 7.3 we show the graph of $J(\omega)$ against $\omega\tau_c$ for fixed τ_c as calculated for the present model of translational diffusion. We also show, for comparison, the corresponding curve for the Lorentzian approximation. Qualitatively, the two curves are not dissimilar. However, it should be emphasised that the scales are logarithmic which tends to suppress the differences; at $\omega\tau_c \sim 0.5$ observe that the Lorentzian approximation overestimates the value of the diffusion model by a factor of approximately 2.

7.4.4 Low frequency behaviour

The nature of the spectral density function, and the consequent behaviour of the relaxation times, has particular importance at low frequencies. Very generally low frequencies correspond to long times. In other words, low frequency measurements, particularly of T_1, should give information about the *hydrodynamic* processes going on in a liquid (and indeed sometimes in solids as well). We shall see how this is so.

Let us consider experiments where the correlation time τ_c is kept fixed while relaxation times are measured at different Larmor frequencies. Here the graph of

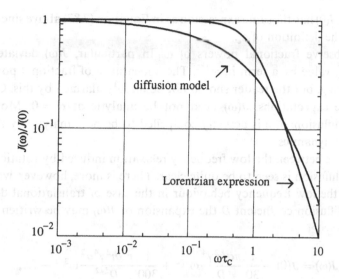

Figure 7.3 Spectral density function for the diffusion model compared with the Lorentzian approximation.

Figure 7.3 is relevant rather than that of Figure 7.1, which applies to variations of τ_c at a given frequency. From a cursory inspection of Figure 7.3 it is clear that the low frequency behaviour for rotational diffusion and that for translational diffusion are rather different. We shall now quantify this.

For the rotational diffusion model, which has a Lorentzian spectral density, we can expand Equation (7.70) to give the low frequency behaviour

$$J(\omega) = J(0) - \frac{2}{3}M_2\tau_c^3\omega^2 + \frac{2}{3}M_2\tau_c^5\omega^4 - \ldots, \tag{7.86}$$

where

$$J(0) = \frac{2}{3}M_2\tau_c. \tag{7.87}$$

The low frequency relaxation rate is seen, from equation (7.86), to start deviating from its constant $\omega = 0$ value by a term in ω^2. It is likely that the reader will evince no surprise at this result; $J(\omega)$ is smoothly varying and an even function of ω. Things are not so simple, however, for the case of translational diffusion.

Taking the translational diffusion model of the previous section, upon expanding Equation (7.79) we obtain

$$J(\omega) = J(0) - \frac{5}{18}\sqrt{5M_2}\tau_c^{3/2}\omega^{1/2} + \frac{5}{36}\sqrt{5M_2}\tau_c^{5/2}\omega^{3/2} - \ldots \tag{7.88}$$

and, of course, $J(0)$ has the same expression as in Equation (7.87) above since this really follows from the definition of τ_c.

Here we observe fractional powers of ω. In particular, $J(\omega)$ deviates from its zero-frequency value by a term in $\omega^{1/2}$. The appearance of fractional powers might appear surprising, but the reader should not be unduly alarmed by this. Contrary to common-sense expectations, $J(\omega)$ need not be analytic at $\omega = 0$. Moreover, for translational diffusion, $J(0)$ is actually compelled to be non-analytic by the requirements of hydrodynamics.

The difference between the low frequency relaxation induced by rotational and by translational diffusion is seen to be quite clear. There is more, however, which can be inferred from the low frequency behaviour in the case of translational diffusion. In terms of the diffusion coefficient D the expansion of $J(\omega)$ may be written as

$$J(\omega) = J(0) - \frac{1}{30} \frac{\pi \alpha \hbar^2 \gamma^4}{D} \omega^{1/2} + \frac{1}{300} \frac{\pi \alpha \hbar^2 \gamma^4 a^2}{D^{5/2}} \omega^{3/2} - \ldots, \qquad (7.89)$$

where α is the spin density and a is the hard core diameter, as previously defined. Observe that the term in $\omega^{1/2}$ contains only constants and the macroscopic quantities α and D. By contrast the term in $\omega^{3/2}$ (and the other powers) contains the microscopic dimension a. It is no accident that the $\omega^{1/2}$ term has this special property. Indeed for any model of the motion which leads to diffusion at long times, the term in $\omega^{1/2}$ will be identical, whereas the other terms in the expansion will differ according to the precise details of the particles' motion. Unfortunately the demonstration of this fact is beyond the scope of this book; the interested reader is referred to Cowan and Fardis (1991).

To summarise then, for translational diffusion at low frequencies the relaxation rates start to deviate from their constant value by a term in $\omega^{1/2}$, which depends on the diffusion coefficient but on no microscopic details of the system.

In principle, we see that so long as the relaxation is dipolar in origin this is a powerful technique for the study of diffusion, complementing the field gradient method discussed in Section 4.5. Here there is no need for a gradient field with the attendant calibration problems. However, there are other instrumentational problems which should be mentioned. Firstly one requires measurements of relaxation times over a range of Larmor frequencies. Thus the magnetic field has to be varied and the spectrometer must be operated over a range of frequencies. Unfortunately many commercial spectrometers operate at a fixed frequency. Secondly, we require measurements specifically at low frequencies. From the discussions of Chapter 3 we see that there can be sensitivity problems at low frequencies so that the relaxation signals may well not be observable at the required frequencies. The technique of relaxation measurement in the rotating frame may here be used to advantage.

7.4.5 High frequency limiting behaviour

We shall, for completeness, make a brief mention of the limiting behaviour of the spin relaxation at high frequencies. As discussed at the end of Section 7.4.2, the diffusive description of motion ceases to have validity at very high frequencies. Furthermore, it is precisely in this régime that the microscopic details of the motion have particular influence on the relaxation. Thus the results we shall now derive must be treated with caution.

For the model of rotational diffusion, whose dipolar spectral density is given by Equation (7.70), expansion in inverse powers of ω gives

$$J(\omega) \sim \frac{2M_2\tau_c}{3}[(\omega\tau_c)^{-2} - (\omega\tau_c)^{-4} + (\omega\tau_c)^{-6} - \ldots]. \qquad (7.90)$$

This indicates that T_1 will become proportional to ω^2 at high frequencies. The translational diffusion model we have been considering leads to an asymptotic behaviour for $J(\omega)$ as follows

$$J(\omega) \sim \frac{M^2\tau_c}{\sqrt{5}}\left[(\omega\tau_c)^{-3/2} - \frac{2}{5}(\omega\tau_c)^{-5/2} + \ldots\right], \qquad (7.91)$$

implying that T_1 will increase as $\omega^{3/2}$ at high frequencies. Once again we note the occurrence of fractional powers in the *translational* diffusion case. Lenk (1977) has observed T_1 to be proportional to $\omega^{3/2}$ in a number of fluid polymers.

It has been pointed out by Ayant, Belorizky, Alzion and Gallice (1975) that the simple translational diffusion model, such as that in Section 7.4.3, has certain unphysical features. In particular, the short time behaviour of the diffusion propagator $P(r,r_0,t)$ does not correctly describe collisions between the particles. The expression used ignores collisions, permitting the particles to intersect and actually to pass through each other. The intersection is only forbidden for the initial and the final states. The precise details of the particles' collisions are clearly unimportant in the long time/low frequency limit; this is not so at short times/high frequencies. Ayant *et al.* used a more realistic treatment of diffusion whereby the boundary condition on the particles colliding was respected at all times. The result is that at high frequencies $J(\omega)$ is predicted to vary as ω^{-2} rather than the $\omega^{-3/2}$ of the simpler model. They quote experimental data which supports such behaviour. There is clearly a paradox here.

While it is not our intention to discuss these phenomena in detail, a few points are worthy of mention. We must consider the different time/frequency scales involved in these systems. We can talk of diffusion only for times longer than the interval between particle collisions. Thus the diffusive description is appropriate only at frequencies below the inverse collision time. Now the high frequency asymptotic expansion of $J(\omega)$ is valid for frequencies significantly greater than the inverse correlation time. The collision time is essentially the time to travel a distance of a mean free path, whereas we recall that the correlation time estimates the time for a particle to travel a distance equivalent to the size of a particle. We must therefore consider the relative magnitudes of the mean free path and the particle size. In particular, when the mean free path is very much smaller than the size of the particles, as in a dense liquid of large molecules, then the diffusion description can remain appropriate concurrently with the asymptotic expansion of $J(\omega)$ so that the $\omega^{-3/2}$ behaviour will be expected. Conversely for a fluid of mobile atoms or small molecules it can be seen that the details of the particles' collisions will be important and then the ω^{-2} law will obtain.

7.5 Some general results

The following subsections contain discussions of various topics of an advanced nature; they can be omitted on a first reading of the book.

7.5.1 Scaling treatment of relaxation

In each of the various relaxation models discussed in the previous sections there was a single microscopic time τ which characterised the dynamical behaviour of the system. Thus in Equations (7.66) and (7.75) the characteristic times for rotational and for translational diffusion were defined. On the assumption that the microscopic dynamics *can* be described in terms of a single such time, a number of general inferences may be made about the relaxation in these systems.

Let us write the autocorrelation function $G(t)$ as the product of its initial value $G(0)$ and a normalised shape function g:

$$G(t) = G(0)g(t/\tau). \tag{7.92}$$

The shape function $g(t/\tau)$ is unity at $t = 0$ and its dimensionless argument t/τ indicates that τ is the characteristic time of this system; it is the natural time unit in terms of which the dynamical behaviour of the system scales. Note that for simplicity we are considering a rotationally invariant system so that $G(t)$ need not be encumbered by a spin-flip subscript.

The spectral density function $J(\omega)$ is found from the Fourier transform of Equation (7.92):

$$J(\omega) = \int_{-\infty}^{\infty} G(0)g(t/\tau)\exp(i\omega t)\mathrm{d}t,$$

which, on changing variables of integration through $x = t/\tau$, may be written

$$J(\omega) = G(0)\tau j(\omega\tau), \tag{7.93}$$

where $j(z)$ is the Fourier transform of the shape function $g(x)$:

$$j(z) = \int_{-\infty}^{\infty} g(x)\exp(ixz)dx.$$

It is instructive to write the expression for the adiabatic T_2 in terms of this formalism. Recall that T_2 is related to the zero-frequency value of $J(\omega)$ and since we now have

$$J(0) = G(0)\tau j(0)$$

this gives us

$$1/T_2 = \frac{3}{2} G(0)\tau j(0)$$

$$= M_2 \tau \frac{j(0)}{2}. \qquad (7.94)$$

This result bears a similarity to the conventional expression for T_2:

$$1/T_2 = M_2 \tau_c.$$

particularly since $j(0)$ is simply a number of the order of unity. The difference between the present and the traditional expression is in the definition of the characteristic time used. The approaches are, however, quite consistent since while the correlation time τ_c has a precise definition, as in Box 7.3, the scaling time used in this section is only specified within a multiplicative constant.

Turning now to longitudinal relaxation let us start, for simplicity, with the expression for T_1:

$$1/T_1 = J(\omega).$$

This is correct for relaxation through external magnetic field inhomogeneities (Appendix E), but for dipolar relaxation it neglects the effect of the double frequency term. This does not detract from the validity of our general conclusions; we shall explicitly show this at the end of the section.

Using the expression for $J(\omega)$ as in Equation (7.93) we have

$$1/T_1 = G(0)\tau j(\omega\tau).$$

If we now divide this expression by τ or multiply it by ω then the resulting expressions $1/T_1\tau$ and ω/T_1 both depend on ω and τ only through the product $\omega\tau$:

$$1/T_1\tau = G(0)j(\omega\tau), \qquad (7.95)$$

$$\omega/T_1 = G(0)\omega\tau j(\omega\tau). \qquad (7.96)$$

This means that T_1 measured at fixed τ – varying ω – (Equation (7.95)) and T_1 measured at fixed ω – varying τ – (Equation (7.96)) will both be, for a given system, universal functions of $\omega\tau$. Thus data plotted in this way will fall on single curves. However, we shall have to wait until Section 8.3 for a more impressive demonstration of this.

Finally, as promised, we must demonstrate that the above conclusions are not invalidated on the inclusion of the double frequency term in the dipolar T_1. Therefore we now write

$$1/T_1 = J(\omega) + 4J(2\omega).$$

Using $J(\omega)$ from Equation (7.93) this gives

$$1/T_1 = G(0)\tau[j(\omega\tau) + 4j(2\omega\tau)],$$

and since the expression in the square brackets depends on ω and τ only through the product $\omega\tau$, we may write

$$[j(\omega\tau) + 4j(2\omega\tau)] = \tilde{j}(\omega\tau) \tag{7.97}$$

so that, as before, T_1 takes the form

$$G(0)\tau\tilde{j}(\omega\tau)$$

and thus the previously established results still hold.

7.5.2 The T_1 minimum

As in the previous section, we shall denote the Fourier transform of the normalised autocorrelation shape function $g(t/\tau)$ by $j(z)$. We also recall that where necessary the double frequency term is subsumed into a composite $\tilde{j}(z)$ function. Now generally (except in very unusual circumstances where $j(0)$ diverges) $j(z)$ starts from its initial value $j(0)$ and it is a decreasing function of its argument. The product $zj(z)$ starts from zero when z is zero, initially growing linearly. However, for large z the decay of $j(z)$ will outweigh the linear growth of the z pre-factor and the product will decrease. So somewhere inbetween, when z is of the order of unity, at $z = z'$ say, $zj(z)$ must have a maximum.

Looking at the expression for T_1 as a function of τ (at fixed frequency), Equation (7.95), we see that the existence of a maximum in $zj(z)$ tells us that there will necessarily be a minimum in T_1 as the correlation time is varied, such as that indicated in Figure 7.1 and the experimental examples, Figures 8.3, 8.5 and 8.9. Furthermore, since we also saw that $zj(z)$ initially grows linearly, this tells us that $1/T_1$ is proportional to the correlation time on the 'fast' side of the T_1 minimum – in the $T_1 = T_2$ region.

The 'position' of the T_1 minimum is given by $\omega_0\tau = z'$ or

$$\tau_{min} = z'/\omega_0. \tag{7.98}$$

Thus when a minimum is observed, to within a constant of the order of unity, the characteristic time of the motion may be estimated by the Larmor period.

The value of T_1 at the minimum may be written, from Equation (7.96), as

$$\frac{1}{T_1^{\min}} = \frac{G(0)z'j(z')}{\omega_0} \tag{7.99}$$

or, since $G(0) = M_2/3$, and denoting the number $z'j(z')$ by K^{-1}:

$$T_1^{\min} = \frac{\omega_0}{M_2} \times 3K. \tag{7.100}$$

We see that at the minimum the value of T_1 is proportional to the Larmor frequency. As the reader will see in Problem 7.3 for a Gaussian correlation function $3K = 1.97$, while for an exponential $3K = 3$.

The conclusion, from this discussion, is that observation of minima in T_1 gives both an estimate for the characteristic time of the system and an indication of the origin of the local fields: the mechanism for relaxation.

7.5.3 *Moments and transverse relaxation – revisited*

Knowledge of all moments permits a complete description of the transverse relaxation profile; recall that the moments are essentially the coefficients of the time expansion of the transverse relaxation function $F(t)$. We saw that while it is possible to obtain *formal* expressions for the various moments, evaluation of the higher order moments becomes increasingly difficult so that a full calculation of the relaxation is not feasible in this way. However, in the case of fluids we explained, at the beginning of this chapter, that the averaging effects of the atomic motion permitted certain approximations to be made. In particular, we saw in Equation (7.23) that to leading order in τ^{-1} the general behaviour of all higher order moments could be expressed in terms of M_2 and τ. We are now in a position to sharpen that relation in the light of the cumulant expansion method that was used in this chapter.

The transverse relaxation function in the adiabatic limit was found to be identical (apart from the factor of 3) with that obtained in the classical model of Chapter 4. As obtained in this chapter, in Equation (7.41) we have

$$F(t) = \exp\left[-3\int_0^t (t - \tau)G(\tau)d\tau\right]. \tag{7.101}$$

One of the conclusions drawn, particularly from the discussions in Chapter 4, was that the precise form of $G(t)$ is not so important for determining the relaxation profile. In particular, in the fast motion limit only its initial height and area are required. Thus for the purpose of this discussion we shall assume a simple Gaussian form for $G(t)$ which we shall write as

$$G(t) = M_2 \exp\left[-\frac{\pi}{4}\left(\frac{t}{\tau_c}\right)^2\right]. \tag{7.102}$$

The numerical factors here inside the exponential ensure that τ_c is the correctly defined correlation time.

Substituting this expression into Equation (7.101) for $F(t)$ and repeatedly differentiating, we obtain a power series expansion for the relaxation profile

$$F(t) = 1$$

$$- t^2 \frac{M_2}{2}$$

$$+ t^4 \left(\frac{\pi M_2}{48} \tau_c^{-2} + \frac{M_2^2}{8} \right)$$

$$- t^6 \left(\frac{\omega^2 M_2}{960} \tau_c^{-4} + \frac{\omega M_2^2}{96} \tau_c^{-2} + \frac{M_2^3}{48} \right)$$

$$+ t^8 \left(\frac{\pi^3 M_2}{21\,504} \tau_c^{-6} + \frac{\pi^2 M_2^2}{23\,040} \tau_c^{-4} + \frac{\pi M_2^3}{384} \tau_c^{-2} + \frac{M_2^4}{384} \right) \tag{7.103}$$

$$- \cdots$$

When the motion is fast the leading term in each of the powers of t will dominate. In that case, from the coefficients of these terms we find the moments, which may be expressed in a general form as

$$M_{2(n+1)} = \frac{M_2}{\tau_c^{2n}} \left(\frac{\pi}{4} \right)^n \frac{(2n)!}{n!}, \tag{7.104}$$

which does indeed have the form asserted in Equation (7.23), but now we have the numerical coefficients. The derivation of Equation (7.104) will be taken up in Problem 7.4. For the present its plausibility may be checked by substituting values for n and comparing with the relevant terms in Equation (7.103).

It is of interest to explore the relation between the time dependence of $F(t)$ and $G(t)$ a little more deeply. If Equation (7.101) is inverted then $G(t)$ may be expanded as

$$3G(t) = M_2 + \frac{t^2}{2} 3M_2^2 + \frac{t^4}{4!} 30M_2^3 + \frac{t^6}{6!} 630M_2^4 + \frac{t^8}{8!} 22\,680M_2^5 + \cdots$$

$$- \frac{t^2}{2} M_4 - \frac{t^4}{4!} 15M_4 M_2 - \frac{t^6}{6!} 420M_4 M_2^2 + \frac{t^8}{8!} 18\,900M_4 M_2^3 - \cdots$$

$$+ \frac{t^4}{4!} M^6 + \frac{t^6}{6!} (28M_6 M_2 + 35M_4^2) + \frac{t^8}{8!} (1260M^6 M_2^2 + 3150M_4^2 M_2) + \cdots$$

$$- \frac{t^6}{6!} M_8 - \frac{t^8}{8!} (45M_8 M_2 + 210M_6 M_4) - \cdots$$

$$+ \frac{t_8}{8!} M_{10} + \cdots$$

$$\tag{7.105}$$

The terms have been arranged in such a way that each column contains a given power of time and each row contains a given power of the motion Hamiltonian. When the motion is fast the highest power of the motion Hamiltonian will dominate in each power of time. These are the terms along the diagonal in Equation (7.105). It will be observed that these diagonal terms are the expansion of the second derivative of $F(t)$. Thus one may re-sum these leading terms of the expansion, to give the simple result

$$G(t) = -F''(t). \tag{7.106}$$

Box 7.4 A paradox

The observant reader will note a strange paradox: the simple relation between $G(t)$ and $F(t)$ implied by Equation (7.106) is impossible! Consider the first derivative of $F(t)$ at very long times. The relation gives

$$F'(\infty) = -\int_0^\infty G(t)\mathrm{d}t.$$

The integral on the right hand side is recognised, from Equation (4.37), as the transverse relaxation rate, T_2^{-1}. This is obviously a finite quantity, whereas $F'(\infty)$, the left hand side, must clearly go to zero. To resolve the paradox let us retrace the steps by which we arrived at the offending equation.

We started with a respectable expression for the relaxation $F(t)$, in the form of Equation (7.101). We then inverted it and expanded as a power series in time, the various terms being shown in Equations (7.105). Then we identified the dominant terms in the series, rejected the others, and then resummed the series. Although we took the leading term in each power of t, the problem is that at larger times some of the neglected terms in the higher powers of t may, in fact, be greater than the retained terms in the lower powers. Thus Equation (7.106) breaks down at long times. Nevertheless, as an indicator of the short time behaviour the relation is quite useful.

Recall that the precise form of the autocorrelation function is not so important; all (or at least most) will lead to exponential relaxation in the fast motion limit. The main use for relations of the form of Equation (7.106) is in obtaining approximate expressions for $G(t)$ which may be used both for constructing improved expressions for $F(t)$ and the transverse relaxation time, as well as evaluating the longitudinal relaxation time.

7.5.4 T_1 sum rules and moments

Moments were introduced specifically in the study of *transverse* relaxation. However, since, as we have just seen, there is a close connection between the moments and the autocorrelation function $G(t)$, and furthermore since the spin lattice relaxation is related to $J(\omega)$ which is the Fourier transform of $G(t)$, it follows that T_1 and the moments must be related. We shall investigate such a relationship in this section and we will thereby establish a number of potentially useful results.

Starting from the expression for the dipolar T_1:

$$1/T_1 = J(\omega) + 4J(2\omega),$$

by a change of variables in the Fourier integral for the double frequency term, we can write

$$1/T_1 = \int\limits_{-\infty}^{\infty} [G(t) + 2G(t/2)]\exp(i\omega t)\mathrm{d}t. \tag{7.107}$$

Again we have assumed rotational invariance for simplicity but the generalisation to distinct $G_1(t)$ and $G_2(t)$ functions is straightforward.

If we now invert the Fourier integral of Equation (7.107), giving

$$G(t) + 2G(t/2) = \frac{1}{2\pi} \int\limits_{-\infty}^{\infty} \frac{\exp(-i\omega t)}{T_1}\mathrm{d}\omega,$$

then expansion in powers of time leads to

$$\sum_{n=0}^{\infty} \frac{G^{(n)}(0)}{n!} t^n + 2\sum_{n=0}^{\infty} \frac{G^{(n)}(0)}{n!}\left(\frac{t}{2}\right)^n = \frac{1}{2\pi} \int\limits_{-\infty}^{\infty} \sum_{n=0}^{\infty} \frac{(-i)^n \omega \pi t^n}{n!T_1}\mathrm{d}\omega.$$

Equating powers of time then gives

$$\left(1 + \frac{1}{2^{n-1}}\right) G^{(n)}(0) = \frac{(-i)^n}{2\pi} \int\limits_{-\infty}^{\infty} \frac{\omega^n}{T_1}\mathrm{d}\omega$$

or, since for even n the integral is symmetric,

$$\int\limits_0^{\infty} \frac{\omega^n}{T_1}\mathrm{d}\omega = (-i)^n \pi \left(1 + \frac{1}{2^{n-1}}\right) G^{(n)}(0), \tag{7.108}$$

which leads to a set of frequency sum rule expressions for T_1 for increasing n. The sum rules can be expressed in terms of moments, using the relations established in Equation (7.105). In this way, for $n = 0, 2, 4$, for example we obtain

$$n = 0: \int_0^\infty \frac{1}{T_1} d\omega = 3\pi G(0) - \pi M_2;$$

$$n = 2: \int_0^\infty \frac{\omega^2}{T_1} d\omega = -\frac{3}{2}\pi G^{(2)}(0) = \frac{\pi}{2}(M_4 - 3M_2^2); \qquad (7.109)$$

$$n = 4: \int_0^\infty \frac{\omega^4}{T_1} d\omega = \frac{9}{8}\pi G^{(4)}(0) = \frac{3\pi}{8}(M_6 - 15M_4 M_2 + 30M_2^3).$$

Again we emphasise that in the case of rapid motion it is the leading moment in the expression which dominates. Moreover, we also recover conditions on the moments that the motion be regarded as rapid, such as $M_4 \gg M_2^2$.

If one has a set of T_1 data taken over a range of Larmor frequencies then the natural inclination is to interpolate smoothly between the points while extrapolating in an intelligent way beyond the end points. Relations such as the sum rules above provide a test of the validity of such procedures. One can immediately tell if all the area of $J(\omega)$ has been exhausted or if there is some unforeseen behaviour hiding between or beyond the experimental points.

8

Some case studies

8.1 Calcium fluoride lineshape

8.1.1 Why calcium fluoride?

Although the calculation of the NMR absorption lineshape in a solid is a well-defined problem, the discussions of Chapter 6 have indicated that a complete and general solution is difficult and indeed unlikely. From the practical point of view one would like to explain/understand the characteristic features of transverse decays and lineshape as reflecting details of internal structure and interactions, while from the theoretical point of view the interest is the possibility of treating a 'relatively' simple many-body dissipative system.

In this respect it is worthwhile to consider the solvable models considered in Chapter 6. The model of Section 6.2 actually arose from discussions in the seminal paper on calcium fluoride by Lowe and Norberg in 1957. The essential point was that the time evolution generated by the dipole interaction is complicated because of non-commutation of the various spin operators. In the solvable models there is only an $I_z I_z$ part of the interspin interaction. Since the Zeeman interaction also involves only I_z this means that all operators commute and the time evolution can be calculated purely classically. It is only in this case that the evolution of each spin can be factored giving a separate and independent contribution from every other spin. In other words it is only in this case that each spin can be regarded as precessing in its own static local field – and the problem is solved trivially. Each many-body eigenstate is simply a product of single-particle eigenstates. However, once transverse components of the interspin interaction are admitted then everything becomes coupled together and a simple solution is no longer possible. The many-body eigenstates no longer factorise. In the spirit of the local field picture it may be regarded as if there were a time variation induced in the local fields.

The model system used for many lineshape/relaxation studies has been the ^{19}F resonance in crystalline calcium fluoride. Calcium fluoride is an ideal material for a

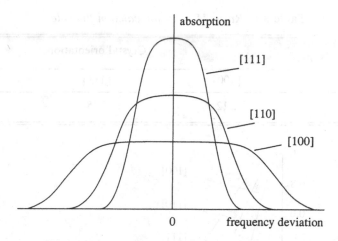

Figure 8.1 Calcium fluoride lineshapes for various crystal orientations.

number of reasons. The ^{19}F isotope has 100% abundance and its nucleus has spin $\frac{1}{2}$ so it has no quadrupole moment. Thus quadrupole broadening of the resonance line does not occur. Furthermore the fluorine dipole moment is quite large; it is the third largest of all naturally occurring nuclei, ensuring a comfortably large signal-to-noise ratio, and in CaF_2 the fluorine atoms form a simple cubic lattice, the most amenable to calculations. Furthermore the calcium atoms are relatively inert. The abundant isotope has no magnetic moment while only 0.13% comprises ^{43}Ca which has a small but non-zero magnetic moment. The effect of this on the fluorine resonance can be ignored. The Debye temperature of the crystal is high, at 474 K, so that phonon motion of the nuclei has negligible effect, particularly since the lattice is cubic. The system may thus be regarded as a simple cubic lattice of pure dipoles.

8.1.2 Overview

The earliest NMR investigations of calcium fluoride studied the CW absorption lineshape. Figure 8.1 shows ^{19}F lineshapes for a single crystal of calcium fluoride for three crystal orientations. The crystallographic directions [100], [110] and [111] indicate the direction of the static magnetic field. One sees that the resonances are unremarkable 'bell shaped' curves with no discernible structure although the width is clearly strongly dependent on the crystal orientation.

The second and fourth moments of these lineshapes are in good agreement with those calculated from Equations (6.66) and (6.67). In particular, the (calculated) ratio M_4/M_2^2 for the three orientations is shown in Table 8.1. Since this ratio is *of the order of* 3 the temptation, following the discussion in Section 6.3.4, is to infer that the lines are *approximately* Gaussian. However, closer inspection shows that the lines are distinctly

Table 8.1. *Ratio* M_4/M_2^2 *for calcium fluoride.*

	Crystal orientation		
	[100]	[111]	[110]
M_4/M_2^2	2.12	2.38	2.25

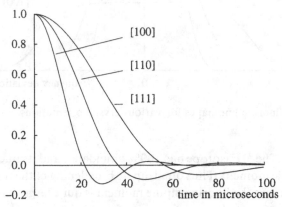

Figure 8.2 Calcium fluoride FID for various crystal orientations.

flatter than Gaussian; clearly this is related to the ratio M_4/M_2^2 being nearer to 2 than to 3.

The deviation from Gaussian behaviour is much more dramatic in the time domain, as shown in Figure 8.2. If the resonance line were truly Gaussian then the FID, the Fourier transform of the line profile, would also be Gaussian – a non-negative function. It was therefore a surprise when Lowe and Norberg in 1957 observed 'A beat structure has been found on free-induction decays . . .'. Although for the reasons discussed above their main interest was in calcium fluoride, they discovered that the characteristic beat structure was a universal phenomenon found in all solids they investigated at 77 K, including paraffin, ice, zirconium hydride, perspex and various (solid) alcohols. Note also the transverse relaxation in solid hydrogen, shown in Figure 6.9, which shows this behaviour.

The ubiquity of the beat structure in solid FIDs is an indication that there is something fundamentally inadequate about the Gaussian approximation for absorption lineshapes; presumably in three dimensions each atomic site does not have enough neighbours for the Central Limit theorem to be applicable.

In the following few subsections we shall follow some of the arguments expounded to explain the lineshape and FID of calcium fluoride. In doing so we shall make plentiful references to the concepts developed in Chapter 6.

8.1.3 Moment calculations

The moments are essentially the derivatives of the FID at time $t = 0$. If one knew all moments then they could be assembled into an infinite power series to give the FID. The problem in calculating the observed relaxation behaviour is a consequence of having only a limited amount of information. From this point of view all 'theoretical' procedures for calculating lineshapes or relaxation profiles are fundamentally methods for using the *known* moments to make reasonable guesses for the *unknown* moments.

Before proceeding to a discussion of the various theoretical methods, we shall pause to summarise precisely what moment information has been calculated for calcium fluoride. Formal expressions for the dipolar M_2 and M_4 were given in Equations (6.66) and (6.67) for the case of spin $\frac{1}{2}$. In order to evaluate these moments the various lattice sums must be performed. This is relatively trivial for M_2 and Van Vleck gave numerical expressions in his 1948 paper. For the cubic lattice Bruce (1957) evaluated M_4 by summing the contributions from the 26 nearest neighbours surrounding a given spin. He gave a summary of the then state of knowledge with values of M_2 and M_4 for the static magnetic field aligned along the three crystallographic axes [100], [110] and [111]. There was an attempted evaluation of M_6 by Glebashev in 1957, but his results were incorrect. Bruce's M_4 calculation was refined by Canters and Johnson (1972) who made a more accurate evaluation of the lattice sums. Because of the rapidly increasing difficulty of evaluating the higher moments (even the formal expressions are prohibitive), very little progress was then made until computer based techniques were developed. In 1973 Jensen and Hansen completed symbolic evaluation of the various commutators and traces together with numerical calculations of the lattice sums for the sixth and eighth moments. To give some idea of the complexity of these calculations they state that the evaluation of M_8 used some 1200 times the computing time of M_6. Table 8.2 summarises the results of the various moment calculations. The numbers in Table 8.2 are given in units of gauss (10^{-4} T). The nearest neighbour spacing has been taken as 5.4465 Å, corresponding to a temperature of 4.2 K. This temperature, the boiling point of liquid helium, has been chosen for comparison with some of the most sensitive measurements made.

8.1.4 Abragam's approximation function

Most approaches to the FID/lineshape problem involve *deriving* expressions for the relaxation or the resonance line. By contrast to this, Abragam (1961) 'simply' made the observation that the function

Table 8.2. *Calculated moments of calcium fluoride.*

	Crystal orientation		
	[100]	[111]	[110]
$M_2^{1/2}$	3.603	1.493	2.218
$M_3^{1/4}$	4.349	1.854	2.715
$M_6^{1/6}$	4.85	2.11	3.07
$M_8^{1/8}$	5.37	2.36	3.42

$$F(t) = \exp\left(-\frac{a^2 t^2}{2}\right)\frac{\sin(bt)}{bt} \qquad (8.1)$$

is a very good approximation to the observed FID when the constants a and b are chosen to be consistent with the correct (calculated) values of M_2 and M_4. In other words, one identifies the expansion of this function:

$$F(t) = 1 - \frac{t^2}{2}\left(a^2 + \frac{b^2}{3}\right) + \frac{t^4}{4!}\left(3a^4 + 2a^2 b^2 + \frac{b^4}{5}\right) + \cdots$$

with the moment series

$$F(t) = 1 - \frac{t^2}{2}M_2 + \frac{t^4}{4!}M_4 - \cdots$$

so that a and b are given by

$$a = \left[\frac{6M_2 - \sqrt{(30)(3M_2^2 - M_4)^{\frac{1}{2}}}}{6}\right]^{\frac{1}{2}},$$

$$b = \left[\frac{\sqrt{(30)(3M_2^2 - M_4)^{\frac{1}{2}}}}{2}\right]^{\frac{1}{2}}.$$

One sees from the expressions for a and b that Abragam's method will break down in the 'super-Gaussian' case, i.e. when $M_4 > 3M_2^2$. Essentially this method accommodates 'sub-Gaussian' lines by incorporating *equally spaced* zeros into the free induction decay.

Since $\sin(bt)/bt$ is the Fourier transform of a rectangular box of width $2b$, it follows that the lineshape is a convolution of a Gaussian of RMS width $2a$ with a rectangle of width $2b$. Values for the ratios of these widths are given in Table 8.3. As Abragam pointed out, the width of the box is considerably greater than the width of the Gaussian it envelopes. He wrote in 1961 that 'No entirely convincing physical model explaining

Table 8.3. *Ratio of width of box to width of Gaussian.*

	Crystal orientation		
	[100]	[111]	[110]
b/a	4.20	2.78	3.39

these very simple features has been suggested so far.' It is now known that the features are not quite so very simple as he suggested, as will be discussed in the next section. But the sentiments he expressed then are still largely applicable today.

The Fourier transform of Abragam's function, Equation (8.1), may be evaluated, giving the absorption lineshape $I(\Delta)$:

$$I(\Delta) = \frac{1}{4b}\left[\text{erf}\left(\frac{b + \Delta}{a\sqrt{2}}\right) + \text{erf}\left(\frac{b - \Delta}{a\sqrt{2}}\right)\right],$$

which is the sum of a rising and a falling error function. This may be pictured physically by regarding the rectangular box as the sum of a rising and a falling step function. Each of these integrates the Gaussian to an error function and thus we obtain the sum of a rising and a falling error function. The parameter b simply determines the separation of the left and right sides, the pre-factor $1/b$ then normalising the area. The slope of the sides of the resonance is determined by the Gaussian parameter a.

The zeros of the relaxation function can be considered in a direct manner in the following way. If $F(t)$ has zeros at t_i then it can be written as the product

$$F(t) = f(t)\prod_i (1 - t/t_i),$$

where $f(t)$ is a smooth function without any zeros. Then if $F(t)$ is even in t then the positive and negative zeros can be combined, giving

$$F(t) = f(t)\prod_i [(1 - t/t_i)^2]. \tag{8.2}$$

From this perspective Abragam's function embodies the following two points:

(1) The zeros in $F(t)$ are equally spaced.
   If $t_n = n\pi/b$ then the infinite product of roots evaluates to the $\sin(bt)/bt$ factor:

$$\prod_{n=1}^{\infty}\left(1 - \frac{b^2 t^2}{(n\pi)^2}\right) = \frac{\sin(bt)}{bt}. \tag{8.3}$$

This may be shown fairly simply, as discussed in Problem 8.1.

(2) The smooth function $f(t)$ is a Gaussian curve.

Certainly $f(t)$ should be an even function of t and since it must be smooth at the origin, the first derivative (and all odd ones) will vanish at the origin. So a Gaussian is a reasonable guess – but nothing more.

In fact both these points are not quite correct when considering the extensive and precise measurements of Lowe and Engelsberg (1974), which we report in Section 8.1.7. Although at long times the zeros of $F(t)$ tend to an equal spacing, in accordance with the memory function discussion of Section 6.4.6, at short times this is not quite so: the earlier zeros are more widely spaced. Also although the envelope profile $f(t)$ is approximately Gaussian at short times, it tends to an exponential at long times, again in accordance with the memory function arguments of Section 6.4.6.

Accepting that Abragam's function is not the whole story, Parker in 1970 showed how it could be regarded as the first term of an expansion in spherical Bessel functions:

$$F(t) = \exp\left(-\frac{a^2 t^2}{2}\right) \sum_{n=0}^{\infty} c_{2n} j_{2n}(bt). \tag{8.4}$$

This is a convergent series, and the widths a and b are taken to be parameters which may be chosen to speed the convergence. That is they are adjusted so that the second term of the expansion is very much less than the first and succeeding terms similarly smaller still. Clearly Abragam's choice for a and b is ideal when only M_2 and M_4 are known.

The coefficients c_0, c_2, c_4 depend only on M_2 and M_4. Using these two moments the fastest convergence of the series is obtained with Abragam's a and b whereupon one obtains $c_0 = 1$ and $c_2 = c_4 = 0$, recovering Abragam's original expression. Knowledge of higher moments permits the evaluation of more of the coefficients c_n; M_6 permits evaluation of c_6 and from M_8 one can obtain c_8. Thus with more available moments one can successively improve on the Abragam formula. In such cases the a and b coefficients must be recalculated according to the moments available. Using the available values for M_6 and M_8 does not seem to make much improvement.

8.1.5 *Experimental measurements on calcium fluoride*

The first measurements on calcium fluoride were performed by Purcell, Bloembergen and Pound in 1946. They used a single crystal in the form of a cylinder with the [110] direction along the cylinder axis. They made CW measurements at a Larmor frequency of 29.1 MHz, observing the variation of the linewidth and peak height with crystal orientation.

Pake and Purcell in 1948 used a new crystal. This also was in the form of a cylinder with the [110] direction along the cylinder axis. From their measurements they extracted values for M_2 and M_4 for the three crystal orientations. In 1957 Bruce refined these results. He used Pake and Purcell's crystal, and he made very much more accurate measurements. The curves of absorption in Figure 8.1 are essentially as published by Bruce. Using his improved measurements Bruce was able to obtain better estimates for M_2 and M_4 for comparison with Van Vleck's theoretical values. The agreement was good.

1957 also saw publication of the seminal work by Lowe and Norberg. They used the same crystal as Bruce and working at the lower temperature of 77 K for signal enhancement, they made the first pulsed NMR measurements on calcium fluoride. They were able to observe the FID relaxation over two orders of magnitude. This covered a time duration of approximately 70 μs for the [100] and [110] orientations and 100 μs for the [111] direction. In this way they observed three zeros for [100] and two zeros for both [110] and [111]. In that paper Lowe and Norberg derived the Fourier transform relation between the lineshape and the FID. For confirmation of this they performed a Fourier transformation of Bruce's lineshape data, showing that the result was equivalent to their FID measurements.

New measurements were reported by Barnaal and Lowe in 1966. Their experiments were performed at a Larmor frequency of 10 MHz, at a temperature of 47 K for improved signal-to-noise, using a new crystal of calcium fluoride. The time range covered was essentially similar to that of Lowe and Norberg. The main claim of this work was greater precision, particularly with regard to the crystal orientation. This seems to be borne out, considering their improved agreement with M_2 and M_4 calculations.

Lowe, Vollmers and Punkkinen (1973) developed a technique for measuring the relaxing magnetisation at arbitrarily short times. This zero-time resolution (ZTR) method, discussed in Chapter 9, overcomes the usual difficulty in pulsed NMR whereby the receiver takes a finite time to recover from the overloads caused by the transmitter pulses. As the authors point out in their paper, it is knowledge of this very short time behaviour which leads most directly to evaluation of moments.

A major paper was published by Engelsberg and Lowe in 1974. They had performed extensive measurements using a new crystal. At short times, below 4.5 μs, they used the ZTR method, working at 77 K. Conventional FID methods were used for times longer than 3 μs at both 77 K and 4.2 K. Using the lower temperature together with signal averaging enabled measurements to be made at longer times. Since there was considerable overlap between the time ranges of the various methods, the three sets of data could be scaled to join smoothly. In this way measurements were made over four orders of magnitude, so that eight zeros were observed for the [100] orientation up to 160 μs, five zeros for the [110] orientation up to about 170 μs and four zeros for [111] up to 220 μs.

It became quite apparent, from the new data, that Abragam's expression (Equation (8.1)) was failing to give a good fit. As mentioned in the previous section, the zeros were *not* equally spaced at short times although they tended to equal spacing at long times, and the envelope of the relaxation, which started decaying in a Gaussian manner, slowed down and tended to an exponential at longer times. Engelsberg and Lowe fitted their extensive data to a rather complex function. We shall take up their story again in Section 8.1.7. However, at this point we must backtrack and consider some of the theoretical attempts at explaining the observed relaxation and lineshape data.

8.1.6 Theories of lineshape and relaxation

The first attempt at the explanation of the calcium fluoride FID was made in the 1957 paper of Lowe and Norberg. They appreciated that the difficulty in making a complete calculation originated in the details of the (truncated) dipolar Hamiltonian. Writing this as

$$\mathscr{H}'_1 = \alpha + \beta, \tag{8.5}$$

where

$$\alpha = \sum_{j<k} A_{jk} S_j S_k,$$

$$\beta = \sum_{j<k} B_{jk} S_j^z S_k^z,$$

the problem is that the α term and the β term do not commute. If they did commute, then the time evolution operator could be factorised as

$$\exp[i(\alpha + \beta)t] = \exp(i\alpha t)\exp(i\beta t).$$

Now since α commutes with the total spin operators, it means that this term cancels in the evolution equation and the time dependence of the relaxation would be generated entirely by the b part and this can be calculated exactly, as we showed in Section 6.2. But α and β don't commute, so what can be done? Since α and β have similar magnitudes there is no obvious small parameter which could be used for a perturbation parameter. Lowe and Norberg decided to account for the non-commutativity of α and β by forcibly factorising the time evolution operator as

$$\exp[i(\alpha + \beta)t] = \exp(i\alpha t)\chi(t)\exp(i\beta t).$$

The operator $\chi(t)$ represents the effect of α and β not commuting. Obviously $\chi(t)$ cannot be evaluated, but Lowe and Norberg performed an expansion of it in powers of time. They were able to complete the calculations up to t^4, expressing $F(t)$ as the sum

$$F(t) = F_0(t) + F_2(t)\frac{t^2}{2!} + F_3(t)\frac{t^3}{3!} + F_4(t)\frac{t^4}{4!}.$$

Clearly this expansion will diverge for sufficiently long times, but they did find good agreement with their experimental data to times beyond the first few zeros.

There have been various attempts to improve upon Lowe and Norberg's result. It is, after all, only a short time approximation and the (time varying) coefficients of t^n are by no means unique. Other ways of separating and treating the interaction Hamiltonian could be equally valid.

Mansfield (1966) used a many-body Green's function method whereby he derived a coupled set of equations involving successively higher order Green's functions. The equations were solved by approximating higher order functions in terms of products of lower order ones. Mansfield did this using a frequency dependent decoupling procedure justified on intuitive grounds. Mansfield's results are equivalent to the memory function method and his decoupling scheme is equivalent to taking a rectangular box for the memory function. His fit to the FID data for calcium fluoride is *not* better than the calculation of Lowe and Norberg over the time range of the first few zeros, but it does have the merit of converging at long times.

Evans and Powles (1967) developed a convergent approximation for the FID in a different way. They started from the truncated dipolar Hamiltonian, Equation (8.5), and they developed an expansion in powers of α, the $\mathbf{I}_j \cdot \mathbf{I}_k$ part. They factored the time evolution operator as

$$\exp[i(\alpha + \beta)t] = \exp\left[i\int_0^t \beta(\tau)d\tau\right]\exp(i\alpha t),$$

where

$$\beta(\tau) = \exp(i\alpha\tau\beta)\exp(-i\alpha\tau)$$

(Appendix C.3), and then they expanded the time-ordered exponential. In this way they obtain a series for $F(t)$ as

$$F(t) = F_0(t) + F_1(t) + F_2(t) + \ldots,$$

where at each stage the approximation is convergent.

As with Lowe and Norberg, this method requires calculated quantities other than M_2 and M_4. Evans and Powles evaluated $F_0(t)$ and $F_1(t)$ for a lattice of $5 \times 5 \times 5$ spins, and they found that the approximation $F_0(t) + F_1(t)$ gave a good fit to FID data covering the first few zeros. However, as with Mansfield's method, over that range it was not a significant improvement over the Lowe and Norberg approximation.

This is no surprise. It is a perturbation expansion, in powers of α, the $\mathbf{I}_j \cdot \mathbf{I}_k$ part of the truncated dipolar Hamiltonian. The condition of validity, which allows the perturbation expansion to be terminated at low order, is that α should be much less than β. Now it is true that A_{jk} is *one third* the magnitude of B_{jk}, but that can hardly be regarded as justification for terminating the perturbation series at the first order term.

A memory function method was used by Borckmans and Walgraef (1968). They showed, from first principles, that the FID function obeys a memory equation and thus they had an explicit (if intractable) expression for the memory kernel. Although they decided to approximate the memory function by a Gaussian, they then developed the more general argument we presented at the end of Section 6.4.6 to show that an exponentially damped sine curve was the expected long time behaviour. This is independent of the form of the memory function, but they did give values for the period and the time constant appropriate to the Gaussian memory kernel.

Parker and Lado (1973) applied the memory function method to the calculation of NMR lineshapes in a variety of systems. In considering the calcium fluoride case they started by using the calculated M_6 values, together with the known M_2 and M_4, to support the use of a Gaussian memory function. They argued that the ratio 'M_4/M_2^2' for the memory function is approximately 3. Now from Equation (6.90) this ratio is:

$$\frac{K_4/K_0}{(K_2/K_0)^2} = \frac{M_6/M_2^3 - 2M_4/M_2^2 + 1}{(M_4/M_2^2 - 1)^2},$$

which has values 2.16, 2.21 and 2.26 for the [100], [111] and [110] orientations.

In fact they obtained a dip in the centre of their calculated lines, as is shown in Figure 6.15. This is understandable since the ratio M_4/M_2^2 for the resonance line is known to be (Table 8.1) closer to 2 than 3. The conclusion they drew was that a Gaussian shaped memory function, while adequate for certain asymptotic considerations, was not suitable for describing the general features of the resonance line (or the relaxation). So they decided to see what the memory function *should* be like. They calculated (numerically) the shape of the memory function which corresponded to Abragam's FID function. The result was a memory function quite similar in shape to the FID but decaying more rapidly. In particular, it exhibited some oscillation as it decayed.

Based on this observation Parker and Lado used for the *memory function* a generalisation of the Abragam function with four adjustable parameters. They used values of M_2–M_8 to determine the parameters. Using this procedure they obtained very good agreement with experimental FID measurements, out to the fifth zero in the [100] orientation.

The good result is hardly surprising, and one might argue that you only get out what you put in. However, it was an important discovery that the memory function decayed more rapidly than the relaxation function. It certainly cannot decay *more slowly*, but it could have decayed on a similar time scale. If that were the case then there would be absolutely no use whatsoever in using the memory function method. Since the memory function decays faster than the relaxation this implies that the relaxation is *somewhat* insensitive to the details of the memory function and so a modest approximation to the memory function gives a rather better approximation to the relaxation function.

The square box approximation of Mansfield and the Gaussian approximation of others thus gave qualitatively reasonable relaxation profiles, while the *good* memory function of Parker and Lado led to the *very good* relaxation profile.

8.1.7 Engelsberg and Lowe's data

As we explained in Section 8.1.5, Engelsberg and Lowe published extensive relaxation measurements in 1974 covering over four orders of magnitude. By this stage in the story it is becoming clear that there is no simple mathematical function which will fit the data so there is no point in trying to derive one. Thus the main concern of the authors was to extract from their measurements values for the moments up to relatively high order and to compare their determinations of M_2, M_4, M_6 and M_8 with the available calculated values. For each of the three usual crystal orientations they determined moments up to M_{14}, including error bounds, and the agreement of their M_2–M_8 was extremely good.

The procedure adopted by Engelsberg and Lowe was to fit their data to a carefully chosen function which could then be differentiated to obtain the moments. In this way data over the entire experimentally measured time range contributed to the moment evaluation. We saw in the discussion surrounding Equation (8.2) that a function with zeros could be factored into a function with no zeros multiplied by a product explicitly giving the zeros. Engelsberg and Lowe thus fitted their data to a function of the form

$$F(t) = f(t)(1 - R^2 t^2) \prod_{i=1}^{60} [1 - (t/t_i)^2],$$

where $f(t)$ is the smooth function and the product indicated incorporates the first 60 zeros. Into this product function were inserted the experimentally measured zeros and the other zeros were assumed equally spaced. The second factor takes approximate account of the zeros above the sixtieth, R being given by

$$R^2 = \sum_{i=61}^{\infty} t_i^{-2}.$$

The approximation is valid for those zeros occurring far beyond the region of measurement. It is known that the envelope function $f(t)$ decays as a Gaussian at short times, but it tends towards an exponential at long times. Thus they found it convenient to use as a fit function

$$f(t) = \exp\{C[A - (A^2 + t^2)^{\frac{1}{2}}]\}, \tag{8.6}$$

which has the correct limiting behaviour. When the zeros in the data were factored out the appropriate $f(t)$ was found to fit the reduced data extremely well over the entire experimental range, through six zeros in the [100] case.

It can be seen that this whole procedure is an empirical process to enable a smooth function to be fitted through the data. No other claims are made for the chosen functional form.

8.1.8 Postscript

We should mention briefly two further approaches to the calcium fluoride lineshape problem which do not fit logically into the previous discussion.

Powles and Carazza (1970) adopted an information theory approach to obtaining the 'most likely' profile for the absorption line consistent with the moment information available. Essentially they define an 'entropy' function S for a given normalised line profile $I(\omega)$ as

$$S = - \int I(\omega)\ln I(\omega)d\omega$$

and a 'trial' function $I(\omega)$ is varied so as to maximise the entropy subject to the constraints of the known moments. Knowing only M_2 and M_4 the functional form for $I(\omega)$ is

$$I(\omega) = \exp(-\alpha\omega^2 - \beta\omega^4)I(0),$$

where α and β are Lagrange multipliers which are adjusted to give the correct values for M_2 and M_4. Thus α and β are functions of M_2 and M_4.

The line profiles so found give a reasonable approximation to the experimental $I(\omega)$. However, for the broader lines, where the magnetic field is along the [110] and the [100] directions, there is a slight dip in the centre of the 'theoretical' curves. Powles and Carazza claim that including the theoretical value of M_6 'would give a virtually perfect fit.' It should, however, be pointed out that the functional form obtained for $I(\omega)$ is rather intractable. In particular it does not permit analytic Fourier transformation. Furthermore, there is the fundamental point that although it is positive definite, $I(\omega)$ is *not* a probability function so that in reality it does not make sense to define an entropy function from it. All that can be said is that this does provide a systematic procedure for finding line profiles.

Gordon (1968) looked at the question of placing error bounds on the likely line profile, from the knowledge of a limited number of moments. In this method one must use a confidence interval or specified resolution. Then upper and lower bounds on the functional form of $I(\omega)$ are found, which may be expressed as Padé approximants. As with the entropy procedure, a small dip is found in the centre of the line, not observed experimentally.

This completes the survey of the calcium fluoride question. Although an enormous effort has gone into the problem it is clear that no complete solution has been found. In particular no *simple* solution to this simply-posed question is available.

8.2 Glycerol

8.2.1 Motivation

Since the earliest days of NMR glycerol, $CH_2OH \cdot CHOH \cdot CH_2OH$, has been cited as a prototypical example for motion-induced relaxation. The viscosity of pure glycerol is some 1500 times that of water at room temperature and it is a rapidly decreasing function of temperature. Although the melting point of solid glycerol is around 290 K, it can be supercooled to as low as 185 K. The boiling point is 563 K. Thus there is a large temperature range available. Also, the viscosity can be decreased through dilution with water. The result is that the motion of the constituent molecules can be varied with ease over a very large range so that it is possible to pass from the fast motion ($\omega_0 \tau_c \ll 1$) limit to the slow motion limit ($\omega_0 \tau_c \gg 1$), traversing the region of the T_1 minimum with ease.

8.2.2 Relaxation mechanisms

Following the pioneering work of Bloembergen (1948) it was believed that the mechanism for proton relaxation in glycerol was through the rotational modulation of the internuclear dipolar fields – the *intra*-molecular interaction. This type of model had been very successful in explaining the dielectric relaxation in glycerol and it seemed a natural extension of that description to the magnetic case. More recent studies, however, have indicated that it is the translational diffusion of the molecules: modulation of the *inter*-molecular interactions, which is the dominant source of relaxation. We shall be able to deduce this quite clearly using some of the results of the previous chapter.

Let us start by following arguments in the spirit of Bloembergen. The assumption was that the relaxation had its origin in *rotational* diffusion of the glycerol molecules. In that case the spectral density function $J(\omega)$ has the Lorentzian form, as is often adopted empirically. For rotational diffusion, from Equations (7.66) and (7.62), the correlation time τ_c is proportional to η/T where η is the viscosity and T is the absolute temperature. The shape of the graph of $\ln T_1$ against $\ln(\eta/T)$ will follow a curve which may be found from the form of $J(\omega)$. To the left of the T_1 minimum the slope of the curve will be -1 (independent of the shape of $J(\omega)$ – see Section 7.5.1). To the right of the minimum the slope of the curve is predicted to be $+1$ for the Lorentzian $J(\omega)$.

8.2.3 Experimental measurements

Bloembergen's originally published relaxation measurements for pure glycerol at various temperatures are shown in Figure 8.3 for Larmor frequencies of 4.8 MHz and 29 MHz. The solid lines give the theoretical T_1 as calculated from the Lorentzian $J(\omega)$.

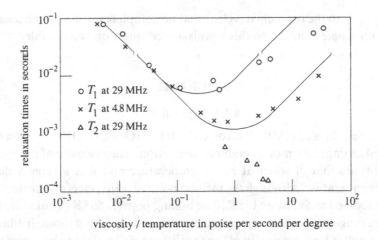

Figure 8.3 Bloembergen's 1948 proton relaxation in glycerol.

It will be observed that there are some small discrepancies between the T_1 data and the theoretical curves. There is a broadening of the minima and, moreover, the data to the right of the minima do not seem to approach the expected slope of $+1$. Bloembergen attributed these discrepancies to a possible distribution of correlation times.

A very extensive programme of NMR measurements on glycerol was undertaken by Luszczynski, Kail and Powles (1960). They took particular care to ensure their specimens were dry; on the basis of viscosity measurements they had concluded that Bloembergen's specimens must have contained some 2% of water. This new study confirmed that it was impossible to fit the data to a single Lorentzian $J(\omega)$. As is often done in such cases, it was assumed that there was a range of correlation times, resulting in a superposition of Lorentzians. These workers also made comparisons with the correlation times inferred from dielectric and ultrasonic measurements. One of the main conclusions of this work was that the translational motion of the molecules is faster than the rotational motion. This surprising result could be understood in terms of the hydrogen bonding between molecules. As suggested by Luszczynski *et al.* it could be that reorientation of the glycerol molecules occurs when allowed by the translational motion, but on average several translational jumps are required to permit appreciable molecular reorientation. Clearly then, the relative effects of these two types of motion on the relaxation must be considered. However, at that stage the Lorentzian form for $J(\omega)$ was still being adopted for the translational relaxation.

8.2.4 *High and low frequencies*

Measurements of the proton T_1 in glycerol at high frequencies were reported by Lenk (1971). By that time the difference in the functional form of $J(\omega)$ for translational and

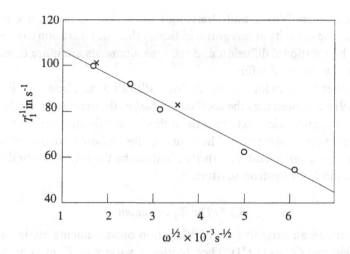

Figure 8.4 Proton relaxation in glycerol at low frequencies.

rotational motion was appreciated. Lenk found that the high frequency behaviour could be described by the power law expression

$$T_1 \propto \omega_0^\alpha,$$

where the exponent α was approximately 1.4. For the rotational model – Lorentzian $J(\omega)$ – an exponent of 2 would be expected. The value of 1.4 is closer to the 3/2 predicted for the translational diffusion (of large molecules; recall the discussion at the end of Section 7.4.5). Lenk attributed the small difference of 0.1 between the theoretical and the experimental values of the exponent to the small influence of processes other than isotropic translation.

If one now looks back at the original data of Bloembergen, in Figure 8.3, there is an indication of this very behaviour. To the right of the T_1 minima we have already noted that data points did not join a line of slope $+1$ as expected from the rotational relaxation model. A slope of $+\frac{1}{2}$ follows from the translational diffusion model and the data certainly seem more consistent with this. Problems 7.1 and 7.2 consider explicitly the question of the slopes of these asymptotic lines.

We saw in Section 7.4.4 that low frequency relaxation measurements were able to distinguish between rotational and translational diffusion in a most direct manner: $T_1(\omega) - T_1(0)$ is proportional to ω_0^2 in the former case and proportional to $\omega_0^{\frac{3}{2}}$ in the latter. Furthermore, this is independent of the microscopic details of the diffusive processes, unlike the situation pertaining at high frequencies. Low frequency proton T_1 measurements on glycerol at a temperature of 22.9 °C have been performed by Harmon (1970). In Figure 8.4 we show Harmon's data (circles) together with two points (crosses) which he quotes from Noack and Preissing (1967). The T_1 points are plotted as a function of $\omega_0^{\frac{1}{2}}$. The linearity of this graph indicates that the relaxation can certainly not be explained by rotational diffusion and the measurements are quite consistent with relaxation by translational diffusion.

There is further evidence for the translational diffusion case, however. Recall that in Equation (7.89) we showed that the coefficient of $\omega_0^{\frac{1}{2}}$ in the expression for $J(\omega)$ involved only macroscopic quantities, including the diffusion coefficient; there were no adjustable model dependent parameters. The values of the diffusion coefficient obtained in this way are in good agreement with those obtained by the spin-echo method (Section 4.5) and as found from neutron scattering.

8.2.5 *The* T_1 *minimum*

Finally we mention an extensive set of relaxation measurements made by Preissing, Noack, Kosfeld and Gross (1971). They plotted a variety of T_1 minima and using a translational diffusion model the fits to the data, even in the region of the minima, are very good; certainly they are far better than the traditional Lorentzian expression for $J(\omega)$. A typical set of data, taken at a Larmor frequency of 80 MHz is shown in Figure 8.5. The solid line is derived from Equation (7.79) together with the known temperature dependence of the self-diffusion coefficient. The fit of the experimental data to the theoretical curve is most impressive as the minimum is the area where discrepancies between different spectral density functions will be most pronounced. Recall that on the 'fast' side of the minimum (the left hand side on our figures) we stated that the behaviour was essentially independent of the shape of $J(\omega)$, while to the other side the function will tend to a power of ω depending on the details of the 'tail' of the function. It is the region of the minimum, however, which is particularly sensitive to the precise form of the function. Thus the fit of the data to the curve in Figure 8.5 may be taken as very firm evidence indeed that it is the modulation of the internuclear dipolar interaction by translational diffusion which is responsible for the relaxation. It thus appears, so many years since the first observation of NMR in glycerol, that the correct mechanism for the proton relaxation has finally been ascertained.

8.3 Exchange in solid helium-3

8.3.1 *Introduction*

Helium gas is rare, comprising only some 5.2 parts per million by volume of the earth's atmosphere, and of that approximately 2% is the light isotope ^{3}He. However, in contrast to its heavier brother, ^{3}He has a magnetic moment and therefore it can be studied using NMR. The magnetogyric ratio of ^{3}He is quite large at 32 MHz/T so that NMR detection is relatively easy, and since ^{3}He has spin $\frac{1}{2}$ there are no quadrupole

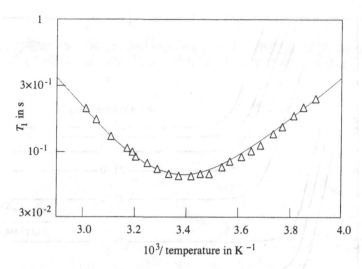

Figure 8.5 Temperature variation of proton relaxation in glycerol at 80 MHz.

interactions to consider; as with calcium fluoride one is dealing with a pure dipole system. Because helium is an inert gas there are no chemical compounds and it remains monatomic. So helium may be regarded as the lightest, simplest substance. In reality, however, its very lightness leads to quantum mechanical effects, causing a richness and diversity of phenomena not to be found in other materials. Solid ^{3}He comprises a regular lattice of spins $\frac{1}{2}$, as does calcium fluoride. We will see, however, that its NMR behaviour is very different indeed (Landesman 1973).

8.3.2 Measurement of T_2

The transverse relaxation in solid ^{3}He is unusual (for a solid) in two ways. The magnetisation decays over a time scale very much longer than the 40 µs or so estimated as the rigid lattice value from M_2. Furthermore the relaxation is observed to be exponential, with none of the structure ordinarily characteristic of transverse relaxation in solids. Of course, these are both indicative of rapid motion of the atoms. Since the transverse relaxation is exponential it follows that a relaxation time T_2 fully describes the relaxation process.

In Figure 8.6 we show measurements of T_2 as a function of temperature for a range of densities, covering both the high pressure hcp and the low pressure bcc phases. Following conventional practice we express the density in terms of the molar volume. Two distinct régimes may be discerned. At high temperatures there is a rapid temperature variation of T_2. Here the atoms are moving by changing place with thermally activated vacancies. It is, however, the lower temperature region which is of particular interest. Here T_2, and thus the atomic motion, is independent of tempera-

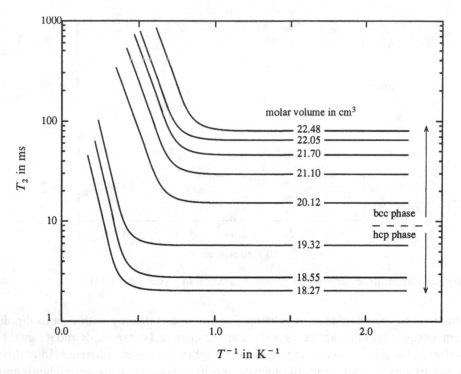

Figure 8.6 T_2 in solid ^{3}He as a function of inverse temperature for various molar volumes.

ture, but it has a very rapid dependence on the interparticle spacing. It is believed that the 'motion' responsible for this relaxation is a consequence of the quantum mechanical delocalisation of the particles.

The temperature independent transverse relaxation measurements may be interpreted using the motional averaging expression $T_2^{-1} = M_2 \tau_c$, to give a correlation time τ_c which varies with density. It is found that τ_c varies very rapidly, as the eighteenth power of the molar volume, changing by a factor of 100 while the molar volume increases from 18 to 22.

In the following subsections we shall see how this quantum motion can be given a microscopic description in terms of an 'effective' Hamiltonian and how the parameters of the Hamiltonian can be inferred from relaxation measurements. Although the discussion above was based upon a consideration of T_2, following a consideration of the origin of the quantum motion it will be expedient to treat transverse and longitudinal relaxation together.

8.3.3 Zero point motion

The atoms in a solid are confined to a volume whose magnitude is approximately the cube of the interatomic spacing. Heisenberg's uncertainty principle,

$$\Delta x \Delta p \geq \hbar,$$

implies that a particle confined within a distance Δx will have a momentum uncertainty Δp greater than or equal to $\hbar/\Delta x$. This momentum uncertainty will result in a minimum kinetic energy $E = (\Delta p)^2/2m$. Thus we see that confining a particle, as in a solid, gives it kinetic energy so that it can no longer be regarded as stationary. The lighter the particle the greater is this zero point energy. In solid helium the zero point energy is comparable with the van der Waals binding energy. For this reason helium is the only substance which will not solidify unless subjected to an external pressure and even in the solid there is considerable motion of the atoms. In solid ^{3}He this will have serious implications for NMR and other properties. So far as NMR is concerned solid ^{3}He may be regarded as a fluid-like system.

This may be looked at in another way. In a solid there are lattice vibrations which may be quantised as simple harmonic oscillators. The energy of such an oscillator is $(n + \frac{1}{2})\hbar\omega$, where n is the number of quanta, depending on temperature and given by the Bose–Einstein distribution function. At low temperatures n will be very small and the $\frac{1}{2}$ will dominate the energy; these are the zero point phonons. Solid helium is unique in that the amplitude of the zero point oscillations can be a large fraction, perhaps 30%, of the interatomic spacing and there is thus considerable delocalisation of the atoms. The direct effect of the lattice vibrations may be accounted for by a simple renormalisation of calculated lattice sums such as

$$\left\langle \sum_i \frac{1}{r_{ij}^6} \right\rangle = \xi \sum_i \frac{1}{\langle r_{ij}^6 \rangle^6},$$

where renormalisation factor ξ has been found to be approximately 0.75.

There is, however, a more important consequence of the delocalisation of the atoms. With such large excursions from their equilibrium positions there is no guarantee that particles will return to the site whence they came. Adjacent particles will move in a correlated way due to their mutual interactions, and they might well move from site to site in the process. This may be considered in the following way.

8.3.4 Exchange

For simplicity let us take two particles on adjacent lattice sites, constrained by the mean field exerted by the other particles in the solid. Since we shall be concerned specifically with the positions of the particles we will describe the states by the trial wavefunctions $\psi_a(r_1)$ and $\psi_b(r_2)$. Here

$\psi_a(r_1)$ means particle 1 is in the state centred on site a,
$\psi_b(r_2)$ means particle 2 is in the state centred on site b.

When there is an interaction between the particles – the van der Waals interaction in solid helium – then the product state $\psi_a(r_1)\psi_b(r_2)$ cannot be an eigenstate of the Hamiltonian. In other words the state where particle 1 is located at site a and particle 2 is located at site b is not a stationary state. The trial eigenstates will be the symmetric and antisymmetric combinations

$$\psi_+(r_1,r_2) = \frac{1}{\sqrt{2}}[\psi_a(r_1)\psi_b(r_2) + \psi_a(r_2)\psi_b(r_1)]f(r_2 - r_1),$$

$$\psi_-(r_1,r_2) = \frac{1}{\sqrt{2}}\{\psi_a(r_1)\psi_b(r_2) - \psi_a(r_2)\psi_b(r_1)\}f(r_2 - r_1),$$

where the function $f(r_2 - r_1)$ accounts for the hard core repulsion of the atoms at small separations; it goes to zero rapidly as $r_2 - r_1$ approaches the hard core dimension.

The expectation value for the pair energy in these two states is

$$E_\pm = \int \psi_\pm^*(r_1,r_2)\mathcal{H}_{12}\psi_\pm(r_1,r_2)dr_1dr_2,$$

where the Hamiltonian $\mathcal{H}_{12}$ is the sum of the interaction energy $V(r_1 - r_2)$ and the kinetic energies of the particles (we have ignored the change in normalisation of the states due to the correlation $f(r_2 - r_1)$). We can express the trial energy eigenvalues in terms of the (angular) frequencies I and J

$$E_\pm = \hbar(I \pm J),$$

where, incorporating the correlation $f(r_2 - r_1)$ into an effective Hamiltonian,

$$\hbar I = \int \psi_a^*(r_1)\psi_b^*(r_2)\mathcal{H}_{12}^{\text{eff}}\psi_a(r_1)\psi_b(r_2)dr_1dr_2,$$

$$\hbar J = \int \psi_a^*(r_1)\psi_b^*(r_2)\mathcal{H}_{12}^{\text{eff}}\psi_a(r_2)\psi_b(r_1)dr_1dr_2.$$

Here $\hbar I$ is the *direct* energy, the diagonal matrix element of the interaction energy, evaluated between identical states. This is simply the mean energy of the interaction. And $\hbar J$ is the matrix element of the interaction energy evaluated between states where the particles have exchanged places. This is referred to as the *exchange* energy.

Since the energy eigenstates have the intrinsic time evolution

$$\psi_+(t) = \psi_+(0)\exp[i(I + J)t],$$
$$\psi_-(t) = \psi_-(0)\exp[i(I - J)t],$$

we can use this to study the behaviour of the localised state $\Psi = \psi_a(r_1)\psi_b(r_2)$, where particle 1 is located at site a and particle 2 is located at site b. This state is given in terms of the eigenstates as

$$\Psi = \frac{1}{\sqrt{2}}[\psi_+(r_1,r_2) + \psi_-(r_1,r_2)].$$

Then the time evolution of the localised state is

$$\Psi(t) = \frac{1}{\sqrt{2}}\{\psi_+(0)\exp[i(I + J)t] + \psi_-(0)\exp[i(I - J)t]\}$$

$$= \exp(iIt)[\psi_a(r_1)\psi_b(r_2)\cos(Jt) + \psi_a(r_2)\psi_b(r_1)\sin(Jt)].$$

We see that the initial state oscillates between the state $\psi_a(r_1)\psi_b(r_2)$ and the state $\psi_a(r_2)\psi_b(r_1)$ with frequency J. In other words the two particles continually exchange places at this frequency.

Within the limitation of considering only two particles, the arguments here are quite general; they should be applicable to all solids. However, usually the exchange frequency is negligible and the effect can be ignored. The magnitude of the exchange frequency depends on two things. The product $\psi_a^*(r_1)\psi_b(r_1)$ is the product of wavefunctions where particle 1 is on site a and on site b. For strongly localised particles either one or other element of the product will be zero. It will only be non-zero if there is an overlap of the wavefunctions of the particles on adjacent sites. As well as the overlap, the other thing which determines the exchange frequency is the interaction energy V. The van der Waals interaction for helium is the weakest of any element. However, the delocalisation of the atoms means that the wavefunctions are broad; for helium alone is there appreciable overlap of neighbouring wavefunctions. This more than compensates, and as a consequence the exchange in helium is significant, with a frequency of the order of megahertz.

Since ^{3}He has a nuclear spin the exchange motion of the particles will involve a corresponding change of spin on the lattice sites. Furthermore, so far as the magnetic properties of the system are concerned it is the *spins* on the sites which are important. We would therefore like to shift attention from the bodily motion of the particles to the variations of spin on the lattice sites. This is achieved directly by considering the symmetry of the helium two-particle wavefunction, which in the general case is the product of a spatial part and a spin part. Since ^{3}He is a fermion there is a requirement that the overall wavefunction be antisymmetric. This means that a symmetric spatial wavefunction, ψ_+ must be accompanied by an antisymmetric spin part and *vice versa*. Thus

symmetric ψ_+ has spin part $|-\rangle = \dfrac{1}{\sqrt{2}}\{|\uparrow\downarrow\rangle - |\downarrow\uparrow\rangle\}$,

antisymmetric ψ_- has spin part $|+\rangle = \dfrac{1}{\sqrt{2}}\{|\uparrow\downarrow\rangle + |\downarrow\uparrow\rangle\}$,

and arguments identical to those above lead us to infer that the states $|\uparrow\downarrow\rangle$ and $|\downarrow\uparrow\rangle$ oscillate between each other with angular frequency J.

This behaviour, which is fundamentally a consequence of the (spin independent) van der Waals interaction, may now be interpreted as occurring under the influence of an effective spin Hamiltonian. This makes for a much simpler description, as we shall see. Thus we introduced an operator $\mathscr{H}_x$ which depends on spin variables, which has the same eigenvalues E_+ and E_- as does the original space-dependent V. In other words

$$\begin{aligned} E_\pm &= \langle \pm | \mathscr{H}_x | \pm \rangle \\ &= \langle \uparrow\downarrow | \mathscr{H}_x | \uparrow\downarrow \rangle \pm \langle \uparrow\downarrow | \mathscr{H}_x | \downarrow\uparrow \rangle \\ &= \hbar I \pm \hbar J. \end{aligned}$$

The effect of $\mathscr{H}_x$ in the second term is to flip both spins, turning the state $|\uparrow\downarrow\rangle$ into the state $|\downarrow\uparrow\rangle$; this is achieved by an operator of the form $I_+^a I_-^b + I_-^a I_+^b$. The effect of $\mathscr{H}_x$ in the first term is to leave the spin orientations unchanged; this is achieved by the unit operator or by an operator of the form $I_z^a I_z^b$. Therefore we may choose the general form for $\mathscr{H}_x$ as

$$\mathscr{H}_x = \alpha + \beta \mathbf{I}^a \cdot \mathbf{I}^b.$$

We then have

$$\hbar I = \alpha + \beta \langle \uparrow\downarrow | I_z^a I_z^b | \uparrow\downarrow \rangle = \alpha - \beta/4,$$

$$\hbar J = -\frac{\beta}{2} \langle \uparrow\downarrow | I_+^a I_-^b + I_-^a I_+^b | \uparrow\downarrow \rangle = -\beta/2,$$

so that in terms of the direct integral I and the exchange integral J the effective spin Hamiltonian is

$$\mathscr{H}_x = \hbar(I - J/2) - 2\hbar J \mathbf{I}^a \cdot \mathbf{I}^b. \tag{8.7}$$

This has the same eigenvalues and eigenstates as the original interaction operator $V(r_1 - r_2)$ and therefore $\mathscr{H}_x$ induces precisely the same evolution of the fermionic wavefunctions. The first term of Equation (8.7) is a constant, so it may be ignored. The resultant expression is called the Heisenberg exchange interaction. We must emphasise that the origin of this interaction is the spin independent van der Waals force; it is *not* a magnetic interaction. It is just that the magnetic effects of the interaction are the same as those form the spin Hamiltonian $\mathcal{H}_x$, and this is a much easier way to model spin/magnetic behaviour. The treatment of electron exchange leading to ferromagnetism is a classic example of this; there it was the electrostatic Coulomb repulsion which Heisenberg cast as an effective spin or magnetic interaction.

In solid ^{3}He it is not entirely appropriate to consider only pairwise exchange. In reality one should find the stationary states of the many-body wavefunction under the influence of the interparticle van der Waals interactions. This is a formidable problem, but by analogy with the two-particle case considered above, it is clear that the stationary states will involve superpositions of permutations of different particles on different sites and when the time evolution of a given localised configuration of atoms is considered, they will continually change between different permutations. There will be pairwise permutations, three-particle interchanges, cycles of four particles, five-particle etc. etc. Thus when expressed as an effective spin Hamiltonian the consequences of the interactions between the particles result in

$$\mathcal{H}_x = -\sum_n (-1)^n J_n P_n,$$

where n is the number of particles in the cycle, J_n is the exchange frequency for the n particle exchange, and P_n is the generator of permutations of n particles.

It is believed that in the bcc phase of ^{3}He the cycles comprising two, three and four spins are dominant while in the hcp phase two-spin and three-spin cycles are particularly important. There is a further simplification which follows since three-spin exchanges can be expressed as a superposition to two-spin exchanges. Thus, certainly in the hcp case, the exchange Hamiltonian may be written simply as a pairwise Heisenberg exchange Hamiltonian:

$$\mathcal{H}_x = -2\sum_{i<j} J_{ij} \mathbf{I}^i \cdot \mathbf{I}^j,$$

where we have considered the exchange between all pairs and the restriction $i < j$ ensures that each pair is counted only once.

The exchange coefficients J_{ij} decrease rapidly with distance, they are only significant for next neighbours. Assuming they are the same for all directions (at least *approximately* true for hcp ^{3}He) we may then take out the single coefficient J, where the sum is restricted to next neighbours only

$$\mathcal{H}_x = -2J \sum_{i<j,\mathrm{nn}} \mathbf{I}^i \cdot \mathbf{I}^j.$$

Even in the cases where isotropic pairwise exchange is not an entirely correct description of the system, it is often a reasonable approximation, and the exchange coefficient J may then be regarded as an effective exchange frequency, a single parameter characterising the rate of the motion. In this respect we expect that J^{-1} will be a measure of the correlation time τ_c for the variation of the dipolar interaction.

We emphasise that the exchange Hamiltonian is a *model* for the NMR behaviour of solid ^{3}He, applicable in the temperature range of the flat region of Figure 8.6. Above about 1 K thermally excited vacancies make an increasingly important contribution to the relaxation, while for significantly lower temperatures the full complexity of the system must be considered and at millikelvin temperatures the solid undergoes a phase transition to a spin-ordered phase.

In the intermediate temperature range, however, the pairwise exchange Hamiltonian provides an effective and simple description of the spin dynamics in terms of which the NMR behaviour may be understood.

8.3.5 Relaxation

We learned in Chapter 7 that the relaxation times T_1 and T_2 for a 'fluid' system can be expressed as

$$
\left.
\begin{aligned}
1/T_1 &= J_1(\omega) + 4J_2(2\omega) \\[2mm]
1/T_2 &= \frac{3}{2}J_0(0) + \frac{5}{2}J_1(\omega) + J_2(2\omega),
\end{aligned}
\right\}
\tag{8.8}
$$

where the spectral density functions $J_m(\omega)$ are, for the case of the dipole–dipole interaction, the Fourier transforms of the corresponding dipolar correlation functions $G_m(t)$. These were given in Equation (7.40) as

$$
G_m(t) = \frac{\mathrm{Tr}\{D_m(t)D_{-m}(0)\}}{\hbar^2 \mathrm{Tr}\{I_z^2\}}
$$

and the dipolar components are given in Equation (5.26).

We now come to the important distinction between the treatment of ^{3}He and that of the fluid system of Chapter 7. Real bodily motion of the particles means that a spin's spatial coordinates x, y and z are varying with time. Recall the discussion in Section 7.1.4 where we indicated that a motion Hamiltonian has the effect of operating on the coordinate operator **r** causing it to evolve with time and we also saw that a semi-classical description of the motion was possible where the co-ordinates, taken as classical variables, were simply assumed to vary with time. In the case of ^{3}He where we are using a spin Hamiltonian such as in Equation (8.7) to describe exchange, the time variation in the dipole interaction is now in the spin operators. Just as in Equation (7.20) we express the time dependence of the $D_m(t)$ as

$$D_m(t) = \exp(i\mathcal{H}_x t/\hbar)D_m\exp(-i\mathcal{H}_x t/\hbar),$$

but now

$$D_m(t) = \frac{\mu_0\hbar^2\gamma^2}{4\pi\sqrt{5}}\sum_{i\neq j}(-1)^m\frac{Y_2^{-m}(\Omega_{ij})}{r_{ij}^3}T_{ij}^m(t),$$

where the spin flip operators T_{ij}^m are defined in Equation (5.24) and their time evolution is generated by $\mathcal{H}_x$. We thus obtain

$$G_m(t) = \left(\frac{\mu_0}{4\pi}\right)^2\frac{8\pi\hbar^2\gamma^4}{5N}\sum_{i\neq j, k\neq l}\frac{Y_2^{m*}(\Omega_{ij})Y_2^m(\Omega_{kl})}{r_{ij}^3 r_{kl}^3}\Gamma_{ijkl}(t), \tag{8.9}$$

which may be compared with Equation (7.35). Here we now have the four spin correlation function

$$\Gamma_{ijkl}(t) = \frac{N\text{Tr}\{T_{ij}^m(0)T_{kl}^{-m}(t)\}}{\text{Tr}\{I_z^2\}}, \tag{8.10}$$

which contains the time dependence. Fourier transformation of the $G_m(t)$ then gives the spectral density functions $J_m(\omega)$ from which the relaxation times can be found, Equation (8.8).

The exchange-induced relaxation behaviour would thus be known if one had expressions for the autocorrelation functions, knowing the time variation of the $\Gamma_{ijkl}(t)$ and combining them in the lattice sums of Equation (8.9). Of course, in practice, just as in the case of the FID relaxation, it is impossible to perform such calculations and resort must be made to approximations.

Equation (8.9) may be expanded in powers of time as

$$G(t) = \tfrac{1}{3}\{M_2 - \tfrac{1}{2}M_4 t^2 + \ldots\}.$$

The coefficients expressed in this way are the moments associated with the transverse relaxation – compare Equation (7.105). Also, we have invoked rotational invariance so that the index m may be discarded. To calculate the moments the derivatives of $\Gamma_{ijkl}(t)$ are found by forming commutators with the exchange Hamiltonian and taking traces. Then the lattice sums implicit in Equation (8.9) are calculated.

The second and fourth moments have been evaluated. The second moment is

$$M_2 = \frac{9C}{20}\frac{\gamma^2\hbar^2}{a^6},$$

where C is a constant depending on the crystal lattice:

$$C = 12.25 \text{ for bcc lattice,}$$
$$C = 14.45 \text{ for hcp lattice.}$$

Note that, as expected, the second moment is independent of J. The fourth moments are conveniently expressed as

$$M_4 = 22.8M_2J^2 \text{ for bcc lattice,}$$
$$M_4 = 42.0M_2J^2 \text{ for hcp lattice.}$$

It is then possible to choose an appropriate functional form for $G(t)$ which is consistent with these values of M_2 and M_4.

8.3.6 Traditional treatment

Experimentally it is observed, from T_1 measurements, that the spectral density function $J(\omega)$ is reasonably approximated by a Gaussian in the hcp phase and by a decaying exponential in the bcc phase. Thus the hcp correlation function $G(t)$ should be approximately Gaussian, while that for the bcc phase should have a Lorentzian profile. In the conventional treatment of exchange induced relaxation in solid ^{3}He the calculated M_2 and M_4 are fitted to these functions. Taking the Fourier transform then yields expressions for the spectral density functions from which the relaxation times may be found as a function of J:

bcc   hcp

$$G(t) = \frac{M_2}{3} \frac{1}{1 + M_4t^2/2M_2} \qquad\qquad G(t) = \frac{M_2}{3} \exp\left(-\frac{M_4}{2M_2}t^2\right).$$

But since

$$M_4 = 22.8M_2J^2 \qquad\qquad\qquad M_4 = 42.0M_2J^2$$

one obtains

$$G(t) = \frac{M_2}{3} \frac{1}{1 + 11.4J^2t^2} \qquad\qquad G(t) = \frac{M_2}{3} \exp(-21J^2t^2).$$

And taking the Fourier transform yields

$$J(\omega) = \frac{0.31M_2}{J} \exp\left(-\frac{\omega}{3.38J}\right) \qquad\qquad J(\omega) = \frac{0.05M_2}{J} \exp\left(-\frac{\omega^2}{84J^2}\right)$$

giving the relaxation times as a function of exchange frequency J and Larmor frequency ω. In particular the adiabatic T_2 is given by

$$\frac{1}{T_2} = 0.465\frac{M_2}{J} \quad \text{bcc lattice,}$$

$$\frac{1}{T_2} = 0.075\frac{M_2}{J} \quad \text{hcp lattice,} \qquad\qquad (8.11)$$

which may be compared with the qualitative discussion of Section 8.3.2, while T_1 is given by

$$\frac{1}{T_1} = \frac{0.31M_2}{J}\exp\left(-\frac{\omega}{3.38J}\right) + \frac{1.24M_2}{J}\exp\left(-\frac{\omega}{1.69J}\right) \quad \text{bcc lattice,}$$

$$\frac{1}{T_1} = \frac{0.05M_2}{J}\exp\left(-\frac{\omega^2}{84J^2}\right) + \frac{0.25M_2}{J}\exp\left(-\frac{\omega^2}{21J^2}\right) \quad \text{hcp lattice.}$$

Observe, particularly from the T_2 expressions, that J^{-1} does indeed give a measure of the dipolar correlation time τ_c, as expected.

These spectral density functions make definite predictions about the frequency dependence of the relaxation times, in particular that of T_1. In practice these predictions are not quite consistent with the extensive experimental data available. Since these expressions have a single frequency parameter J and the coefficient of ω/J is fixed by the short time expansion, while the *functional* form of the high frequency behaviour might be correct it is not possible to ensure *numerical* agreement as there is no free parameter with which to scale the frequency. This casts doubt on the values of J inferred from these analyses. In Section 8.3.11 onwards we shall describe a procedure for obtaining more realistic approximations to the spectral density functions, which rectifies this deficiency. But first we must pause to consider another phenomenon for which the exchange interaction is responsible.

8.3.7 Spin diffusion

The flip-flop nature of the exchange interaction means that any excess magnetisation in part of a specimen will gradually become distributed uniformly. This diffusive process is very much slower than that found in fluids, but it may still be observed by the technique of spin echoes in a field gradient, discussed in Section 4.5.

Since J is the rate at which the $|\uparrow\downarrow\rangle$ configuration changes to the $|\downarrow\uparrow\rangle$ configuration, a simple counting argument implies that the order of magnitude of the diffusion coefficient for the magnetisation will be

$$D \approx Ja^2,$$

where a is the nearest neighbour separation. In this section we shall show that the exchange interaction does indeed lead to diffusion of the spins and that this results in the characteristic cubic decay of spin echoes in a field gradient, familiar from the discussions of Section 4.5. We shall obtain a more precise expression for the relationship between J and D, and this will be used in the examination of some experimental measurements on solid ^{3}He.

Let us start with a precise specification of what is meant by *diffusion*. Fundamentally, of course, this is any process which obeys the diffusion equation. We shall, however, adopt a more appropriate, but equivalent, statement. As may be seen from the diffusion equation, the effect of diffusion is to cause a (spatially) sinusoidal inhomogeneity of wavevector q to decay exponentially as

$$\exp(-Dq^2t).$$

So if the inhomogeneity decays exponentially and if the inverse time constant is proportional to q^2 then the coefficient of q^2 is the diffusion coefficient. With these points in mind we can make connection with the experimental procedure.

Following the discussions of Section 4.5 we consider a small sinusoidal field applied in addition to the static field B_0. The extra magnetic field experienced at the ith site is written as

$$b_i = \frac{\sqrt{2G}}{q}\sin(qz_i),$$

where z_i is the z coordinate of the ith site. In Section 4.5 we were considering a classical treatment where we had to calculate the autocorrelation function for the local magnetic field variations. In Chapter 7 we saw that in the quantum mechanical generalisation it was the interaction energy of this 'local' field which was required. There we were interested in the autocorrelation function of the dipolar interaction.

In this case the interaction is that between the spins and the field inhomogeneity:

$$\mathcal{H}_1 = -\hbar\gamma\sum_i I_z^i b_i, \tag{8.12}$$

so that the corresponding autocorrelation function is then

$$G(t) = \frac{2\gamma^2 G^2}{q^2}\sum_{i<j,\text{nn}} b_i b_j \frac{\text{Tr}\{I_z^i(t)I_z^j(0)\}}{\text{Tr}\{I_z^2\}}.$$

The time evolution of the spin operators I_z is generated by the exchange interaction

$$I_z^i(t) = \exp\left(i\frac{\mathcal{H}_x t}{\hbar}\right)I_z^i(0)\exp\left(-i\frac{\mathcal{H}_x t}{\hbar}\right).$$

8.3.8 Moment calculations

Once again we are confronted with the canonical problem of relaxation theory. We *expect/hope* that $G(t)$ will decay exponentially, at least away from the vicinity of $t = 0$. However, a full calculation is impossible and resort must be made to traditional moment expansion methods. Thus we write

$$G(t) = G(0)\left\{1 - \frac{m_2 t^2}{2} + \frac{m_4 t^4}{4!} - \cdots\right\},$$

where here the moments (different from the moments previously considered) are given by

$$m_{2n} = (-1)^n \frac{\Sigma_{ij}\sin(\mathbf{q}\cdot\mathbf{r}_i)\sin(\mathbf{q}\cdot\mathbf{r}_j)K_{ij}^{(2n)}}{\Sigma_i\sin^2\mathbf{q}\cdot\mathbf{r}_i}$$

and

$$K_{ij}^{(2)} = \text{Tr}\{[\mathscr{H}_x, I_z^i][I_z^j]\}/\hbar^2\text{Tr}\{I_z^{i2}\}$$
$$K_{ij}^{(4)} = \text{Tr}\{[\mathscr{H}_x,[\mathscr{H}_x, I_z^i]][\mathscr{H}_x,\mathscr{H}_x, I_z^j]]\}/\hbar^2\text{Tr}\{I_z^{i2}\}.$$

For the bcc and the hcp lattices these moments have been evaluated (Redfield 1959) to leading order in q:

$$\left.\begin{array}{l} m_2 = (8/3)J^2a^2q^2, \\ m_4 = (208/3)J^4a^2q^2, \end{array}\right\} \text{bcc lattice,}$$

$$\left.\begin{array}{l} m_2 = 4J^2a^2q^2, \\ m_4 = 136J^4a^2q^2. \end{array}\right\} \text{hcp lattice.}$$

The question is now how to incorporate these moments into a suitable expression for $G(t)$, and in particular to see whether a decaying exponential is a suitable function. At this stage it is helpful to recall the dictum of Section 6.3.4, that an (approximately) exponential function is appropriate when the ratio m_4/m_2^2 is very large. Well here we have

$$m_4/m_2^2 \approx (aq)^{-2},$$

which becomes indefinitely large as the wavevector q tends to zero. From the practical point of view we know this is the appropriate limit, since it is then that the sinusoidal field variation becomes a good approximation to the experimentally-applied linear field gradient.

8.3.9 The hydrodynamic limit

Encouraged by the magnitude of the ratio m_4/m_2^2 we shall approximate the Fourier transform of $G(t)$ by a truncated Lorentzian curve. As discussed in Section 6.3.4, this permits us to respect the very short time behaviour, where exponential behaviour is absolutely forbidden, while giving an exponential decay at longer times. In this case we shall adopt a Gaussian curve for the truncation function. See Problem 6.5 and Cowan, Mullin and Nelson (1989) for an expansion of this point. Then the relaxation time is given by

$$\frac{1}{\tau} = \left(\frac{\pi}{2}\right)^{\frac{1}{2}} \frac{m_2^{3/2}}{m_4^{1/2}}. \tag{8.13}$$

(Compare Equation (6.74) for the box truncation function.) Let us examine the q dependence of the relaxation time, where we must consider the higher order terms in m_2 and m_4. The general form for τ^{-1} will be

$$\frac{1}{\tau}\text{const} \times Ja^2q^2 + O(q^4).$$

This is encouraging. We know that diffusion is a *hydrodynamic* process; it occurs at long times, over large distance, where the atomic structure of the system is irrelevant. This is exactly the $q \rightarrow 0$ limit discussed above. We see, then, that in this limit the inverse relaxation time is proportional to q^2, which is precisely the criterion that the process *is* diffusive, as discussed in Section 8.3.7.

The symmetry of the interaction Hamiltonian, Equation (8.12), is such that it changes sign when $z \rightarrow -z$. This together with the exponential decay of the correlation function is sufficient, from the considerations of Section 4.3.5, to ensure that the spin-echo height follows the (exponential) cubic decay characteristic of diffusive systems.

Using Equation (8.13), and taking the coefficient of q^2, we obtain the spin diffusion coefficient for the two lattices as

$$D = 0.655Ja^2 \text{ bcc lattice,}$$
$$D = 0.860Ja^2 \text{ hcp lattice.}$$

The qualitative arguments of Section 8.3.7, that D is proportional to and of the order of Ja^2, are thus vindicated and the numerical coefficients for the expressions have been found. These may be compared with experimental measurements.

8.3.10 Spin diffusion measurements

Spin diffusion data from a number of sources for the bcc phase of solid ^{3}He are shown in Figure 8.7, plotted as a function of Ja^2. The straight line has the 'theoretical' slope of 0.655 and the fit through the data is convincing. This gives considerable support to the discussions of the previous sections. But things are not quite so straightforward. What is measured is the spin diffusion coefficient as a function of molar volume. So the crucial question is: how is the exchange frequency inferred from the molar volume?

The traditional answer is that J values are used that are reasonably consistent with measurements of T_1, T_2, magnetic susceptibility and thermal capacity. This points to the weakness of the procedure outlined in Section 8.3.6 where functional forms for the spectral density function were adopted on a rather arbitrary basis. Values of J obtained from those procedures alone would be questionable, and there would be inconsistencies between values derived from T_1 and T_2 values. An improved method for interpreting relaxation measurements is presented in the following subsections.

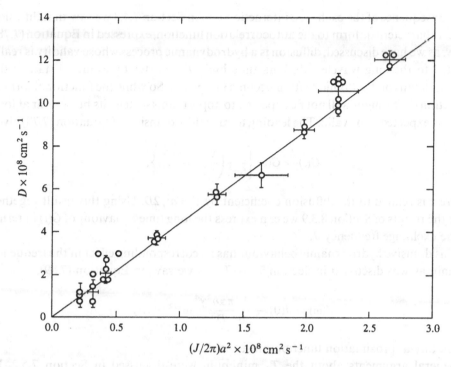

Figure 8.7 Spin diffusion coefficient in the bcc phase of ^{3}He.

8.3.11 Approximating the dipolar correlation function

In the following subsections we shall describe the construction of approximations to the dipolar autocorrelation function $G(t)$, and thus the corresponding spectral density $J(\omega)$, which are substantially better than the simple expressions used in Section 8.3.6. We will consider specifically the bcc phase of solid ^{3}He. The procedure of Section 8.3.6 involved combining the rigorously evaluated short time behaviour, the moment expansion, into an expression for $G(t)$ whose Fourier transform *tries* to reflect the correct high frequency behaviour of $J(\omega)$ as inferred from T_1 measurements.

The procedure we shall describe here involves the use of extra information. We shall augment knowledge of the short time moment expansion of $G(t)$ and the observed high frequency form of $J(\omega)$ with two new things. We shall require the long time behaviour of $G(t)$ and the short frequency behaviour of $J(\omega)$ to be consistent with the spin diffusion discussed in the previous section. Also we will ensure that the 'K value' of Section 7.5.2, associated with the observed T_1 minimum data, is reflected in the chosen approximation functions. Essentially this gives information about $J(\omega)$ in its mid range.

We start with spin diffusion. We have seen in Section 8.3.7–8.3.10 that spin diffusion is a consequence of the exchange interaction and in Section 7.4.3 we saw that diffusion gave a characteristic form to the autocorrelation function, expressed in Equation (7.78). Now, as we have discussed, diffusion is a hydrodynamic process whose validity is really limited to the long wavelength, long time limit. At shorter times and distances the precise details of the atomic motion become important. So while the functional form of Equation (7.78) might well not be expected to apply to this system, its behaviour at long times *is* expected to be valid. The leading term in the expansion of Equation (7.78) gives

$$G(t) \sim G(0) \left\{ \frac{1}{6\pi^{\frac{1}{2}}} \left(\frac{\tau}{t} \right)^{3/2} - \cdots \right\},$$

where τ is related to the diffusion coefficient by $\tau = a^2/2D$. Using this result together with the results of Section 8.3.9 we can express the long time behaviour of $G(t)$ in terms of the exchange frequency J.

This diffusive hydrodynamic behaviour has its corresponding effect in the frequency domain as was discussed in Section 7.4.4. There we saw, in Equation (7.89),

$$J(\omega) = J(0) - \frac{1}{30} \frac{\pi a \hbar^2 \gamma^4}{D^{3/2}} \omega^{1/2} + \cdots,$$

which any approximation function must satisfy.

General arguments about the T_1 minimum were discussed in Section 7.5.2. In particular, we saw that the Larmor frequency at which a minimum occurred (as the correlation time is varied) gave an order of magnitude estimate for the inverse correlation time and that the value of T_1 at the minimum is proportional to the Larmor frequency at which the minimum is observed. Equation (7.100) encapsulated this as:

$$T_1^{\min} = \frac{\omega_0}{M_2} \times 3K,$$

where K is a number characteristic of the *shape* of the spectral density function.

We have plotted values of T_1 at the minimum against Larmor frequency in Figure 8.8. The T_1 data have been divided by the square of the molar volume V_m to remove the dependence on M_2 which is also present when the density is changed. The straight line through the data is given by

$$\frac{T_1^{\min} \text{ (in ms)}}{V_m^2 \text{ (in cm}^6)} = 2.826 \times 10^{-2} \frac{\omega_0}{2\pi} \text{ (in MHz)}.$$

This 'fit' to the data takes account of the larger experimental error in the data point at 50 MHz.

The functional form chosen to approximate $G(t)$ is given by

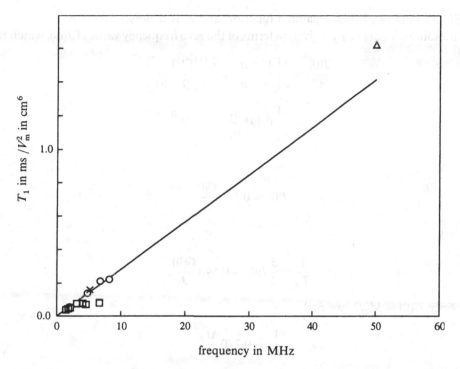

y-axis: T_1 in ms $/V_{\mathrm{m}}^2$ in cm^6

x-axis: frequency in MHz

Figure 8.8 T_1 minima at different Larmor frequencies for bcc ^{3}He. Data points: open circles, Richardson, Hunt and Meyer (1965); open squares, Richards, Hatton and Giffard (1965); crosses, Reich (1963); open triangles, Chapellier *et al.* (1985).

$$\frac{G(t)}{G(0)} = \frac{a_1}{(1 + b_1^2 J^2 t^2)^4} + \frac{a_2}{(1 + b_2^2 J^2 t^2)^{3/4}},$$

where the parameters a_1, a_2, b_1, b_2 are chosen to satisfy the above-mentioned criteria. The second term of this expression gives the correct long time behaviour, while its Fourier transform leads to the corresponding short frequency form. The expression for the spectral density function can be written in analytic form as

$$J(\omega) = \frac{a_1 \pi}{96 b_1 J}\left[\left(\frac{\omega}{b_1 J}\right)^3 + 6\left(\frac{\omega}{b_1 J}\right)^2 + 15\left(\frac{\omega}{b_1 J}\right) + 15\right]\exp\left(-\frac{\omega}{b_1 J}\right)$$

$$+ \frac{a_2}{2^{1/4}}\left(\frac{2}{\pi}\right)^{1/2}\Gamma\left(\frac{1}{4}\right)\frac{1}{b_2 J}\left(\frac{\omega}{b_2 J}\right)^{1/4} K_{1/4}\left(\frac{\omega}{b_2 J}\right), \qquad (8.14)$$

while the values for the parameters are

$$a_1 = 0.84, \quad a_2 = 0.16,$$
$$b_1 = 1.768, \quad b_2 = 2.736.$$

In Equation (8.14), Γ is the gamma function and K is a Bessel function.

The adiabatic part of T_2 is given in terms of the zero frequency value of $J(\omega)$, which is

$$\frac{J(0)}{G(0)} = \frac{1}{J}\left[\frac{15\pi}{96}\frac{a_1}{b_1} + \frac{\pi^{1/2}\Gamma(\frac{1}{4})}{\Gamma(\frac{3}{4})}\frac{a_2}{b_2}\right]$$

$$= \frac{1}{J}\left[0.491\frac{a_1}{b_1} + 5.244\frac{a_2}{b_2}\right]$$

or

$$J(0) = 0.540\frac{G(0)}{J},$$

so that

$$\frac{1}{T_2} = \frac{3}{2}J(0) = 0.540\frac{G(0)}{J},$$

but since $G(0) = M_2/3$, we find

$$\frac{1}{T_2} = 0.270\frac{M_2}{J},$$

which is the quantitative manifestation of the qualitative result $T_2^{-1} = M_2\tau_c$, quoted in Section 8.3.2. In other words $\tau_c = 0.270/J$; again we see that J^{-1} is a measure of the correlation time for the exchange, but now with an improved pre-factor. This should be compared with Equation (8.11) which was based on using the simple Lorentzian $G(t)$/exponential $J(\omega)$. That gave a numerical pre-factor of 0.465 so there is a ratio of 0.58 between that and the present 'improved' result.

8.3.12 *Comparison with experiment*

In Figure 8.9 we show all published T_1 measurements on ^{3}He in the bcc phase plotted in 'reduced' form. We know, in general, from the discussions of Section 7.5.1 and in particular from the discussion of the previous section that for a given system T_1/J is a single-valued function of ω/J. Now since the exchange frequency J is proportional to the frequency of the T_1 minimum, if follows that the same universal behaviour will be displayed when plotting T_1 divided by the frequency of the minimum against Larmor frequency divided by the frequency of the minimum. It is this that we have plotted in Figure 8.9, where the T_1 values have also been divided by the square of the molar volume to account for the variation of M_2 (and thus $G(0)$) with density. The data points are seen to fall very well on a single curve. At higher frequencies there is some discernible deviation from the universal behaviour, which we ascribe to differing crystal

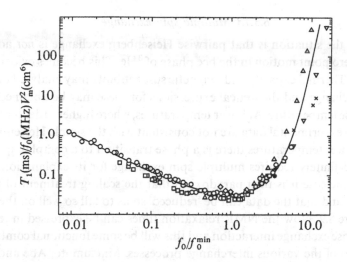

Figure 8.9 Reduced plot of T_1 data for bcc ^{3}He together with 'theoretical' curve. Data points: open circles, Richardson *et al.* (1965); open squares, Richards *et al.* (1965); open triangles, Thomlinson *et al.* (1972); inverted triangles, Bernier and Guerrier (1983); diamonds, Chapellier *et al.* (1985); crosses, Kirk and Adams (1972); pluses, Beal *et al.* (1964).

orientations; the analysis makes the assumption of a polycrystalline sample, taking averages over crystal orientation.

Having the relaxation data presented in this way facilitates the testing of proposed 'theoretical' forms for the spectral density function $J(\omega)$. We have the proposed function in Equation (8.14). The relation between the exchange frequency and the frequency of the minimum is found from Equation (7.98) and the functional form of $J(\omega)$:

$$\omega_0^{\min} = 2.42J.$$

Use of this enables Equation (8.14) to be plotted on the graph of experimental data. The solid line in Figure 8.9 shows this. We must conclude that Equation (8.14) does indeed provide a good approximation to the functional form of the spectral density function.

The result of all these considerations is that the exchange frequency can be deduced in a consistent manner from the measurements made at each molar volume. The best fit to our analysis yields

$$\frac{J}{2\pi} = 14.08 \left(\frac{V_m \ (\text{in cm}^3)}{24} \right)^{18.3} \text{MHz.} \tag{8.15}$$

This compares favourably with the results obtained by other means.

8.3.13 Multiple spin exchange

The reality of the situation is that pairwise Heisenberg exchange is *not* adequate to describe the zero point motion in the bcc phase of ^{3}He. This becomes evident at lower temperatures. Thermal capacity and magnetic susceptibility may analysed on the basis of pairwise exchange and theoretical expressions for these may be obtained as power series in inverse temperature. At lower temperatures, where higher order terms become important, the experimental data are not consistent with the simple Heisenberg model and at even lower temperatures there is a phase transition to a complex spin-ordered phase which definitely requires multiple spin exchange for its explanation.

This being the case it is then paradoxical that the scaling treatment of relaxation works so well, and that the data can be 'reduced' so as to fall so well on the universal curve of Figure 8.9. Now the NMR relaxation times can be expressed in terms of an *effective* pairwise exchange interaction and this will be some functional combination of the frequencies of the various interchange processes. Matsumoto, Abe and Izuyama (1989) have shown that when considering a restricted subset of two- three- and four-particle exchange, the effective pairwise exchange frequency determining the NMR relaxation times is the combination

$$J_{\text{eff}}^2 = J_{\text{nn}}^2 - 14 J_{\text{nn}} J_{\text{t}} + 6.7 J_{\text{nn}} K_{\text{p}} + 61 J_{\text{t}}^2 - 54 J_{\text{t}} K_{\text{p}} + 18 K_{\text{p}}^2,$$

where J_{nn} is the pairwise exchange frequency, J_{t} is the frequency of the three-particle process and K_{p} is the four-particle exchange frequency for interchanges in a plane.

The high quality of the reduced data plot, Figure 8.9, implies that so far as NMR relaxation is concerned, the system *can* be understood in terms of single time parameter characterising the atomic motion. This means that the interchange frequencies of the various exchange cycles all scale with density in a similar manner, as Equation (8.15), in fact. Such behaviour has been obtained by the path integral calculations of Ceperley and Jacucci (1988). There is, however, no *a priori* reason to expect these exchange processes to vary with density in the same way; indeed it is a surprise that they do. One is led to wonder if there may be a more elementary physical origin of the atomic motion, and that multiple spin exchange might be a descriptive construction rather like the planetary epicycles of Ptolemy. Currently there is no satisfactory first-principles theoretical explanation, but it is likely that vacancy-interstitial formation is the fundamental process out of which all exchange cycles are built. This would give the single energy or time scale for all exchange processes. The elucidation of this remains one of the unsolved problems in the theory of solid helium.

9

The density operator and applications

9.1 Introduction to the density operator

9.1.1 Motivation and definition

It is often inconvenient, for one reason or another, to keep track of all the variables of a complex system. For example, in a many-body system it would be impracticable to consider the co-ordinates of each particle. Furthermore, such information is actually not of much interest. Thus in a spin system the behaviour of the individual magnetic moments is unimportant; the components of the total magnetisation are the variables of primary interest. Very generally, for a system with $\sim 10^{23}$ co-ordinates, one is unlikely to need more than, say, ten variables to describe its observable properties. However, it is quite clear that this reduced amount of information is no longer sufficient to write down a wavefunction and, therefore, it is no longer possible to calculate the evolution of such a system using the usual methods of quantum mechanics.

In Chapter 5 we saw that it was possible, by the introduction of probabilistic arguments, to calculate the evolution of the total magnetisation of a spin system. The calculations were performed using the machinery of quantum mechanics and meaningful and useful results were thereby obtained. In this chapter we will look at things from a rather different point of view. We will not, initially, be concerned with the expectation values of certain specified observables, rather we direct our attention to the general description of the 'state' of the system and the way it evolves with time – possibly towards thermal equilibrium. One consequence of this will be a formalisation of the methods of Chapter 5. However, we will also be able to extend those results to see how the evolution of a general variable or set of variables may be calculated from quantum mechanical first principles.

A quantum mechanical system for which one knows the values of all its co-ordinates can be represented by a wavefunction. In general, with full information a quantum system may be represented by a state vector or ket. This description is not possible, however, in the absence of full information. We will see, in this chapter, that in such

291

cases it is still possible to make a quantum mechanical specification of the state of a system in terms of a *density operator*. For conventional treatments of the density operator see Fano (1957), tu Haar (1961), Landau and Litshitz (1969) and von Neumann (1955). We shall start the discussion by introducing the density operator for a system where we do have complete information – when we can write down a wavefunction. Such a density operator is said to describe a *pure state*. After examining some of the properties of such a pure state density operator, including its equation of motion, we shall consider the case of less than maximal information when we will introduce what is known as a *reduced* density operator. The state of a system specified in this way, lacking full description, is called a *mixed* state.

To introduce the full density operator, describing a pure state, let us consider a system in the quantum state s, and the expectation value A of an operator $\mathcal{A}$ in this state. In Chapter 5, in Equation (5.1) we saw that the value of A could be expressed as an integral involving the system wavefunction ψ_s. If the system is represented by the ket $|s\rangle$ then the expectation value A can be written in bra-ket notation as

$$A = \langle s|\mathcal{A}|s\rangle. \tag{9.1}$$

We shall express this result in a slightly different form in terms of the operator defined by the ket-bra $|q\rangle\langle p|$. To do so we make use of the following identity

Identity: $$\langle p|q\rangle = \mathrm{Tr}\{|q\rangle\langle p|\} \tag{9.2}$$

This relation holds for all bras and kets $\langle p|$ and $|q\rangle$; its proof is taken up in Problem 9.1. In terms of this result we may identify the bra $\langle p|$ with $\langle s|\mathcal{A}$ and the ket $|q\rangle$ with $|s\rangle$ so that Equation (9.1) may be written as:

$$A = \mathrm{Tr}\{|s\rangle\langle s|\mathcal{A}\}.$$

If we now define an operator ρ (the density operator) by

$$\rho = |s\rangle\langle s|, \tag{9.3}$$

then the expectation value for any observable operator $\mathcal{A}$ takes the simple form

$$A = \mathrm{Tr}\{\rho\mathcal{A}\}. \tag{9.4}$$

Thus we see that for a given system in a specified quantum state we can form the operator ρ; then the expectation value of any operator $\mathcal{A}$ representing an observable A is given by the trace of the product of the operator $\mathcal{A}$ with the density operator ρ.

Box 9.1 Explanation of terminology

The operator ρ introduced in this chapter has its analogue in classical statistical mechanics. A classical observable A may be expressed as a function of the generalised co-ordinate variables $\{q_i\}$ and momentum variables $\{p_i\}$ of a many particle assembly. The average value of this observable may be found as an integral over a probability density

$$\langle A \rangle = \int \rho(p,q) A(p,q) dp dq.$$

Here $\rho(p,q)$ represents the density of points in a phase space ensemble. Now it is the integral over p and q of classical mechanics, which becomes the sum over states, or trace, of quantum mechanics. Therefore the above equation is the classical counterpart of the quantum expectation value expression

$$A = \mathrm{Tr}\{\rho \mathscr{A}\}$$

and thus the classical density function in phase space goes over to the quantum density *operator*.

In the literature the operator ρ is sometimes loosely referred to as a density *matrix*. Strictly speaking, in this context a matrix is a representation of an operator in a given basis. Thus once a complete orthonormal set of states $|i\rangle$ has been specified, the elements $\langle i|\rho|j\rangle = \rho_{ij}$ constitute a density matrix. We shall make use of the energy representation in some of our discussions. When we refer to a density *matrix* we will usually mean the matrix in the energy representation.

9.1.2 Properties of the full density operator

We have defined the density operator ρ for the system in the state s, represented by the ket $|s\rangle$, by Equation (9.3)

$$\rho = |s\rangle\langle s|.$$

This, according to the terminology introduced in the previous section, is thus a full density operator describing a pure state.

There are a number of properties of the density operator which we now demonstrate – for the pure state density operator.

(1) Normalisation: $\mathrm{Tr}\rho = 1$.

This follows from the normalisation of the kets $|s\rangle$ and the identity, Equation (9.2),

$$\text{Tr}\rho = \text{Tr}\{|s\rangle\langle s|\} = \langle s|s\rangle = 1.$$

This property is equivalent to the requirement that the expectation value of the unit operator is unity.

(2) The i, i diagonal matrix element of the density operator gives the probability that the system is found in the state i.

If we expand the ket $|s\rangle$ in terms of a complete orthonormal set of kets:

$$|s\rangle = \sum_j a_j |j\rangle,$$

then ρ may be written as

$$\rho = |s\rangle\langle s| = \sum_{j,k} a_j a_k^* |j\rangle\langle k|.$$

Now taking the i, i matrix element of this we have

$$\rho_{ii} = \sum_{j,k} a_j a_k^* \langle i|j\rangle\langle k|i\rangle$$

$$= \sum_{j,k} a_j a_k^* \delta_{ij}\delta_{ki} = a_i a_i^*.$$

So we see that $\rho_{ii} = |a_i|^2$ which is indeed the probability of being in (strictly, of observing) the ith state.

(3) ρ is Hermitian: $\rho^+ = \rho$.

This must be so since, from property (2), the diagonal elements must always be real. However, we can prove Hermiticity as follows:

$$\rho_{ij} = \langle i|s\rangle\langle s|j\rangle$$
$$= \langle s|j\rangle\langle i|s\rangle$$
$$= (\langle j|s\rangle\langle s|i\rangle)^*$$

$$= \rho_{ji}^*$$

and therefore

$$\rho = \rho^+,$$

so that the density operator is necessarily Hermitian.

(4) ρ is idempotent: $\rho^2 = \rho$.

This property is true only for a pure state density operator, since then

$$\rho^2 = |s\rangle\langle s|s\rangle\langle s|,$$

however, by normalisation

$$\langle s|s\rangle = 1$$

and therefore

$$\rho^2 = |s\rangle\langle s| = \rho.$$

We shall see that the first three of these properties are always obeyed by a density operator. However, the last property, idempotence, is only true for a pure state. It is thus a useful indicator of whether a state is pure or not.

9.1.3 Equation of motion – von Neumann's equation

Most of the discussions of the time variation of quantum mechanical systems have been carried out so far in this book from the Heisenberg point of view. That is, the time variation is ascribed to the operators representing the observables. However, as we explained in Chapter 5, there is an alternative approach, in which the observables' operators remain constant while the quantum states evolve with the passage of time: the Schrödinger picture. Very often, in one's introduction to quantum mechanics the Schrödinger picture is used, and the fundamental equation of motion is then the time dependent Schrödinger equation. This brings out the wave–particle duality most directly. The Heisenberg picture does, however, have a number of important advantages. We saw in Chapter 2 how the connection between classical and quantum descriptions of nature is most directly made from that standpoint.

Since the density operator is constructed from the kets which describe the system's state, we see that it will only have a time variation in the Schrödinger picture; in the Heisenberg picture it remains constant. As the material of this chapter unfolds it should become apparent to the reader that in the discussions of Chapter 5 the density operator, in its time-independent Heisenberg form, was actually introduced 'through the back door'. The discussion in this chapter is mainly from the Schrödinger point of view, so that a consideration of the equation of motion obeyed by the density operator then becomes of central importance.

In the Schrödinger picture the kets have a time dependence as given by Schrödinger's equation of motion

$$\frac{\partial}{\partial t}|s\rangle = -\frac{i}{\hbar}\mathcal{H}|s\rangle,$$

where $\mathcal{H}$ is the Hamiltonian for the system. The density operator is defined as the product of a ket with a bra. One must thus differentiate the product:

$$\frac{\partial}{\partial t}|s\rangle\langle s| = \left(\frac{\partial}{\partial t}|s\rangle\right)\langle s| + |s\rangle\left(\frac{\partial}{\partial t}\langle s|\right).$$

The Schrödinger equation above gives the time derivative of $|s\rangle$, while its Hermitian conjugate equation gives the derivative of the bra $\langle s|$,

$$\frac{\partial}{\partial t}\langle s| = \frac{i}{\hbar}\langle s|\mathcal{H}.$$

Combining these we obtain

$$\frac{\partial\rho}{\partial t} = -\frac{i}{\hbar}\mathcal{H}|s\rangle\langle s| + \frac{i}{\hbar}|s\rangle\langle s|\mathcal{H}$$

$$= -\frac{i}{\hbar}(\mathcal{H}\rho - \rho\mathcal{H})$$

or, since we recognise the commutator in this last equation, we can write

$$\frac{\partial\rho}{\partial t} = -\frac{i}{\hbar}[\mathcal{H},\rho], \tag{9.5}$$

which is known as the von Neumann equation. This equation bears a startling resemblance to the Heisenberg equation for the time dependence of an observable's operator, but with opposite sign. This is to be expected. In the Schrödinger picture the operators representing observables have no time dependence; time variation is contained in the kets or the density operator. In the Heisenberg picture the observable operators have the usual Heisenberg time variation, while ρ, like the wavefunctions or kets, does not depend on time. But in both cases expectation values must be the same. The solution to the von Neumann equation is

$$\rho(t) = \exp\left(-i\frac{\mathcal{H}}{\hbar}t\right)\rho(0)\exp\left(i\frac{\mathcal{H}}{\hbar}t\right), \tag{9.6}$$

so the expectation value of the operator $\mathcal{A}$ may be written

$$\langle\mathcal{A}(t)\rangle = \text{Tr}\{\rho(t)\mathcal{A}\}$$

$$= \text{Tr}\left\{\exp\left(-i\frac{\mathcal{H}}{\hbar}t\right)\rho(0)\exp\left(i\frac{\mathcal{H}}{\hbar}t\right)\mathcal{A}\right\}. \tag{9.7}$$

Now by cyclic permutation within the trace this may also be expressed as

$$\langle\mathcal{A}(t)\rangle = \text{Tr}\left\{\rho(0)\exp\left(i\frac{\mathcal{H}}{\hbar}t\right)\mathcal{A}\exp\left(-i\frac{\mathcal{H}}{\hbar}t\right)\right\}$$

$$= \text{Tr}\{\rho\mathcal{A}(t)\}. \tag{9.8}$$

This is exactly the time evolution in the Heisenberg picture; here $\mathscr{A}$ varies with time while the density operator ρ, or equivalently the state vectors, is constant. Careful examination of Equations (9.7) and (9.8) should convince you that the Schrödinger time evolution for ρ and the Heisenberg time evolution for $\mathscr{A}$ have opposite senses.

9.1.4 The reduced density operator

Thus far, in the discussion of the density operator, there have been no new concepts introduced, only some definitions. The full density operator simply provides a new way of expressing some of the old-established relations of quantum mechanics. The real importance of the density operator, however, is connected with the use of the *reduced* density operator. This we now introduce.

We continue to be interested in the expectation value of an operator $\mathscr{A}$, but now we consider the case where $\mathscr{A}$ is a function of only a few of the many variables (coordinates) of the entire system. For instance, in the NMR context, $\mathscr{A}$ could depend on spin but not lattice variables. In this case let us denote by a the variables upon which $\mathscr{A}$ depends, and b the remaining variables, of which $\mathscr{A}$ is independent.

The expectation value of $\mathscr{A}$ is given by

$$\langle \mathscr{A} \rangle = \mathrm{Tr}_{a,b}\{\rho(a,b)\mathscr{A}(a)\},$$

where the trace is explicitly expressed as being evaluated over the two sets of variables denoted by a and b. Now the trace is simply a diagonal sum, and since the operator $\mathscr{A}$ is independent of the b variables, this trace may be evaluated first, i.e.

$$\langle \mathscr{A} \rangle = \mathrm{Tr}_{a}\{\mathrm{Tr}_{b}\{\rho(a,b)\}\mathscr{A}(a)\}.$$

In this expression we see that the operator $\mathrm{Tr}_{b}\{\rho(a,b)\}$ contains all the information necessary to evaluate the expectation value of $\mathscr{A}$. Thus we see that if the operator σ is defined by

$$\sigma(a) = \mathrm{Tr}_{b}\{\rho(a,b)\}, \tag{9.9}$$

then the expectation value of $\mathscr{A}$ may be written

$$\langle \mathscr{A} \rangle = \mathrm{Tr}\{\sigma\mathscr{A}\},$$

so that σ is used here just as we would use a density operator. This is a *reduced* density operator. It depends on fewer variables than the full density operator and it is therefore a simpler object.

Reduced density operators are more useful than the full density operator as first defined. Clearly a reduced density operator does not contain complete information about a system; such a density operator cannot, in general, be expressed as a ket-bra product $|s\rangle\langle s|$. For a macroscopic many-body system containing $\sim 10^{23}$ particles we should require some 10^{23} variables to specify its state. Contrasting this, a suitable reduced density operator would have just sufficient dimension to give values for the observables of interest.

9.1.5 Physical meaning of the reduced density operator

The reduced density operator, as we have seen, contains less than complete information to specify a quantum state fully. From the operational point of view we may say that a reduced density operator simply contains sufficient information to give values for a given set of observables. However, from the definition of σ, in Equation (9.9) we shall see that the reduced density operator has a direct and simple interpretation. We will now investigate the connection between the full and the reduced density operator. The reduced density operator is defined by

$$\sigma(a) = \text{Tr}_b \rho(a,b).$$

where, as before, a represents variables of interest and b variables not of interest. The full density operator ρ is a function of the variables a and b. The ket and bra of which ρ is composed may be expanded in a complete orthonormal set of the a and b eigenkets and eigenbras:

$$|s\rangle = \sum_{i,j} A_{ij} |a_i\rangle |b_j\rangle,$$

so that the full density operator may be written as

$$\rho = |s\rangle\langle s|$$

$$= \sum_{i,j,i',j'} A_{ij} A_{i'j'}^* |a_i\rangle |b_j\rangle\langle a_{i'}|\langle b_{j'}|$$

$$= \sum_{i,j,i',j'} A_{ij} A_{i'j'}^* |a_i\rangle\langle a_{i'}| |b_j\rangle\langle b_{j'}|.$$

Now let us take the trace over the b variables. In doing so we shall use the orthonormality of the b kets to obtain

$$\sigma = \text{Tr}_b \rho = \sum_{i,j,i'} A_{ij} A_{i'j}^* |a_i\rangle\langle a_{i'}|.$$

We shall now specify the a kets to be chosen so as to make ρ diagonal in that variable. In this representation the reduced density operator σ will be diagonal and then

$$\sigma = \sum_{i,j} A_{ij}A_{ij}^* |a_i\rangle\langle a_i|$$

$$= \sum_i \left\{ \sum_j |A_{ij}|^2 \right\} |a_i\rangle\langle a_i|.$$

However, $|A_{ij}|^2$ is the probability that the system is in the state having the a variable a_i and the b variable b_j, so summing over all j we obtain the overall probability that the a variable is a_i regardless of the value of the b variable – i.e. the probability that a system specified by a variables only is in the a_i state. Thus we can write

$$\sigma = \sum_i p_i \sigma_i, \tag{9.10}$$

where

$$\sigma_i = |a_i\rangle\langle a_i|,$$

the (full) density operator which would describe a system of only a variables in the a_i state, and p_i is the probability that the system is in that state.

Thus finally we arrive at the physical interpretation of the reduced density operator. If we consider only the variables of interest then the reduced density operator is simply an average over the various possible pure state density operators.

In the treatment followed we have expressions for the various probabilities occurring in the average. However, it is equally possible to consider the reduced density operator in ignorance of he other variables and to define the reduced density operator as an ensemble average; this is the way to proceed when one does not know for certain which state the system is in. In other words we see that when we don't know exactly which state the system is in, while we can't perform an average over the wavefunctions or kets, the lack of knowledge can be accommodated by making the average over the density operators of the respective states. The reduced density operator provides a convenient and systematic way to do this.

9.1.6 Properties of the reduced density operator

We shall now examine how the properties of the full density operator, derived in Section 9.1.2, apply in the present case. Most properties will be seen to carry over. However, we shall see that some properties will be different for the reduced density operator.

(1) Normalisation: $\text{Tr}\rho = 1$.

This follows from the normalisation of the kets $|s\rangle$ and the identity, Equation (9.2),

$$\text{Tr}\rho = \text{Tr}\{|s\rangle\langle s|\} = \langle s|s\rangle = 1.$$

It remains true since we are simply completing the trace operation already begun in the definition of σ.

(2) The i, i diagonal matrix element clearly remains the probability that the system is in the (less well specified) state i.

(3) σ remains Hermitian.

This follows obviously; nothing in the process of constructing a reduced density operator from a full density operator destroys hermiticity.

(4) σ is *not* idempotent: $\sigma^2 \neq \sigma$.

This property of the reduced density operator is different for a mixed state. We can demonstrate this by writing

$$\sigma = \sum p_i |i\rangle\langle i|$$

so that

$$\sigma^2 = \sum p_i p_j |i\rangle\langle i||j\rangle\langle j|.$$

Orthonormality then gives

$$\sigma^2 = \sum p_i^2 |i\rangle\langle i|.$$

Now since the probabilities must sum to unity, a p_i cannot be greater than 1. It then follows that p_i^2 cannot be greater than p_i. Thus in general

$$p_i^2 \leq p_i$$

so that we conclude that

$$\sigma^2 \leq \sigma,$$

the equality holding when one of the p_i is unity (and the rest are therefore zero), i.e. for a pure state.

Idempotence is thus seen to provide a simple test of whether a density operator is describing a pure or a mixed state.

9.1.7 *Equation of motion for the reduced density operator*

We know the equation of motion obeyed by a full density operator: this is the von Neumann equation, Equation (9.5),

$$\frac{\partial \rho}{\partial t} = -\frac{i}{\hbar}[\mathscr{H}, \rho].$$

We also know that it is much more convenient to work with a reduced density operator, discarding that information superfluous to our needs. The question then arises as to whether the reduced density operator obeys a similar equation of motion or not. To be precise, we shall ask two questions. Firstly, under what circumstances does a reduced density operator obey an equation like Equation (9.5)? Secondly, if/when a reduced density operator does not obey an equation of that form, what equation of motion does it obey? The first question will be answered in this section. The answering of the second question will be taken up in Section 9.5. The reduced density operator is given by

$$\sigma(a) = \text{Tr}_b\rho(a,b).$$

The time derivative is then

$$\frac{\partial\sigma}{\partial t} = \text{Tr}_b\left\{\frac{\partial\rho}{\partial t}\right\},$$

which, from the von Neumann equation, may be expressed as

$$\frac{\partial\sigma}{\partial t} = -\frac{i}{\hbar}\text{Tr}_b[\mathcal{H},\rho].$$

If the system Hamiltonian $\mathcal{H}$ does *not* depend on the b variables then we can take the trace straight through to the ρ, giving the same von Neumann equation for the reduced density operator

$$\frac{\partial\sigma}{\partial t} = -\frac{i}{\hbar}[\mathcal{H},\sigma]. \tag{9.11}$$

However, if the Hamiltonian *does* depend on some of the b variables then the trace cannot be carried through and a rather more complicated equation of motion ensues. The solution of the von Neumann equation

$$\sigma(t) = \exp\left(-i\frac{\mathcal{H}}{\hbar}t\right)\sigma(0)\exp\left(i\frac{\mathcal{H}}{\hbar}t\right) \tag{9.12}$$

thus gives the time development of the reduced density operator *only* when σ covers all the variables of $\mathcal{H}$. Otherwise the evolution of σ is not generated by a unitary transformation.

It must be emphasised at this stage that the time variation considered here is in the Schrödinger picture. It is within this picture that the time evolution of σ may be complicated (non-unitary). However, if one chooses to work within the Heisenberg picture then time development remains a unitary transformation – of the observable's operator. For this reason it is usually simpler, formally, to use the Heisenberg approach. This was the method of Chapter 5. The Schrödinger approach is, however, of use when considering the *general* behaviour of a system, not concentrating on a particular variable. An example is the approach of a system to thermal equilibrium. We are not quite at the stage of considering the *approach* to equilibrium but we shall, in the next section but one, examine the form of the density operator which describes equilibrium states. Before that we pause to consider a simple example of the use of the density operator.

9.1.8 Example: Larmor precession

The Hamiltonian for an assembly of non-interacting spins in a magnetic field, the Zeeman Hamiltonian, is given by:

$$\mathcal{H}_z = -B_0\gamma\hbar I_z = \hbar\omega_0 I_z,$$

where I_z is the operator for the total z component of spin, summed over all particles:

$$I_z = \sum_i I_z^i.$$

Since the Hamiltonian separates as a sum of identical single-particle Hamiltonians, it follows from the discussion of the previous section that a von Neumann equation of motion can be written for the *single-particle* reduced density operator. For a system of spins $\frac{1}{2}$, which we consider, this means the reduced density operator may be represented as a 2×2 matrix:

$$\sigma = \begin{pmatrix} a & b + ic \\ b - ic & 1 - a \end{pmatrix}. \tag{9.13}$$

We have chosen to express σ in the above form so that normalisation and hermiticity are automatically guaranteed. In the usual representation the operators for the three components of spin I_x, I_y and I_z are given by

$$I_x = \begin{pmatrix} 0 & 1/2 \\ 1/2 & 0 \end{pmatrix}, I_y = \begin{pmatrix} 0 & -i/2 \\ i/2 & 0 \end{pmatrix}, I_z = \begin{pmatrix} 1/2 & 0 \\ 0 & -1/2 \end{pmatrix}.$$

Then their expectation values may be found as

$$\langle I_x \rangle = \text{Tr}\{\sigma I_x\} = \text{Tr}\left\{ \begin{pmatrix} a & b+ic \\ b-ic & 1-a \end{pmatrix} \begin{pmatrix} 0 & 1/2 \\ 1/2 & 0 \end{pmatrix} \right\} = b,$$

$$\langle I_y \rangle = \text{Tr}\{\sigma I_y\} = \text{Tr}\left\{ \begin{pmatrix} a & b+ic \\ b-ic & 1-a \end{pmatrix} \begin{pmatrix} 0 & -i/2 \\ i/2 & 0 \end{pmatrix} \right\} = c,$$

$$\langle I_z \rangle = \text{Tr}\{\sigma I_z\} = \text{Tr}\left\{ \begin{pmatrix} a & b+ic \\ b-ic & 1-a \end{pmatrix} \begin{pmatrix} 1/2 & 0 \\ 0 & -1/2 \end{pmatrix} \right\} = a - 1/2,$$

so that the values of $\langle I_x \rangle$, $\langle I_y \rangle$ and $\langle I_z \rangle$ are determined uniquely by the parameters b, c and a. Alternatively, this argument may be turned around so that the density operator may be expressed in terms of the expectation values $\langle I_x \rangle$, $\langle I_y \rangle$ and $\langle I_z \rangle$. Thus we may write σ as

$$\sigma = \begin{pmatrix} \frac{1}{2} + \langle I_z \rangle & \langle I_x \rangle + i\langle I_y \rangle \\ \langle I_x \rangle - i\langle I_y \rangle & \frac{1}{2} - \langle I_z \rangle \end{pmatrix}. \tag{9.14}$$

We see that this σ is completely specified in terms of the expectation values of the three components of spin. A larger density operator, for instance, one describing spin 1 particles, would require knowledge of extra quantities. This is considered in Problem 9.2.

The spin $\frac{1}{2}$ density operator of Equation (9.14) obeys the von Neumann equation of motion, Equation (9.11), when the Hamiltonian is the one-particle Zeeman Hamiltonian

$$\mathcal{H}_Z = \hbar\omega_0 I_z$$

so that I_z refers to a single spin. The equation of motion is then

$$\frac{\partial \sigma}{\partial t} = i\omega_0 [I_z, \sigma],$$

which has solution

$$\sigma(t) = \exp(i\omega_0 I_z t)\sigma(0)\exp(-i\omega_0 I_z t). \tag{9.15}$$

In the representation where I_z is diagonal it is a simple matter to evaluate the exponential:

$$\exp(i\omega_0 I_z t) = \exp\begin{pmatrix} i\omega_0 I_z t/2 & 0 \\ 0 & -i\omega_0 I_z t/2 \end{pmatrix} = \begin{pmatrix} \exp(i\omega_0 I_z t/2) & 0 \\ 0 & \exp(-i\omega_0 I_z t/2) \end{pmatrix}.$$

The time evolution of σ is then

$$\sigma(t) = \begin{pmatrix} \frac{1}{2} + \langle I_z(t) \rangle & \langle I_x(t) \rangle + i\langle I_y(t) \rangle \\ \langle I_x(t) \rangle - i\langle I_y(t) \rangle & \frac{1}{2} - \langle I_z(t) \rangle \end{pmatrix}$$

$$= \begin{pmatrix} \exp(i\omega_0 I_z t/2) & 0 \\ 0 & \exp(-i\omega_0 I_z t/2) \end{pmatrix}\begin{pmatrix} \frac{1}{2} + \langle I_z(0) \rangle & \langle I_x(0) \rangle + i\langle I_y(0) \rangle \\ \langle I_x(0) \rangle - i\langle I_y(0) \rangle & \frac{1}{2} - \langle I_z(0) \rangle \end{pmatrix}\begin{pmatrix} \exp(-i\omega_0 I_z t/2) & 0 \\ 0 & \exp(i\omega_0 I_z t/2) \end{pmatrix}.$$

On working through the matrix multiplications we obtain:

$$\sigma(t) = \begin{pmatrix} \frac{1}{2} + \langle I_z(0) \rangle & (\langle I_x(0) \rangle + i\langle I_y(0) \rangle)\exp(i\omega_0 t) \\ (\langle I_x(0) \rangle - i\langle I_y(0) \rangle)\exp(-i\omega_0 t) & \frac{1}{2} - \langle I_z(0) \rangle \end{pmatrix}, \tag{9.16}$$

which indicates that the diagonal elements of the matrix remain constant while the off-diagonal elements oscillate at the Larmor frequency. Upon separating the various elements we finally find

$$\langle I_x(t)\rangle = \langle I_x(0)\rangle\cos(\omega_0 t),$$
$$\langle I_y(t)\rangle = \langle I_y(0)\rangle\sin(\omega_0 t),$$
$$\langle I_z(t)\rangle = \langle I_z(0)\rangle,$$

which are immediately recognised as the equations for Larmor precession expressed in terms of the quantum mechanical expectation values.

Clearly we have discovered nothing new in this exercise. However, it provides an illustration of the manipulation of the various matrices involved, and it shows how familiar results may be recovered using the technique.

9.2 Thermal equilibrium

9.2.1 The equilibrium density operator

In Section 5.2.2 we considered the treatment of systems in thermal equilibrium. Much of the material of that section will be recalled here, now presented from a different perspective. The discussion there was based on the idea that an equilibrium system could exist in any of its energy eigenstates, each with a probability given by the appropriate Boltzmann factor. We now recognise this description, in density operator language, as a *mixed* state.

To find the density operator which describes a system in equilibrium we note the following two properties of the state of thermal equilibrium:

(1) Observable quantities are constant in time.
(2) $\rho_{ii} \propto \exp(-\mathcal{H}_{ii}/kT)$ in the representation where $\mathcal{H}$ is diagonal.

The first property is really the phenomenological description of an equilibrium state – that which pertains when everything has settled down and stopped changing. However, there is an important consequence. We are saying that the expectation value of the time derivative of any operator is zero:

$$0 = \text{Tr}\left\{\frac{\partial\rho}{\partial t}\mathcal{A}\right\}$$

$$= -\frac{i}{\hbar}\text{Tr}\{[\mathcal{H},\rho]\mathcal{A}\}$$

for any operator $\mathcal{A}$. This can only be so if the commutator of $\mathcal{H}$ with ρ is zero; the equilibrium density operator must commute with the Hamiltonian. Thus ρ is diagonal when $\mathcal{H}$ is diagonal.

However, when $\mathcal{H}$ is diagonal the diagonal elements of ρ are given by the relevant Boltzmann factors: the second property above. Thus we conclude that in the energy representation:

$$\rho_{ij} = \begin{cases} \text{const} \times \exp(-\mathcal{H}_{ij}/kT) & \text{when } i = j, \\ 0 & \text{when } i \neq j. \end{cases} \tag{9.17}$$

In Section 5.2.2 we explained that a function of an operator may be understood in terms of its power series expansion. This is consistent with the more general definition that a function of an operator is another operator, which (a) has the same eigenstates, and (b) whose eigenvalues are the (scalar) function of the eigenvalues of the original operator. This is, in the representation where the operator $\mathcal{A}$ is diagonal,

$$\mathcal{A} = \begin{pmatrix} a_1 & & & 0 \\ & a_2 & & \\ & & \ddots & \\ 0 & & & a_n \end{pmatrix},$$

the function $f(\mathcal{A})$ is given by the operator

$$f(\mathcal{A}) = \begin{pmatrix} f(a_1) & & & 0 \\ & f(a_2) & & \\ & & \ddots & \\ 0 & & & f(a_n) \end{pmatrix}.$$

From this it follows immediately, by Equations (9.17), that the equilibrium ρ is given by:

$$\rho_{eq} = \text{const} \times \exp(-\mathcal{H}/kT).$$

The normalisation condition $\text{Tr}\rho = 1$ gives the constant factor, so that we may write

$$\rho_{eq} = \frac{\exp(-\mathcal{H}/kT)}{\text{Tr}\{\exp(-\mathcal{H}/kT)\}}, \tag{9.18}$$

and the denominator is seen to be the partition function for the system. The expectation value of an operator $\mathcal{A}$ is given by its trace with the density operator:

$$A = \text{Tr}\{\mathcal{A}\exp(-\beta\mathcal{H})\}/\text{Tr}\{\exp(-\beta\mathcal{H})\},$$

where we use the symbol β to represent inverse temperature $1/kT$. We recognise this result as Equation (5.3).

9.2.2 Ensemble average interpretation

The expression for the equilibrium density operator in Equation (9.18) has a direct interpretation in terms of the ensemble average view of the reduced density operator, discussed at the end of Section 9.1.5. If we expand the equilibrium density operator with a complete set of eigenstates $|i\rangle$, using the identity $\Sigma_i |i\rangle\langle i| = 1$, then

$$\rho = \sum_i \frac{\exp(-\beta\mathcal{H})}{\text{Tr}\{\exp(-\beta\mathcal{H}\}} |i\rangle\langle i|.$$

Choosing the states to be the energy eigenstates, so that

$$\mathcal{H}|i\rangle = E_i|i\rangle,$$

then permits the simplification

$$\rho = \sum_i \frac{\exp(-\beta E_i)}{\Sigma_j \text{Tr}\{\exp(-\beta E_j)\}} |i\rangle\langle i|.$$

This can be written in the more suggestive form:

$$\rho = \sum_i p_i \rho_i,$$

where p_i is the Boltzmann factor for the state $|i\rangle$, and ρ_i is $|i\rangle\langle i|$, the pure state density operator for the state $|i\rangle$. This is precisely the way the mixed state density operator was expressed in Equation (9.10). Thus we have an interpretation of the equilibrium density operator as an ensemble average of pure state density operators over a Boltzmann distribution.

9.2.3 A paradox

Common experience tells us that macroscopic systems settle down to states of thermal equilibrium; the discipline of thermodynamics is based on the existence of such states. The state of equilibrium can be attained in two seemingly different ways. On the one hand an isolated system can be disturbed into a non-equilibrium state. In this case its subsequent evolution will return it to thermal equilibrium. On the other hand two bodies (either of which may or may not be in equilibrium) can be brought into contact whereupon they will exchange energy until they have come into equilibrium themselves and with each other. In both cases, in terms of the density operator, we expect an initial state $\rho(0)$ to evolve, with the passage of time, into the equilibrium ρ_{eq}.

The establishment of external equilibrium will be taken up in Section 9.2.6. We start, however, with a consideration of the first case: the isolated system. In this case the time evolution is given by the von Neumann equation so that we expect

$$\exp\left(-i\frac{\mathcal{H}}{\hbar}t\right)\rho(0)\exp\left(i\frac{\mathcal{H}}{\hbar}\right) \rightarrow \frac{\exp(-\beta\mathcal{H})}{\mathrm{Tr}\{\exp(-\beta\mathcal{H})\}}$$

for sufficiently long times t. But this is impossible! Mathematically, the expression may be inverted; we may write $\rho(0)$ as

$$\exp\left(i\frac{\mathcal{H}}{\hbar}t\right)\frac{\exp(-\beta\mathcal{H})}{\mathrm{Tr}\{\exp(-\beta\mathcal{H})\}}\exp\left(-i\frac{\mathcal{H}}{\hbar}t\right) \rightarrow \rho(0),$$

which cannot happen. The time evolution operators commute with the equilibrium density operator, so that ρ_{eq} does not change with time. This is no surprise, since the form of ρ_{eq} was derived specifically to be time independent. As a special case of this problem we note that if the initial $\rho(0)$ were prepared as a pure state then there is no way that a unitary transformation, such as Equation (9.6), can turn this into a mixed state such as ρ_{eq}.

We have therefore reached the worrisome conclusion that no state can evolve into the equilibrium state! This paradox must be resolved. There are actually two aspects to the paradox. Both of the equilibrium properties listed in Section 9.2.1 are impossible. In the energy representation:

(1) the off-diagonal elements of ρ *cannot* become zero;
(2) the diagonal elements of ρ *cannot* approach $\exp(-E_i/kT)$.

We shall demonstrate these explicitly. Let us examine the time evolution of the density operator, as described by Equation (9.6). We consider the behaviour of a matrix element of ρ, in the energy representation. Then, since $\mathcal{H}$ is taken to be diagonal,

$$\rho_{ij}(t) = \exp\left(-\frac{i\mathcal{H}_{ii}t}{\hbar}\right)\rho_{ij}(0)\exp\left(\frac{i\mathcal{H}_{jj}t}{\hbar}\right)$$

$$= \exp\left[\frac{i(\mathcal{H}_{jj}-\mathcal{H}_{ii})t}{\hbar}\right]\rho_{ij}(0). \tag{9.19}$$

This indicates that in the energy representation the off-diagonal elements of ρ oscillate, while the diagonal elements remain constant. In Equation (9.16) we had a special case of this where we saw that the off-diagonal elements of that matrix oscillated at the Larmor frequency.

Our conclusion, which is quite general, is that time evolution causes the off-diagonal elements of the density matrix to oscillate while the diagonal elements remain constant. These are the two aspects of the paradox.

9.2.4 Partial resolution

Let us consider in more detail the question of the off-diagonal elements of ρ. Our argument was that these elements were zero, because ρ had to commute with $\mathcal{H}$, because in equilibrium observable quantities were constant in time, but this is not quite so. In equilibrium there are *fluctuations* in observable quantities. Sometimes these may be seen quite obviously, for example the density fluctuations in a fluid which occur at critical opalescence. Usually, however, the fluctuations are very small or even unobservable. But they always exist. With sufficient magnification the observables fluctuate about their means, as indicated in Figure 4.5. In principle the description of the complete time evolution – fluctuations and all – should be contained in the von Neumann equation. Observables are constant *on average*. And so it is only on average that the off-diagonal elements of the density operator vanish.

A related phenomenon has already been encountered when we considered the dephasing of spins in an inhomogeneous magnetic field and the consequent loss of transverse magnetisation. One might have been inclined to believe that the decay time T_2^* measured the progress towards equilibrium. After all, the equilibrium state has zero transverse magnetisation. But then we learned about the rephasing effect of a 180° pulse; the lost magnetisation could be recovered. What was thought to be an equilibrium state turned out to be nothing of the kind. With a spread of Larmor frequencies the off-diagonal elements of the many-body density matrix will oscillate at different rates. Operators such as the total x component of spin give the sum of various elements of the matrix; thus the dephasing leads to decay and the *appearance* of equilibrium.

From this perspective thermal equilibrium has a subjective, or almost arbitrary, element. The off-diagonal elements of the density matrix oscillate, and an observable such as I_x takes a sum over various elements, and thus exhibits relaxation as phase coherence is lost. However, the system may be manipulated, such as by a 180° pulse to reverse the oscillations, to return the system to coherence from its state of equilibrium. And with more sophisticated ways of manipulating these systems order can be recovered from more incoherent and seemingly random 'equilibrium' states.

9.2.5 Macroscopic systems

The previous section gave a partial resolution of the paradox of the density operator description of equilibrium states. We showed that the off-diagonal elements of the equilibrium density matrix were zero only on average. But there is still the difficulty of the diagonal elements. On the one hand we believe that they should settle down to a Boltzmann distribution, but on the other hand the equation of motion indicates that they remain constant. The origin of this problem may be seen from the von Neumann equation for a diagonal element of ρ expressed here in an arbitrary representation:

$$\frac{\partial \rho_{ii}}{\partial t} = \frac{i}{\hbar} \sum_j (\mathscr{H}_{ji} \rho_{ij} - \mathscr{H}_{ij} \rho_{ji}). \tag{9.20}$$

Since both $\mathscr{H}$ and ρ are Hermitian there is no contribution from the diagonal elements of either $\mathscr{H}$ or ρ. The rate of change of ρ_{ii} depends on the off-diagonal elements of $\mathscr{H}$ and ρ in the ith row and the ith column. Our original discussion was carried out in the energy representation. Then $\mathscr{H}$ was diagonal. We argued that in that representation ρ_{eq} was diagonal although subsequently that requirement was relaxed, but even admitting off-diagonal elements to ρ will not allow its diagonal elements to vary so long as $\mathscr{H}$ is diagonal, i.e. so long as we are in the energy representation. The probability that a system is in a given stationary state is absolutely constant. So to repeat our question, how can the system establish a Boltzmann distribution?

The answer to this will be found through a deeper understanding of the nature of real macroscopic systems. Recall that we are considering a system isolated from its surroundings. That was the justification for using the von Neumann equation to describe the evolution of the system since then the Hamiltonian depended only on the variables of the system and that is also the justification for regarding the eigenstates of the system's Hamiltonian as stationary states. Unfortunately it is impossible to specify a system's Hamiltonian completely. In any real macroscopic system there is a hierarchy of interactions, resulting in a sequence of contributions to the Hamiltonian operating at smaller and smaller energy scales.

For a spin system in an applied magnetic field (greater than 10^{-2} T or so) the largest relevant energy will probably be that of the Zeeman interaction between the dipoles and the field. Considerably smaller than this will be the adiabatic part of the internuclear dipolar interaction. At the next level down is the non-adiabatic part of the dipole–dipole interaction and somewhere on the scale will be interactions between the nuclear spins and bodily atomic motion, interactions with electrons, etc. As a consequence, it is simply inappropriate to talk of the stationary states of a macroscopic body. Whatever approximate Hamiltonian is chosen, its eigenstates will not be truly stationary. The terms of the ignored parts of the Hamiltonian will induce transitions between the approximate eigenstates.

Each term in the Hamiltonian hierarchy of energy scales will cause a frequency modulation of the matrix elements of the term of greater energy. This operates recursively so that after considering a number of levels the transitions induced between the approximate eigenstates can be regarded as being random, within the constraint of energy conservation. As is well known in statistical mechanics, random transitions between energy states which preserve energy conservation will result in a Boltzmann distribution of states.

This resolves the paradox of the establishment of thermal equilibrium *in an isolated system*. In particular we see that in this case thermal equilibrium is a property appropriate to real macroscopic systems.

9.2.6 External equilibrium

We now turn to the case in which thermal equilibrium is established by the interaction of our system of interest with another system. Let us assume that the interactions between the systems are sufficiently weak. That permits us to talk of the energy states of the system of interest, and the effect of the interactions is to induce transitions between the states.

When the interaction is weak it means that there is negligible correlation between the states of the two systems, except that required for energy conservation. As a result a Boltzmann distribution of states will be established in both systems, at a common temperature. If the 'other' system has a sufficiently large thermal capacity then it can be regarded as a 'heat bath' at a constant temperature. In that case the system of interest will settle down to thermal equilibrium at the temperature of the bath. The mathematical treatment of this type of evolution towards equilibrium will be taken up in Section 9.5.

9.2.7 Spin systems

In spin systems there can be separations between levels in the interaction energy hierarchy. Thus there will be a separation between the time scales appropriate for various degrees of 'random' dephasing and apparent equilibrium. As mentioned before, the simplest example of this is the formation of spin echoes following a 180° pulse. Here the approximate stationary states are the eigenstates of the (spatially averaged) Zeeman Hamiltonian. The 'interaction' responsible for the evolution to equilibrium is the magnetic field inhomogeneities leading to a spread of local precession frequencies. The next Hamiltonian in the hierarchy will be the dipole–dipole interaction if the spins are immobile, or the motional modulation of the magnetic field inhomogeneities if the spins are moving very rapidly.

For a range of intermediate motions the third level of interaction can be ignored. Then the second level interactions modulate first level ones at constant frequencies and the equilibrium is 'apparent'. It is then that the 180° pulse can reverse the effect of the second level interactions, which had caused 'thermalisation' of the eigenstates of the first level interaction.

For immobile spins the first level of interaction is the Zeeman Hamiltonian. The second level is the dipole–dipole interaction – only the adiabatic part when the applied magnetic field is large. The third level of interaction can often be ignored, and then it is possible for carefully arranged sequences of RF pulses to reverse the dephasing caused by the dipolar interaction – the so-called magic echo (Abragam and Goldman, 1982, p. 47).

In summary, we can say that spin systems are special because:

(*a*) if there is separation of interaction scales then the random, equilibrium states may not be *so* random;

(b) the tools for manipulating spin states, namely pulses of RF radiation, are more powerful than those available for non-spin systems.

Related to this is the concept of spin temperature, to be introduced in Section 9.3.

We conclude this section with a simple example. In Section 6.2 we considered a realistic model spin system which was soluble, exhibiting irreversible behaviour. The Hamiltonian for that system was

$$\mathcal{H} = -\hbar\omega_0\sum_j I_z^j + \frac{1}{2}\sum_{j\neq k} I_z^j I_z^k B_{jk};$$

the second term represents the interspin interaction. The solution of this system was given in Equation (6.65). We can recast the solution in terms of a density operator. The solution gave the expectation values of I_x, I_y and I_z. Thus we have the information to give the 2×2 reduced density operator for an ensemble of spin $\frac{1}{2}$ particles. The structure of that density operator was given in Equation (9.14), from which we conclude

$$\sigma(t) = \begin{pmatrix} \frac{1}{2} + \langle I_z(0)\rangle & \langle I_+(0)\rangle\exp(i\omega_0 t)\prod_k \cos\left(\frac{B_{jk}t}{2\hbar}\right) \\ \langle I_-(0)\rangle\exp(-i\omega_0 t)\prod_k \cos\left(\frac{B_{jk}t}{2\hbar}\right) & \frac{1}{2} - \langle I_z(0)\rangle \end{pmatrix}.$$

This is a single-particle (reduced) density operator and, since the interaction involves a two spin Hamiltonian, it follows that the time evolution of this density operator is not a unitary transformation.

9.3 Spin dynamics in solids

9.3.1 The concept of spin temperature

Temperature is a concept which is quite familiar to common experience; 'hot' and 'cold' are readily felt by our senses. At the fundamental level, however, the meaning of temperature is tied up with the laws of thermodynamics and its definition, in terms of heat ratios of reversible cycles, is far from intuitive. The aspect of temperature which we have encountered already is connected with the likelihood of a system having a given energy: the Boltzmann principle, which determines the structure of the equilibrium density operator. This principle states that when a system is in thermal equilibrium at a temperature T, the probability that it will be found in a state of energy E is given by

$$p(E) \propto \exp(-E/kT), \tag{9.21}$$

where k is Boltzmann's constant; the corresponding density operator is

$$\sigma = \frac{\exp(-\mathcal{H}/kT)}{\text{Tr}\{\exp(-\mathcal{H}/kT)\}}.$$

We can introduce the idea of spin temperature by inverting Equation (9.21) and defining (for spins) a spin temperature in terms of the probabilities $p(E)$. Although initially this may be regarded simply as a mathematical convenience, we shall see as the arguments unfold that under certain circumstances the spin temperature thus defined may be identified as a respectable thermodynamic quantity. Further discussions will lead to generalisations of the concept of temperature to areas outside the usual remit of classical thermodynamics. Such spin temperature ideas have proved invaluable in the understanding of many NMR phenomena observed in solids.

9.3.2 *Changing the spin temperature*

The concept of spin temperature is clearly best applied in cases where T_1 is much greater than T_2. Then, for a significant portion of the system's evolution, it may be characterised by a spin temperature different from that of the lattice. Let us consider the spin temperature of such a system in various circumstances, when only the Zeeman part of the spin Hamiltonian is considered; interspin interactions are ignored. First, imagine the application of a 90° pulse. When acting on a system in equilibrium, this rotates the magnetisation into the transverse plane where it decays with a time scale T_2. The effect of the 90° is to destroy the longitudinal magnetisation. Translating this into the language of spin temperature, the equilibrium spin temperature starts equal to the lattice temperature. The 90° pulse, destroying the magnetisation, equalises the population of the various spin states. In other words it has created an *infinite* spin temperature.

As if this were not a sufficiently surprising result, now consider the effect of a 180° pulse on a spin system in equilibrium. The magnetisation, initially parallel to the applied magnetic field, becomes reversed. The upper energy states are more occupied than are the lower energy states. Reference to the Boltzmann factor leads us to ascribe to such a state a *negative* spin temperature. Application of a 180° to a system at temperature T results in a spin temperature of $-T$.

Negative temperatures are not normally encountered in realistic physical systems. Ordinarily, states of higher energy are not more populated than those of lower energy. The special feature of a spin system is that it has an upper bound to its energy states. A 'normal' system has an infinite number of states of ever increasing energy. A negative temperature implies higher energy states having higher occupations; infinite energy states would be infinitely populated. Thus negative temperatures for normal systems are nonsensical.

It might be thought that a state where *some* of the energy states were more highly populated than the lower energy states should be considered. Such states are certainly possible, but since no Boltzmann factor could fit such a distribution, no temperature, positive or negative, could characterise it; it is in no sense an equilibrium state.

Starting with a conventional system in thermal equilibrium with its surroundings, and therefore at a positive temperature, the addition of energy alters the population distribution of the states. The temperature increases. An infinite amount of energy would be required to equalise the populations and produce an infinite temperature. For a spin system only a finite energy is needed to produce an infinite spin temperature, simply **M·B**. What happens if more energy is added? Clearly the higher energy states must become more populated, resulting in a negative spin temperature. Negative temperatures correspond to higher energies than positive temperatures. Sometimes this is expressed in the paradoxical statement that negative temperature are *hotter* than positive temperatures! It actually makes more sense to talk in terms of inverse temperatures. This quantity varies smoothly through zero as the temperature goes off to infinity and reappears at minus infinity.

9.3.3 The spin temperature hypothesis

Most of the above discussion of spin temperature is applicable to all systems, solids and fluids. However, as we have said, the real power of the spin temperature idea is in discussing NMR phenomena in solids, when T_1 is long (Redfield, 1955; Goldman, 1970). Then (for times shorter than T_1) the system may be regarded as isolated from its surroundings and the equilibrium established is the *internal* equilibrium discussed in Section 9.2. In the language of that section it is an *apparent* equilibrium. As an example, consider an assembly of stationary non-interacting spins in an inhomogeneous magnetic field. Following a 90° pulse the precessing magnetisation will decay away over a time scale T_2^*, so that for times longer than this the system appears to have reached equilibrium. The off-diagonal elements of the many-body density operator are then oscillating incoherently so that in evaluating the macroscopic magnetisation it is *as if* the off-diagonal elements were zero. In other words the observable quantities have the same values they would have if the system were described by a true equilibrium density operator. However, this example shows the error of the assumption. Application of a 180° pulse will cause a 'time reversal' in this system so that the magnetisation is recovered as a spin echo. In that case the apparent equilibrium was thus not a true equilibrium.

In truth the 'equilibrium' states in solids established in time scales shorter than T_1 are all apparent equilibrium states. Certainly the usual dipole–dipole interaction makes the time evolution much more complex, but even there it is possible to reverse the flow of time – for example the 'magic echo'. Nevertheless, for many applications it is as if the system were in real equilibrium, and thus we are led to the spin temperature hypothesis. This states that after a disturbance a spin system may often be regarded as having evolved to an equilibrium state and the spin temperature of this state is that which gives the correct value for the system's energy. In other words, it is as if the energy were the only constant of the motion.

9.3.4 Energy and entropy

The energy of a spin system may be found by the usual density operator procedure for finding expectation values:

$$E = \mathrm{Tr}\{\mathcal{H}\rho\},$$

which for an equilibrium state becomes

$$E = \frac{\mathrm{Tr}\{\mathcal{H}\exp(-\mathcal{H}/kT)\}}{\mathrm{Tr}\{\exp(-\mathcal{H}/kT)\}}.$$

We saw, in Section 2.1.4, in connection with the derivation of Curie's Law, that frequently the dimensionless quantity $\hbar\gamma B/kT$ is a number much smaller than unity. This permits expansion in powers of this small parameter, a high temperature expansion, and results, among other things, in the linearity between magnetic field and magnetisation. Since it is only for magnetic fields greater than some thousand teslas or for temperatures less than fractions of a millikelvin that this breaks down, we shall, in what follows, generally adopt this high temperature/low polarisation approximation. It is quite possible to perform many of the formal calculations involving spin temperature without making the approximation, but the results can be quite unwieldy. Goldman (1977) has done much work in this area. Thus we are justified in expanding the density operator to first order in $\beta = 1/kT$. In that case

$$\sigma_{eq} = (1 - \beta\mathcal{H})/\mathrm{Tr}1$$

where 1 is the unit operator.

The energy of this system is then

$$E = -\beta\mathrm{Tr}\{\mathcal{H}^2\}/\mathrm{Tr}1, \tag{9.22}$$

since for all systems of interest the Hamiltonian has zero trace. So we see that spin systems with negative energy have positive temperatures, while systems with positive energy have negative temperatures.

The entropy of a system is given by the expression

$$S = -k\sum_j p_j \ln p_j.$$

Now since the density operator is diagonal in the energy representation, and since in the energy representation the diagonal elements are the Boltzmann factors, giving the probabilities p_j, it follows that the entropy can be written in terms of the density operator as

$$S = -kTr\{\sigma\ln\sigma\}$$

or

$$S = -k\langle\ln\sigma\rangle$$

since the trace is just the sum over states. In accordance with the laws of statistical mechanics, the Boltzmann density operator may be found by maximising the entropy. In the high temperature limit the entropy is given by

$$S = k\beta\langle\mathcal{H}\rangle - kTr1$$

$$= k\beta^2Tr\{\mathcal{H}^2\}/Tr1 + \text{const} \qquad (9.23)$$

and we may ignore the constant.

9.3.5 *Adiabatic demagnetisation*

Adiabatic magnetisation is a technique used in many laboratories for the attainment of low temperatures. Spin temperature is the natural way to understand the physics of this cooling method. Nevertheless, our particular interest here is not with the technicalities of low temperature physics. In this section we consider the principles of the adiabatic demagnetisation method.

Imagine a paramagnetic solid, thermally isolated from its surroundings. A magnetic field is applied and we are interested in the effect of varying this field. Since the spin system is thermally isolated, we are considering an *adiabatic* process, i.e. the entropy of the system remains constant. We see that constant entropy implies that the populations of the various energy states remains unchanged and since these populations are given by the Boltzmann factor, this means that its exponent $\hbar\gamma B/kT$ must remain constant. In other words B/T is invariant: as B is varied, T changes proportionately.

Here, then, we have a method for cooling a paramagnetic solid. It is allowed to reach equilibrium in contact with some thermal reservoir and in the presence of a magnetic field. When equilibrium is established the thermal link with the reservoir is broken and the field is reduced. As the field drops, so does the temperature. Before the advent of the dilution refrigerator the standard method of cooling to temperatures below 0.3 K was by the adiabatic demagnetisation of electron paramagnetic salts, particularly cerium magnesium nitrate (CMN). Temperatures down to a few millikelvins may be obtained in this way. With *nuclear* demagnetisation of copper, temperatures below a millikelvin are readily achieved. It would seem that one could get down to the absolute zero of temperature simply by reducing the magnetic field to zero. However, that would be in violation of the Third Law of Thermodynamics. There are various factors which in reality limit the lowest temperatures achievable, and one of these is of particular interest to us.

9.3.6 Dipole fields

Even when the external field is reduced to zero, an individual dipole still sees the dipole fields of its neighbours, which will have a magnitude of a few gauss. For cooling there is no point in reducing the external field below this value. These fields turn out to have a crucial importance for certain NMR experiments carried out in solids. We have already seen, earlier in this chapter, that the dipole fields are responsible for the broadening of the CW absorption line/lifetime of the free precession decay of transverse magnetisation. Here we are concerned with something different.

When the applied field is of the order of the dipole field then the energy splittings due to the Zeeman Hamiltonian and the dipole Hamiltonian will be comparable. In this case spin flip-flops will lead to rapid interchange between Zeeman and dipole energy to maintain the state of equilibrium. The density operator for this system (in the high temperature case) is

$$\sigma_{eq} = [1 - \beta(\mathcal{H}_z + \mathcal{H}_d)]/\text{Tr}1,$$

where the Zeeman Hamiltonian is $-B_0\hbar\gamma I_z$.

Let us assume that the magnetic field is varied sufficiently slowly that the system remains in thermal equilibrium, just as in the previous (high field) case. Then the entropy will remain constant. From Equation (9.23) this means that

$$\beta^2\text{Tr}\{(\mathcal{H}_z + \mathcal{H}_d)^2\} = \text{const.}$$

Now the cross term $\text{Tr}\{\mathcal{H}_z\mathcal{H}_d\} = 0$, as may easily be verified. Thus the adiabatic condition may be written

$$\beta^2 B_0^2\hbar^2\gamma^2\text{Tr}\{I_z^2\} + \beta^2\text{Tr}\{\mathcal{H}_d^2\} = \text{const}$$

or

$$\beta^2 B_0^2\hbar^2\gamma^2\text{Tr}\{I_z^2\} + \beta^2 B_L^2\hbar^2\gamma^2\text{Tr}\{I_z^2\} = \text{const,}$$

where we have introduced the (dipolar) local field B_L:

$$B_L^2 = \frac{1}{\hbar^2\gamma^2}\frac{\text{Tr}\{\mathcal{H}_d^2\}}{\text{Tr}\{\mathcal{H}_z^2\}}.$$

From this we see that the spin temperature varies with the applied magnetic field as

$$T(B_0^2 + B_L^2)^{\frac{1}{2}},$$

which indicates the way in which the temperature fails to go to zero as the applied field is reduced. This is shown in Figure 9.1.

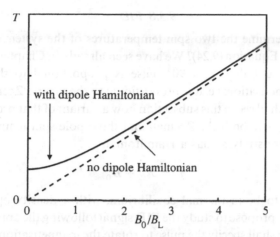

Figure 9.1 Effect of the dipole Hamiltonian on adiabatic demagnetisation.

9.3.7 High field case

When the applied field B_0 is much larger than dipole field B_L the energy splittings due to the Zeeman Hamiltonian are very much greater than those due to the dipole Hamiltonian. In that case there can be no interchange between Zeeman and dipole energy. In other words the Zeeman energy and the dipole energy are separately constants of the motion. The Zeeman Hamiltonian is very much greater than the dipole Hamiltonian. This means that the states of the system are fundamentally Zeeman states, and these are just broadened a little by the dipole interaction. To first order the change of the Zeeman levels is determined by the adiabatic part of the dipole Hamiltonian – that part which commutes with I_z, which we denoted by D_0 in earlier chapters. The effects of the non-adiabatic parts of the dipole interaction are negligible in the high field case.

Since $\langle \mathscr{H}_z \rangle$ and $\langle D_0 \rangle$ are separately constants of the motion the density operator for this system (in the high temperature case) is

$$\sigma_{eq} = (1 - \beta_z \mathscr{H}_z - \beta D_0)\mathrm{Tr1}. \qquad (9.24)$$

Now we have separate spin temperatures for the Zeeman part and the dipole part. Here β_z is the Zeeman inverse temperature and β is the dipole inverse temperature. They are determined from the values of the Zeeman and the dipole energy:

$$E_z = \mathrm{Tr}\{\sigma_{eq}\mathscr{H}_z\} = -\beta_z \mathrm{Tr}\{\mathscr{H}_z^2\}/\mathrm{Tr1}, \Bigg\} \\ E_d = \mathrm{Tr}\{\sigma_{eq}D_0\} = -\beta r\{D_0^2\}/\mathrm{Tr1}, \qquad (9.25)$$

so that E_z determines the Zeeman temperature and E_d determines the dipole temperature.

9.3.8 FID

How could we determine the two spin temperatures of the system described by the density operator of Equation (9.24)? We have seen already in Chapter 5 that the size of the precession signal following a 90° pulse is proportional to the equilibrium z component of magnetisation: i.e. it is proportional to the inverse Zeeman temperature of that system. We shall see in this subsection how a variant of that procedure can, with the present system, give both the Zeeman and the dipole spin temperatures.

The system we are studying has a Hamiltonian

$$\mathcal{H} = \mathcal{H}_z + D_0 \tag{9.26}$$

and it is assumed to be in thermal equilibrium, with a density operator given by Equation (9.24). We propose to study the FID signal following the application of an RF pulse of angle θ. We shall specify the pulse to rotate the magnetisation about the y axis, so it tends to rotate I_z into I_x. By analogy with the procedures of Chapter 5 we shall perform this rotation on the equilibrium density operator to convert the 'Boltzmann distribution of states' into our specially prepared 'non-equilibrium distribution of states'. Then we will use that as the 'weighting factor' in calculating the expectation values of I_x and I_y as they evolve in time under the influence of the system Hamiltonian, Equation (9.26).

We write the equilibrium density operator as

$$\sigma_{eq} = (1 + \beta_z \hbar \omega_0 I_z - \beta D_0)/\text{Tr}1.$$

First we must find the expression for the density operator immediately following the θ pulse, which we denote by $\sigma(0_+)$. The rotation of I_z is easy:

$$I_z \rightarrow I_z \cos\theta + I_x \sin\theta.$$

The rotation of D_0 is a little more complex. Certainly we know that it is bilinear in angular momentum operators and the rotations can be performed on these just as with the I_z above, but it will be helpful to preserve the form of D_0 so far as possible, and to exploit the symmetries of the interaction. Now from Equation (5.26) we have the expression for D_0 given in terms of the spherical harmonic Y and the tensor operator T:

$$D_0 = \frac{\mu_0 \hbar^2 \gamma^2}{4\pi\sqrt{5}} \sum_{i,j} \frac{Y_2^0(\Omega_{ij})}{r_{ij}^3} T_{ij}^0.$$

Rotation of the spin operators will transform the T operator. Rotation by an angle θ may be expressed as

$$T_{ij}^0 = \sum_{m=-2}^{2} T_{ij}^m d_{m,0}^{(2)}(\theta).$$

The coefficients d are a special case of the rotation matrices $\mathscr{D}$ used in Appendix F, where here we perform the rotation about the y axis. The coefficients are given explicitly in Brink and Satchler (1962); the specific ones we require are

$$d^{(2)}_{2,0}(\theta) = d^{(2)}_{-2,0}(\theta) = \sqrt{\frac{3}{8}}\sin^2\theta,$$

$$d^{(2)}_{1,0}(\theta) = -d^{(2)}_{1,0}(\theta) = -\sqrt{\frac{3}{2}}\sin\theta\cos\theta,$$

$$d^{(2)}_{0,0}(\theta) = \frac{1}{2}(3\cos^2\theta - 1).$$

Thus the effect of the rotation on T^0_{ij} is

$$T^0_{ij} \rightarrow \frac{1}{2}(3\cos^2\theta - 1)T^0_{ij} - \sqrt{\frac{3}{2}}\sin\theta\cos\theta(T^1_{ij} - T^{-1}_{ij}) + \sqrt{\frac{3}{8}}\sin^2\theta(T^2_{ij} + T^{-2}_{ij}).$$

An important step is now to transform the middle term containing the difference between the two T operators. From Equation (5.25) we have

$$[I_\pm, T^0_{ij}] = \sqrt{6}T^{\pm 1}_{ij},$$

so that

$$T^1_{ij} - T^{-1}_{ij} = \frac{2i}{\sqrt{6}}[I_y, T^0_{ij}].$$

By multiplying the transformed T by the Y^0_2 we obtain the rotated adiabatic part of the dipolar Hamiltonian. Together with the rotated Zeeman Hamiltonian, we obtain the transformed density operator $\sigma(0_+)$ as (neglecting the denominator)

$$\sigma(0_+) = 1 + \beta_z\hbar\omega_0(I_z\cos\theta + I_x\sin\theta) - \beta\left\{\frac{1}{2}(3\cos^2\theta - 1)D_0 - i\sin\theta\cos\theta[I_y, D_0]\right.$$

$$\left. + \sin^2\theta\frac{\mu_0\hbar^2\gamma^2}{4\pi\sqrt{5}}\sum_{i,j}\frac{Y^0_2(\Omega_{ij})}{r^3_{ij}}\sqrt{\frac{3}{8}}(T^2_{ij} + T^{-2}_{ij})\right\}.$$

Now we want to calculate the expectation values of I_x and I_y as they evolve with time. To do this we simply multiply $I_x(t)$ and $I_y(t)$ by $\sigma(0_+)$ and take the trace. We will use a time evolution generated solely by D_0, which means that we are looking at the magnetisation components in the frame rotating at the Larmor frequency. In taking the traces many of the terms vanish. The only ones which remain are those where I_x multiplies I_x and I_y multiplies I_y, so that

$$\langle I_x(t)\rangle = \text{Tr}\{\sigma(0_+)I_x(t)\} = \beta_z\hbar\omega_0\sin\theta\text{Tr}\{I_x(t)I_x(0)\},$$
$$\langle I_y(t)\rangle = \text{Tr}\{\sigma(0_+)I_y(t)\} = -i\beta\sin\theta\cos\theta\text{Tr}\{I_x(t)[I_x(0), D_0]\}.$$

In the first term we recognise $\text{Tr}\{I_x(t)I_x(0)\}$ as the usual FID envelope which is expected in the absence of dipolar order. We express this as $\text{Tr}\{I_x^2\}F(t)$ where $F(t)$ is the normalised FID. In the second term the operators in the trace can be permuted so that

$$\langle I_y(t)\rangle = i\beta\sin\theta\cos\theta\text{Tr}\{[I_x(t),D_0]I_x(0)\}$$

$$= \beta\hbar\sin\theta\cos\theta\frac{\text{d}}{\text{d}t}\text{Tr}\{I_x(t)I_x(0)\}.$$

So this term is proportional to the *derivative* of $F(t)$. We summarise the results of these calculations as

$$\left.\begin{array}{l}\langle I_x(t)\rangle = \beta_Z\hbar\omega_0\sin\theta\text{Tr}\{I_x^2\}F(t), \\ \langle I_y(t)\rangle = \beta\hbar\sin\theta\cos\theta\text{Tr}\{I_x^2\}F'(t).\end{array}\right\} \tag{9.27}$$

This shows that the in-phase component of the precessing magnetisation is proportional to the Zeeman inverse temperature while the quadrature component is proportional to the dipole inverse temperature. Furthermore we see that the Zeeman signal is maximised using a 90° initiating pulse while the dipole signal is maximised using a 45° pulse. It is interesting to note that the relaxation of both signals is related to the familiar free induction decay function.

9.3.9 Provotorov equations

We have seen that in the high field case a lattice of spins can exist with separate Zeeman and dipole energies. One way of providing an interaction between the Zeeman and the dipole 'bath' is by applying a resonant oscillating or rotating transverse field. When such a field is applied the Hamiltonian for the system is static in the frame rotating with the field, so the equilibrium state will be established in that frame. However, recall we saw in Chapter 2 that in the frame rotating with angular velocity ω the Zeeman field is reduced by an amount ω/γ. Thus when ω is close to the Larmor frequency we enter the 'low field' régime when a common spin temperature will be established. Clearly the rate at which energy is exchanged between the two systems will depend crucially on the distance from resonance. The description of these phenomena is provided by the Provotorov equations (Provotorov, 1961). These provide for solids essentially what the Bloch equations provide for fluids. We shall follow the elegant treatment of Abragam and Goldman (1982).

We can use the results of the last section, in Equation (9.27), to show how the system will respond to the application of a sinusoidal (oscillating or rotating) transverse magnetic field. Those equations give the response to the application of a magnetic field of short duration in the y direction. Alternatively the field could be regarded as being of longer duration, applied in the y direction of the frame rotating at the Larmor frequency. However it is interpreted, the product of the magnitude of the field and its duration are such as to provide the correct rotation angle θ. The region of linear response is when the rotation angle is vanishingly small. In that case the Fourier transform, at frequency Δ, of the precessing magnetisation gives the linear response to a sinusoidal field whose frequency is a distance Δ from resonance.

The linear response to an impulse field is found from Equation (9.27) in the limit of small Δ. We shall combine the in-phase and quadrature components of the solution into a complex combination

$$\langle I_x(t) \rangle + i\langle I_y(t) \rangle = \hbar\theta \text{Tr}\{I_x^2\}[\beta_z\omega_0 F(t) + i\beta F'(t)]. \tag{9.28}$$

The Fourier transform of the FID signal is defined by

$$f(\Delta) = \frac{1}{\pi} \int_0^\infty F(t)\exp(-i\Delta t)dt$$

and from this expression, by integrating by parts, one finds the corresponding Fourier transform of $F'(t)$

$$i\Delta f(\Delta) - \frac{1}{\pi} = \frac{1}{\pi} \int_0^\infty F'(t)\exp(-i\Delta t)dt.$$

By substituting these into Equation (9.28) we find the linear response to the complex sinusoidal field to be

$$\langle I_x(\Delta) \rangle + i\langle I_y(\Delta) \rangle = \omega_1 \text{Tr}\{I_x^2\}[\pi(\beta_z\omega_0 - \beta\Delta)f(\Delta) - i\beta], \tag{9.29}$$

where we have eliminated the angle θ in favour of the magnitude of the rotating field.

On resonance Δ is zero and $f(0)$ is real and at its peak. Then the expression for the magnetisation is real, so it points in the x (rotating frame) direction. Since the applied field is in the y (rotating) direction, we see that on resonance the magnetisation is perpendicular to the field. This is just as predicted by the Bloch equations.

Now let us consider the flow of energy between the Zeeman and the dipole baths. The effective Hamiltonian in the frame rotating with the applied field at a frequency of ω is

$$\mathcal{H}_{\text{eff}} = \hbar\Delta I_z + D_0 + \hbar\omega_1 I_y, \tag{9.30}$$

where Δ is the frequency distance from resonance $\omega_0 - \omega$ and ω_1/γ is the magnitude of the rotating field. The Zeeman energy and the dipole energy are given by

$$E_z = -\alpha\Delta^2\hbar^2\text{Tr}\{I_z^2\}/\text{Tr}1,$$
$$E_d = -\beta D^2\hbar^2\text{Tr}\{I_z^2\}/\text{Tr}1,$$

where

$$D^2 = \text{Tr}\{D_0^2\}/\text{Tr}\{I_z^2\}$$

and $\alpha = \beta_Z \omega_0 / \Delta$, the inverse Zeeman spin temperature in the rotating frame. Then conservation of energy means that $E_Z + E_d$ is constant, or

$$\Delta^2 \frac{d\alpha}{dt} + D^2 \frac{d\beta}{dt} = 0. \tag{9.31}$$

This shows how the variations in the Zeeman and the dipole temperatures are related. First we shall find the time derivative of α. This is approached through

$$\frac{d\alpha}{dt} = -\frac{1}{\Delta \text{Tr}\{I_z^2\}} \frac{d}{dt} \langle I_z \rangle$$

and the derivative of $\langle I_z \rangle$ is

$$\frac{d}{dt} \langle I_z \rangle = \omega_1 \langle I_x \rangle,$$

since only the I_y term in the Hamiltonian has any influence on the change in I_z. We have $\langle I_x \rangle$ in the real part of Equation (9.29), so we finally obtain

$$\frac{d\alpha}{dt} = -\omega_1^2 \pi (\alpha - \beta) f(\Delta) \tag{9.32}$$

and using Equation (9.31) we find the corresponding result for β:

$$\frac{d\beta}{dt} = -\omega_1^2 \pi \frac{\Delta^2}{D^2} (\beta - \alpha) f(\Delta). \tag{9.33}$$

Equations (9.32) and (9.33) are known as the Provotorov equations. They show how energy flow between the Zeeman and the dipole baths may be facilitated by the application of a transverse rotating magnetic field.

9.4 In the rotating frame

9.4.1 The basic ideas

These subsections concern the case in which a rotating or oscillating transverse magnetic field is applied to a system for times much longer than those associated with 90° and 180° pulses. There is the theoretical question associated with this – under these circumstances, what sort of state will the system settle down to? There is also the essentially practical question of what use it might all be. The Provotorov theory gave one answer to these questions; now we consider some others.

Both the questions are answered by viewing such a system from the point of view of the reference frame rotating with the applied field. Such an observer will see a stationary transverse field of magnitude B_1 together with a longitudinal field which will be $B_0 - \omega/\gamma$; the B_0 field is reduced by the effect of the rotation (Section 2.4.3). Thus in this frame a spin simply sees a stationary field. If the rotating field is precisely on-resonance – if the rotation frequency ω is equal to the Larmor frequency – then the longitudinal part of the field will be zero and the spin, in the rotating frame, will experience a stationary transverse magnetic field of magnitude B_1.

In Section 9.3.9 we mentioned the concept of spin temperature in the rotating frame. In truth we used this idea without too much consideration. Let us now look at this in some more detail. Even though the system is being subjected to a time varying magnetic field, so its Hamiltonian is *not* constant, nevertheless in the rotating frame its Hamiltonian *is* time independent. The Zeeman part (interaction with B_0) is zero. There will be an interaction with the transverse field of $-\hbar\gamma I_y B_0$. Also there will be the dipolar interaction, if that is important. Now the adiabatic part D_0, which is the dominant part, will be constant in time as it is invariant under a rotation about the z axis: D_0 commutes with I_z. The non-adiabatic terms will be very small. These will vary with time, at ω_0 and $2\omega_0$, but they can be neglected. Thus the system will experience an effectively constant Hamiltonian in the rotating frame. By analogy with Equation (9.30) the Hamiltonian in the rotating frame will be

$$\mathcal{H}_{\text{eff}} = D_0 + \hbar\omega_1 I_y.$$

In Section 9.3.9 we simply stated that thermal equilibrium would be established in the frame in which the effective Hamiltonian is time independent. This is really an extension of the spin temperature hypothesis. In solids, where the equilibrium states considered are of the 'internal' equilibrium kind, we know that sometimes the spin temperature hypothesis can let us down; the magic echo is an example of this. However, if we consider fluids where there is the strong Hamiltonian of the lattice to be considered, then the equilibrium is of the 'external' kind and the hypothesis of a spin temperature in the rotating frame is on very strong grounds indeed. The following section will examine the rate at which this equilibrium is approached.

9.4.2 Relaxation in the rotating frame

Consider an assembly of spins in an external magnetic field B_0. These are assumed to be in thermal equilibrium with a magnetisation M_0. We apply a 90° pulse to this system to create some transverse magnetisation and we shall study its behaviour as it precesses. Thus far we have described a conventional FID measurement, in which we know that the magnetisation will decay to zero with a time constant T_2. However, this experiment will be different. Following the 90° pulse we shall apply a transverse magnetic field of magnitude B_1 pointing along the magnetisation – i.e. the magnetic field will be rotating at the Larmor frequency. In the rotating frame there will be a stationary transverse magnetisation and a stationary transverse magnetic field B_1. According to the discussion above about the establishment of spin temperature in the rotating frame, we will expect the transverse magnetisation to relax, not to zero, but to the value appropriate to the transverse field.

In the rotating frame we are saying that there is a stationary magnetic field B_1 and the magnetisation, parallel to the field, changes its value to achieve equilibrium. This sounds just like a T_1 process, even though here it is occurring in the transverse plane. Let us see what the time constant will be for such relaxation.

The magnetic field applied to the system is

$$\mathbf{B}(t) = B_0\mathbf{k} + B_1[\cos(\omega_0 t)\mathbf{i} + \sin(\omega_0 t)\mathbf{j}],$$

so that the Hamiltonian is given by

$$\mathcal{H} = \hbar\omega_0 I_z + \hbar\omega_1[I_x\cos(\omega_0 t) + I_y\sin(\omega_0 t)] + \mathcal{H}_d + \mathcal{H}_m, \tag{9.34}$$

where $\mathcal{H}_d$ is the dipolar Hamiltonian and $\mathcal{H}_m$ is the motion Hamiltonian. It is the total Hamiltonian $\mathcal{H}$ which generates the time evolution of the system in the presence of the rotating magnetic field.

To find an expression for the transverse magnetisation we can use the procedures developed in Chapter 5. However, here we shall do the same thing, but using the language of the density operator. In the high temperature/low polarisation case the important (Zeeman) part of the density operator is proportional to I_z. The 90° pulse rotates this to I_x. So this is the important part of the density operator at the start of the relaxation process. We want to study the evolution of $I_x(t)$, where the time dependence is given by the Heisenberg equation. Our recipe of Section 9.1.1 is clear: multiply $I_x(t)$ by the density operator and take the trace. Thus we see that the evolution of the expectation value of I_x is proportional to $\mathrm{Tr}\{I_x(t)I_x(0)\}$, as in Chapter 5, the only difference here being in the Hamiltonian we shall use for the time dependence.

In this case it is not sensible to use the complex form of the relaxation expression. The time evolution of $I_x(t)$ in the rotating frame (where the transverse magnetisation continues to point in the same direction) is generated by the rotating frame version of Equation (9.34). However, as with our treatment of fluids in Chapter 7, it is convenient to transform away both the Zeeman and the motion Hamiltonians. In other words we have an interaction Hamiltonian

$$\mathcal{H} = \hbar\omega_1 I_x + \mathcal{H}_d(t)$$

and the time variation of $\mathcal{H}_d(t)$ is caused by both the Zeeman and the motion Hamiltonians. From Equations (5.26) and (7.29) this is

$$\mathcal{H}_d(t) = \sum_m \frac{\mu_0}{4\pi}\left(\frac{4\pi}{5}\right)^{\frac{1}{2}} \hbar^2\gamma^2 \sum_{i\neq j}(-1)^m \frac{Y_2^{-m}[\Omega_{ij}(t)]}{r_{ij}^3(t)} T_{ij}^m\exp(im\omega_0 t),$$

but in this case there is the transverse field to be considered. This is most easily done by further transforming to a frame rotating around the x axis with frequency ω_1. This will give an extra time dependence to the interaction representation dipole Hamiltonian.

We shall effect this extra rotation in the following way. What is required is a rotation about the x axis through an angle $\omega_1 t$. However, rotations about the z axis are easier to treat. So we use a trick, by rotating by $\pi/2$ about the y axis, then performing the rotation about the z axis through the angle $\omega_1 t$, and finally tipping back to the transverse plane by rotating by $-\pi/2$ about the y axis. We have encountered rotations about the y axis in Section 9.3.8, and Appendix F. Here the T^m operators have the extra time dependence

$$T_{ij}^m(t) = \sum_{m_1,v=-2}^{2} T_{ij}^{m_1} d_{m,v} \tilde{d}_{v,m_1} \exp(iv\omega_1 t),$$

where d is for a rotation of $\pi/2$ about the y axis and $\tilde{d}$ is for the inverse rotation. Consequently the interaction Hamiltonian is given by

$$\mathcal{H}_d(t) = \sum_{m,m_1,v} \frac{\mu_0}{4\pi}\left(\frac{4\pi}{5}\right)^{\frac{1}{2}} \hbar^2\gamma^2 \sum_{i\neq j}(-1)^m \frac{Y_2^{-m}[\Omega_{ij}(t)]}{r_{ij}^3(t)} T_{ij}^{m_1} d_{m,v}\tilde{d}_{v,m_1}\exp[i(m\omega_0 + v\omega_1)t].$$

The cumulant expansion procedure used in Chapter 7 gave an expression for the decay of transverse magnetisation. In terms of the xx correlation function this is

$$F(t) = \exp\left(-\int_0^t (t-\tau)\frac{\mathrm{Tr}\{[I_x,\mathcal{H}_d(\tau)][\mathcal{H}_d(0),I_x]\}}{\hbar^2\mathrm{Tr}\{I_x^2\}}\,d\tau\right).$$

This means that for times longer than the 'correlation time' for the system the relaxation will be exponential with a time constant given by

$$\frac{1}{T} = \sum_0^\infty \frac{\mathrm{Tr}\{[I_x,\mathcal{H}_d(t)][\mathcal{H}_d(0),I_x]\}}{\hbar^2\mathrm{Tr}\{I_x^2\}}\,dt.$$

We can make some very general inferences at this point without detailed calculations. The different parts of the interaction Hamiltonian have the oscillatory time dependence given by $\exp[i(m\omega_0 + v\omega_1)t]$ where m and v range from -2 to $+2$. The integral for T^{-1} then gives Fourier transforms of the correlation functions, resulting in spectral density functions at the frequencies $m\omega_0 + v\omega_1$. We know that in the absence of the transverse field the two d matrices multiply to give the unit matrix and the result reduces to that for conventional transverse relaxation:

$$\frac{1}{T_2} = \frac{3}{2}J_0(0) + \frac{5}{2}J_1(\omega_0) + J_2(2\omega_0).$$

We are particularly interested in the case in which the transverse field B_1 is much smaller than the static field B_0. In that case the effect of adding ω_1 to ω_0 is negligible so that the only place the ω_1 terms can have any importance is in the 'adiabatic' part $J_0(0)$.

A detailed calculation for non-zero ω_1 will then give the $J_0(0)$, $J_0(\omega_1)$ and $J_0(2\omega_1)$ contributions. It can be seen directly from the $d_{m,v}$ elements of the matrix

$$
\mathbf{d} = \begin{pmatrix}
\frac{1}{4} & \frac{1}{2} & \sqrt{\frac{3}{8}} & \frac{1}{2} & \frac{1}{4} \\
-\frac{1}{2} & -\frac{1}{2} & 0 & \frac{1}{2} & \frac{1}{2} \\
\sqrt{\frac{3}{8}} & 0 & -\frac{1}{2} & 0 & \sqrt{\frac{3}{8}} \\
-\frac{1}{2} & \frac{1}{2} & 0 & -\frac{1}{2} & \frac{1}{2} \\
\frac{1}{4} & -\frac{1}{2} & \sqrt{\frac{3}{8}} & -\frac{1}{2} & \frac{1}{4}
\end{pmatrix}
$$

that there is no single frequency term since when $m = 0$ then v must be 0 or ± 2. When the numerical coefficients are found, from taking the commutators, it is found that the $v = 0$ term is also zero. Thus only the $v = 2$ term remains so that the relaxation time *has* to have the form

$$
\frac{1}{T_{1\rho}} = \frac{3}{2} J_0(2\omega_1) + \frac{5}{2} J_1(\omega_0) + J_2(2\omega_0), \tag{9.35}
$$

so as to reduce to the conventional T_2 expression in the limit $\omega_1 \to 0$. We have used the conventional designation $T_{1\rho}$ for this relaxation time, since it is describing *longitudinal* relaxation in the *rotating* frame.

Equation (9.35) indicates the utility of making relaxation measurements in the rotating frame – i.e. in the presence of a transverse rotating field. Assuming that the motion is slower than the Larmor period, so that the ω_0 and the $2\omega_0$ terms can be ignored,

$$
\frac{1}{T_{1\rho}} = \frac{3}{2} J_0(2\omega_1),
$$

indicating that the relaxation rate is measuring the spectral density of the local field fluctuations at frequency $2\omega_1$, but this is just like a T_1 measurement at frequency $2\omega_1$.

The point here is that the experiment is being done at a Larmor frequency of frequency ω_0; it is that (higher) frequency which determines the signal sensitivity etc., but the spectral density is being measured at the low frequency of frequency $2\omega_1$. So there are considerable signal-to-noise advantages. However, there is more. In a conventional measurement scheme, the frequency is varied by changing the B_0 field and retuning the spectrometer – if it is capable of working at different frequencies. This is all a complicated process. In a $T_{1\rho}$ measurement it is only required to change the strength of the transverse field, which is trivial by comparison.

When one considers the hydrodynamic information which is available from low frequency T_1 data as a function of frequency it will be appreciated what a powerful probe $T_{1\rho}$ measurements can provide. Burnett and Harmon (1972) have made such measurements on glycerol, which extend the data of Figure 8.4 down to 'zero frequency'.

9.4.3 Spin locking

If a rotating field B_1 is applied on resonance and if there is some precessing transverse magnetisation M in the same direction of the field, then there will be an energy $-MB_1$ associated with this. It is a Zeeman energy in the rotating frame. Furthermore this energy must change for the transverse magnetisation to relax to its equilibrium value. In this case there is an energy change associated with the transverse relaxation, unlike the case in the absence of a rotating field. Energy must be exchanged between this Zeeman bath and the lattice – hence the designation of $T_{1\rho}$, for T_1 in the rotating frame.

In the absence of motion, when there is no lattice Hamiltonian $\mathscr{H}_m$, in the presence of a resonant transverse field there is nowhere for the Zeeman energy to go and the transverse magnetisation cannot relax. This is just like the previous conventional analysis of the longitudinal relaxation in solids. In reality there will be other processes to effect the relaxation but they will be very slow. In general, we conclude that the relaxation of the transverse magnetisation in the presence of a transverse field B_1 is just like a T_1 process in a longitudinal field of that magnitude. This means that at any time while there is precessing (and relaxing) transverse magnetisation its magnitude may be 'frozen' by applying a transverse rotating field along the direction of the magnetisation. This is known as 'spin locking'.

9.4.4 Zero-time resolution (ZTR)

The technique of spin locking may be exploited for the measurement of transverse magnetisation at very short times following an excitation pulse. This is of particular interest in moment studies, since the moments are most directly related to the coefficients of a time expansion of the FID. We mentioned in Chapter 8, in connection with the extensive measurements which have been made on calcium fluoride, that the zero-time resolution (ZTR) method, developed by Lowe, Vollmers and Punkkinen (1973), had been used to provide short-time data.

There are a number of difficulties associated with making measurements at short times. Of particular importance is the fact that following the intense resonant RF pulses, with magnitudes possibly of kilovolts, the NMR preamplifier will be paralysed for a time while it recovers from the enormous overload. In Chapter 3 we discussed various methods which are used to reduce this dead time and, in principle, this can be reduced to as low as 1 μs, but it is frequently the case that signals cannot be reliably seen for some tens of microseconds following a 90° pulse.

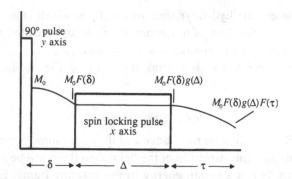

Figure 9.2 Application of a spin locking pulse for the ZTR method.

Now although one might not be able to observe the magnitude of the precession signal at such short times, it should be possible to spin lock the signal and observe it later. The signal must be observed in the absence of a transverse field, but (*a*) since the spin locking field is relatively small, the recovery time might well be shorter, and (*b*) there is no reason why the signal should not be observed at a time significantly after the system has recovered from the spin locking pulse.

In Figure 9.2 we show the pulse sequence for the ZTR method. The initial 90° pulse along the *y* axis tips the equilibrium magnetisation from the *z* axis over to the *x* axis in the rotating frame. This magnetisation is allowed to relax for a time δ after which it will have magnitude $F(\delta)$, this function describing the free decay of transverse magnetisation. The spin locking pulse is then applied for a time Δ, during which there might be some small relaxation $g(\Delta)$. The signal is observed a time τ after the termination of the spin locking pulse. Thus there will be a further attenuation $F(\tau)$ of the magnetisation. The point is that τ is chosen to be long enough so that the receiver has recovered completely, but not so long that the magnetisation has decayed significantly. The times τ and Δ are kept fixed, so these contributions to the signal decay are unchanged. The time interval δ is varied, so that the quantity being measured is $F(\delta)$. In this way the variation of the FID at very short times may be measured. Lowe *et al.* (1973) were able to measure the FID of calcium fluoride for times down to fractions of a microsecond.

9.5 Density operator theory of relaxation

9.5.1 Historical survey

The density operator obeys the Heisenberg equation of motion

$$\frac{\partial \rho}{\partial t} = -\frac{i}{\hbar}[\mathscr{H}, \rho], \tag{9.36}$$

which has solution

$$\rho(t) = \exp\left(-i\frac{\mathcal{H}}{\hbar}t\right)\rho(0)\exp\left(i\frac{\mathcal{H}}{\hbar}t\right),$$

where $\mathcal{H}$ is the Hamiltonian for the system. As we saw in Section 9.1, this describes *reversible* behaviour, and the entropy defined by

$$S = -k\mathrm{Tr}\rho\ln\rho$$

remains a constant.

On the other hand we discussed, in Section 9.2, the fact that real systems *do* behave irreversibly, approaching equilibrium; the entropy of such systems increases.

This paradox was resolved by making a clear distinction between the full density operator and a reduced density operator describing a system. In particular for the sort of systems studied in NMR the reduced density operator would contain only the spin variables, the lattice variables having been averaged over (or the trace was taken). In the case in which there are variables in the Hamiltonian over which the trace has been taken, the von Neumann equation is no longer applicable. In the following section we will examine the sort of equation of motion obeyed by the reduced density operator in this case.

The discussion thus far would appear to have no relevance to solid spin systems, in which there is no motion Hamiltonian to speak of, when T_1 is effectively infinite. Such systems can, however, be accommodated within this framework if we take the trace over spin co-ordinates of all but one of the particles so that we are effectively considering the ensemble average behaviour of a single spin. Since the interaction Hamiltonian (dipolar, for example) involves pairs of spin values, the resultant equation of motion would be irreversible. This was precisely the sort of thing which was seen at the end of Section 9.2.7, where we wrote down the evolution of the spin density operator corresponding to the solvable model considered in Section 6.2. However, most of the following sections will be concerned mainly with motion-induced relaxation.

We shall consider a system described by a Hamiltonian

$$\mathcal{H} = \mathcal{H}_z + \mathcal{H}_d + \mathcal{H}_m,$$

the sum of a Zeeman, dipole and motion parts. Recall that the Zeeman Hamiltonian depends on the spin variables, the motion Hamiltonian depends on the lattice variables and the dipole Hamiltonian depends on both. In accordance with the discussion above, while the full density operator will depend on both spin and lattice variables, the reduced density operator in which we are interested will depend on spin variables only. The time evolution induced by the Zeeman Hamiltonian will thus be of the trivial type, as described by the von Neumann equation. It then makes sense to transform to an interaction picture where this behaviour is removed from the equations of motion leaving the 'interesting' part.

As in Chapter 7, we shall transform to the interaction picture where the direct effects of both the Zeeman and the motion Hamiltonians are removed (*to* the observables *from* the density operator). This leaves only a time dependent dipole Hamiltonian to cause the time dependence of the interaction density operator. Since this Hamiltonian is relatively small, perturbation expansions might be appropriate. The equation of motion for the interaction picture density operator is now

$$\frac{\partial \rho(t)}{\partial t} = -\frac{i}{\hbar}[\mathcal{H}_d(t), \rho(t)], \qquad (9.37)$$

where

$$\mathcal{H}_d(t) = \exp\left[\frac{i}{\hbar}(\mathcal{H}_m + \mathcal{H}_z)t\right] \mathcal{H}_d \exp\left[-\frac{i}{\hbar}(\mathcal{H}_m + \mathcal{H}_z)t\right].$$

For simplicity we shall not make a notational distinction for the density operator in the interaction picture.

The fundamental task in hand is then, starting from Equation (9.37), to find an equation of motion obeyed by the reduced density operator

$$\sigma(t) = \mathrm{Tr}_l \rho(t)$$
$$= \langle \rho(t) \rangle,$$

where the trace or average is taken over the lattice variables. In particular it is to be hoped that at long times $\sigma(t)$ will tend towards the equilibrium form

$$\sigma(t \to \infty) = \sigma_{eq} = \frac{\exp(-\mathcal{H}_z/kT)}{\mathrm{Tr}\{\exp(-\mathcal{H}_z/kT)\}}$$

and the equations should tell us how this state is approached in time.

This general question has had a long history, and a significant boost was given by the interest of the NMR fraternity. The pioneers were Wangsness and Bloch (1953), Bloch (1957) and Redfield (1957). Abragam (1961) gave a good summary of the work to that stage, and finally Hubbard (1961) presented a complete and definitive solution to the problem. Redfield's original paper was somewhat inaccessible and a revised version was subsequently published (Redfield, 1965).

9.5.2 *Master equation for the density operator*

Ideally, the result of such a programme would be an equation of motion for the reduced density operator (in the interaction picture) of the form

$$\frac{\partial \sigma_{\alpha\beta}(t)}{\partial t} = -\sum_{\gamma\delta} R_{\alpha\beta\gamma\delta} [\sigma_{\gamma\delta}(t) - \sigma_{\gamma\delta}^{eq}], \qquad (9.38)$$

at least in the short correlation time limit. Each element of the density operator would relax to its equilibrium value, the decay being a superposition of exponentials. The various works mentioned in the previous section do achieve this to a large measure, within the various approximations invoked.

Equation (9.38) is referred to as the master equation for the density operator. Observe that the relaxation matrix has *four* indices. Thus it cannot be understood as a representation of a conventional Hilbert space operator of quantum mechanics and it is sometimes called a 'superoperator'. There might well be degeneracy among the various elements of R, but ultimately one would expect expressions for the elements in terms of the system properties, and anticipating a little, the expressions will be replaced to correlation functions of the time dependent interaction Hamiltonian.

The conventional derivations of Equation (9.38) have much in common with the relaxation calculations presented in Appendix E in that the 'separation of time scales' is invoked early on so that the relaxing quantity is factored out on the right hand side of the equation of motion. By contrast to this, and within the spirit of the material of the body of this book, we shall use an alternative method based upon a cumulant expansion. As the reader will recall, this has the merit of combining a perturbation expansion in a small quantity (here, again, the dipole Hamiltonian) with some extra input about the nature of the long time behaviour.

9.5.3 Cumulant expansion of the density operator

The formal solution to Equation (9.37) for the time evolution of the interaction picture density operator is given by

$$\rho(t) = \exp\left[-\frac{i}{\hbar}\int_0^t \mathcal{H}_d(\tau)d\tau\right]\rho(0)\exp\left[\frac{i}{\hbar}\int_0^t \mathcal{H}_d(\tau)d\tau\right]$$

in terms of time-ordered exponentials. The equilibrium density operator has no time evolution; when operated on by the time evolution propagator it remains constant. Thus we can subtract ρ_{eq} from both sides of this equation, to give

$$\rho(t) - \rho_{eq} = \exp\left[-\frac{i}{\hbar}\int_0^t \mathcal{H}_d(\tau)d\tau\right][\rho(0) - \rho_{eq}]\exp\left[\frac{i}{\hbar}\int_0^t \mathcal{H}_d(\tau)d\tau\right].$$

We find the corresponding reduced density operator by taking the trace (in the quantum mechanical case) or the average (in the semiclassical) case over the lattice or motion variables. We denote this, for either case, by an angle bracket;

$$\sigma(t) - \sigma_{eq} = \left\langle \exp\left[-\frac{i}{\hbar} \int_0^t \mathcal{H}_d(\tau)d\tau \right] [\rho(0) - \rho_{eq}] \exp\left[\frac{i}{\hbar} \int_0^t \mathcal{H}_d(\tau)d\tau \right] \right\rangle.$$

Here we expect that as time proceeds $\sigma(t) - \sigma_{eq}$ will tend to the null operator. Thus in the spirit of the cumulant expansion we shall introduce an operator $\Psi(t)$ defined by

$$\sigma(t) - \sigma_{eq} = \exp[\Psi(t)]$$

and we expand Ψ in powers of the interaction Hamiltonian $\mathcal{H}_d$ as far as the first non-vanishing term. This is simply the operator equivalent of the procedure described in Sections 7.2.1 and 7.2.2. The result is

$$\sigma(t) - \sigma_{eq} = \exp\left[-\frac{1}{\hbar^2} \int_0^t (t - \tau)\langle[\mathcal{H}_d(0),[\mathcal{H}_d(\tau),(\rho(0) - \rho_{eq})]]\rangle d\tau \right],$$

which is the density operator analogue of Equation (7.27) for the relaxation functions. This equation may be differentiated to give

$$\frac{\partial\sigma(t)}{\partial t} = -\frac{1}{\hbar^2} \int_0^t \langle[\mathcal{H}_d(0),[\mathcal{H}_d(\tau),[\rho(t) - \rho_{eq}]]\rangle d\tau.$$

Although this is a differential equation for the reduced density operator, we have the full density operator on the right hand side. However, we may now invoke the separation of time scales. In other words we are considering the case in which the fluctuations in $\mathcal{H}_d$ occur very rapidly, before σ has had time to change significantly. Then the upper limit of the integral may be extended to infinity and ρ may be replaced by its reduced or averaged counterpart σ so that

$$\frac{\partial\sigma(t)}{\partial t} = -\frac{1}{\hbar^2} \int_0^\infty \langle[\mathcal{H}_d(0),[\mathcal{H}_d(\tau),[\sigma(t) - \sigma_{eq}]]\rangle d\tau, \qquad (9.39)$$

which is indeed an equation solely in σ.

This equation does have the form of the master equation, Equation (9.38), where the relaxation kernel is given by

$$R_{\alpha\beta\gamma\delta} = \frac{1}{\hbar^2}\sum_{\mu}{}'\delta_{\beta\delta}\int_0^\infty \langle \mathcal{H}_{\alpha\mu}(0)\mathcal{H}_{\mu\gamma}(\tau)\rangle d\tau + \frac{1}{\hbar^2}\sum_{\mu}\delta_{\alpha\gamma}\int_0^\infty \langle \mathcal{H}_{\mu\beta}(0)\mathcal{H}_{\delta\mu}(\tau)\rangle d\tau$$

$$-\frac{1}{\hbar^2}\int_0^\infty \langle \mathcal{H}_{\alpha\gamma}(0)\mathcal{H}_{\delta\beta}(\tau)\rangle d\tau - \frac{1}{\hbar^2}\int_0^\infty \langle \mathcal{H}_{\delta\beta}(0)\mathcal{H}_{\alpha\gamma}(\tau)\rangle d\tau,$$

in terms of the matrix elements of the interaction representation Hamiltonian. Since the spin part of the interaction Hamiltonian has oscillatory time variations at frequencies of 0, ω_0 and $2\omega_0$, the integrals in this expression will become Fourier transforms at these frequencies. Thus the elements of the relaxation kernel are expressed, ultimately, in terms of sets of spectral density functions.

9.5.4 An example

We shall finish this section with an example of the use of Equation (9.39) in studying the relaxation of a particular quantity. For simplicity we will consider longitudinal relaxation, from a non-equilibrium initial state. It is convenient to use the language of spin temperature here. We write the spin density operator in the high temperature approximation as

$$\sigma = (1 + \beta\hbar\omega_0 I_z)/\mathrm{Tr}1, \tag{9.40}$$

where the inverse Zeeman spin temperature β describes the system as I_z varies; in the high temperature limit β is proportional to the z component of magnetisation since

$$\langle I_z \rangle = \mathrm{Tr}\{\sigma I_z\} = \beta\hbar\omega_0 \mathrm{Tr}\{I_z^2\}/\mathrm{Tr}1.$$

We are interested in studying how β varies with time. We use the proposed expression for σ in Equation (9.40) which is substituted into Equation (9.39) so that this density operator varies in time as

$$\frac{\partial\sigma(t)}{\partial t} = -\frac{(\beta - \beta_{eq})\omega_0}{\hbar}\int_0^\infty \langle[\mathcal{H}_d(0),[\mathcal{H}_d(\tau),I_z]]\rangle d\tau.$$

Now since β is given by

$$\beta = \frac{\mathrm{Tr}\{\sigma I_z\}\mathrm{Tr}1}{\hbar\omega_0\mathrm{Tr}\{I_z^2\}}$$

its time derivative is

$$\frac{\partial \beta}{\partial t} = \frac{\mathrm{Tr}\left\{\frac{\partial \sigma}{\partial t} I_z\right\} \mathrm{Tr} 1}{\hbar \omega_0 \mathrm{Tr}\{I_z^2\}} .$$

So multiplying the equation for $\partial \sigma / \partial t$ by I_z and taking the trace, we obtain

$$\frac{\partial \beta}{\partial t} = -\frac{(\beta - \beta_{eq})\omega_0}{\hbar^2} \int_0^\infty \frac{\mathrm{Tr}\{I_z[\mathcal{H}_d(0),[\mathcal{H}_d(\tau),I_z]]\}}{\mathrm{Tr}\{I_z^2\}} d\tau,$$

where the trace includes either the average or the trace over lattice variables. This equation indicates an exponential relaxation for the magnetisation, and by a cyclic permutation within the trace we see that the relaxation time is precisely the T_1 as calculated in Chapter 7, the integral of the correlation function of Equation (7.37).

It should be apparent from this discussion that the use of Equation (9.39) for the evolution of the reduced density operator allows the calculation of the relaxation of any quantity from any initial state. The initial state is accommodated in the form chosen for the expression for the reduced density operator σ, as in our Equation (9.40). The operator representing the quantity required is multiplied into the equation of motion for σ and the trace is then taken. This provides a general procedure for the study of relaxation, liberated from the possibly artificial constraints of the methodology of Chapter 5.

10

NMR imaging

10.1 Basic principles

10.1.1 Spatial encoding

Since the NMR resonance frequency of a spin is proportional to the magnetic field it experiences, it follows that in a spatially varying field spins at different positions will resonate at different frequencies. The spectrum from such a system will give an indication of the number of spins experiencing the different fields.

In a uniform magnetic field gradient the precession frequency is directly proportional to displacement in the direction of the gradient; there is a direct linear mapping from the spatial co-ordinates to frequency. Thus the absorption spectrum yields the number of spins in 'slices' perpendicular to the gradient. In Figure 10.1 we show how the spectrum would be built up from such slices.

We have already encountered the concept of spatial encoding of spins in Section 4.5 where we considered diffusion and the way it can be measured using spin echoes. There the important point was that whereas spins in a field gradient with their corresponding spread of precession frequencies suffer a decay of transverse magnetisation, this can be recovered, to a large extent, by the time-reversing effect of a 180° pulse. However, if the particles are defusing then, because of the field gradient, their motion will take them to regions of differing precession frequencies. The resultant additional dephasing cannot be recovered by a 180° pulse, which thus permits the diffusion coefficient to be measured.

In imaging one is concerned with the main dephasing effect of the gradient field. Compared with this the diffusive effects are small, and in our initial treatment we shall assume that the resonating spins are immobile. However, in Section 10.5 we will see how NMR imaging methods can be used to map the spatial variation of the diffusion coefficient as well as the NMR parameters T_1 and T_2. Some good books on NMR imaging, mainly from the medical perspective, include Mansfield and Morris (1982), Morris (1986), Stark and Bradley (1992) and Wehrli (1991).

335

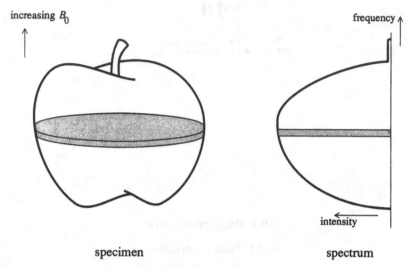

increasing B_0

frequency

intensity

specimen   spectrum

Figure 10.1 Formation of a one-dimensional NMR image.

10.1.2 Field gradients

Central to NMR imaging methods is the application of magnetic field gradients. Now in general the magnetic field gradient is a tensor quantity, having nine elements:

$$\mathbf{G} = \begin{pmatrix} \partial B_x/\partial x & \partial B_x/\partial y & \partial B_x/\partial z \\ \partial B_y/\partial x & \partial B_y/\partial y & \partial B_y/\partial z \\ \partial B_z/\partial x & \partial B_z/\partial y & \partial B_z/\partial z \end{pmatrix}. \tag{10.1}$$

The values of these are not independent, as the magnetic field is constrained to obey Maxwell's equations. The requirement that $\text{div}\mathbf{B} = 0$ implies that the trace of $\mathbf{G}$ must be zero while the static condition $\text{curl}\mathbf{B} = 0$ means that the off-diagonal part of $\mathbf{G}$ is antisymmetric. Thus there are two diagonal and three off-diagonal independent elements; in all, the gradient tensor has five independent elements.

In the presence of the static magnetic field B_0, assumed to be in the z direction, spins will thus see a magnetic field whose x component b_x is made up from the top row of $\mathbf{G}$; whose y component b_y is made up from the middle row of $\mathbf{G}$; and whose z component comprises B_0 together with contribution b_z from the bottom row of $\mathbf{G}$. However, only fields parallel to $\mathbf{B}_0$ have any significant effect. This is none other than the adiabatic approximation which we have seen in earlier chapters where the interaction Hamiltonian was truncated, eliminating those parts which did not conserve m_z. In the present context this may be seen in the following way.

Since the gradient fields are assumed to be very much smaller that the static field the *direction* of the resultant field will be changed negligibly from the z direction. The magnitude of the field is given by

$$B = [(B_0 + b_z)^2 + b_x^2 + b_y^2]^{\frac{1}{2}}$$

$$= (B_0^2 + 2B_0 b_z + b_z^2 + b_x^2 + b_y^2)^{\frac{1}{2}}$$

$$= B_0 \left(1 + \frac{2b_x}{B_0} + \frac{b_z^2 + b_x^2 + b_y^2}{B_0^2}\right)^{\frac{1}{2}}.$$

When the square root is expanded we then find

$$B = B_0 + b_z$$

plus terms of *higher order* in b/B_0, which may therefore be ignored. Thus we neglect fields in the transverse directions, which simplifies considerably the analysis.

In the spirit of this approximation we are only considering the bottom line of **G** which means that the gradient can be regarded as a vector quantity **G**:

$$\mathbf{G} = \frac{\partial B_z}{\partial x}\mathbf{i} + \frac{\partial B_z}{\partial y}\mathbf{j} + \frac{\partial B_z}{\partial z}\mathbf{k}, \tag{10.2}$$

so that the (z component of the) magnetic field at the point **r** can be written as

$$B(\mathbf{r}) = B_0 + \mathbf{G} \cdot \mathbf{r}. \tag{10.3}$$

10.1.3 Viewing shadows

As usual, let us consider a static magnetic field applied in the z direction. Superimposed upon this we will take a magnetic field gradient G in the z direction; this restriction can be relaxed with no difficulty. The z component of the magnetic field at an arbitrary point **r** with coordinates x, y, z is given by

$$B_z(\mathbf{r}) = B_0 + Gz \tag{10.4}$$

so that a spin at this point will precess with an angular frequency

$$\omega = \omega_0 + \gamma Gz. \tag{10.5}$$

If $\rho(\mathbf{r})$ measures the density of spins at the position **r** then the number of spins in the slice at z is given by the integral over x and y multiplied by the slice thickness Δz:

$$\int \rho(x,y,z)dxdy \times \Delta z, \tag{10.6}$$

so that the corresponding lineshape can be expressed as

$$I(\omega - \omega_0) \propto \int \rho \left(x, y, \frac{\omega - \omega_0}{\gamma G} \right) dx dy. \qquad (10.7)$$

There would be similar expressions for the case in which the field gradient is oriented along the x and the y directions.

As it stands, this method enables the observation of a one-dimensional shadowgraph of the specimen. The technique is useful in simple applications, for instance, in determining whether a crystal is growing from the top or the bottom of a sample cell. However, it is fundamentally one-dimensional, giving no information about variations in directions perpendicular to the field gradient. An example of the use of a single gradient in this way is shown in Figure 10.21, where a profile of T_2 has been obtained.

The question naturally arises as to whether it is possible to create a better image than a simple shadowgraph, preferably in two or three dimensions. In other words, can one reconstruct an object from its shadow projections?

10.1.4 Discussion

We shall see in the following sections that there is a whole range of methods for reconstructing images, using NMR, by capturing and processing signals produced in the presence of magnetic field gradients. Important considerations are the efficiency of the process, the time needed to reconstruct an object, and the spatial resolution which can be achieved. Also, discussion so far has been concerned mainly with mapping the density of the spins. It is possible, as we have mentioned, to create images of other NMR-related quantities such as the relaxation times T_1 and T_2 as well as the diffusion coefficient D. These matters will be taken up in Section 10.5.

The method by which a number of projections can be used to create an image of an object is applicable not only to NMR. It was in the early 1970s that Hounsfield developed the idea of imaging using X-rays for medical applications, known as computed tomography (CT). At a similar time both Lauterbur (1973) and Mansfield and Grannell (1973) independently proposed using NMR as a method for imaging. The X-ray methods became a rapid success, but it has taken longer for magnetic resonance imaging (MRI) to become an established medical diagnostic tool. The reason for this was mainly the serious technological problems which had to be overcome with MRI systems, but the potential benefits are enormous.

X-rays are, of course, an ionising radiation. Exposure to X-rays is therefore dangerous and doses must be kept low, a requirement somewhat at variance with the use of multiple exposures needed for image reconstruction. A further serious disadvantage in using X-rays is the limited contrast available in soft tissue. X-ray attenuation is determined by electron density and is essentially proportional to atomic number. This provides a crude and relatively insensitive grey scale, although it is possible to introduce radio-opaque contrast enhancing agents. On the other hand MRI has the ability firstly to determine the density of the chosen nuclei, and secondly to image the spin relaxation times of the nuclei. An early triumph for MRI was in the study of soft tissue, in particular the white matter of the brain. It is now possible to predict with reasonable certainty the development of diseases such as multiple sclerosis before any physical symptoms are displayed.

10.1.5 Spatial resolution

With conventional imaging methods, whether simply viewing by eye or reconstructing from X-rays or ultrasound, one irradiates the object and then studies the scattered or transmitted radiation. The spatial resolution is limited, ultimately, by the wavelength of the radiation; thus electron microscopes are more powerful than their optical counterparts. Although X-rays have a wavelength of a few angstroms, in practice the spatial resolution achievable using them is limited by geometrical considerations (focal spot size) to about a millimetre. With magnetic resonance imaging the wavelength limitation does not apply. This is just as well, since the wavelengths concerned are of the order of metres!

In a conventional scattering experiment the individual scattering centres can be regarded as point emitters and spatial information comes from interference of waves from different centres. Thus the wavelength is fundamental in determining the distance scale involved. In the NMR case, while it is true that the individual nuclei are emitters of RF radiation, the spatial information is dispersed into the frequency domain by magnetic field gradients. It is no longer the wavelength of the radiation which determines the spatial resolution; now it is the magnitude of the gradient, determining the frequency variation which ultimately must be compared with the natural linewidths of the resonance. It is therefore possible to increase the spatial resolution by using larger field gradients and by capturing data over a longer time range. Of course this requires greater signal sensitivity and more computing power/time.

A particular goal at which MRI workers have been aiming is the enhancement of spatial resolution to the state where it may be used as a form of non-invasive microscopy. This requires the application of large amplitude field gradients, with the associated coils and power amplifiers. Current spatial resolution achievable is in the region of $5\,\mu m$.

10.2 Imaging methods

10.2.1 Classification of methods

It is possible to classify the various imaging techniques according to the region of the specimen from which information is captured simultaneously. Thus in the simplest case one would collect data relating only to an individual point – the 'sensitive point'. Then

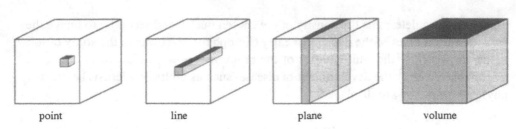

| point | line | plane | volume |

Figure 10.2 Regions of simultaneous data capture.

the image would be built up by moving the sensitive point around the specimen. Alternatively, in increasing order of complexity, one could collect data from a line, a plane, or the entire volume. These possibilities are indicated in Figure 10.2.

While the point method is the slowest to cover an entire object, it does have the advantage of requiring minimal processing to reconstruct the image. In general, increasing the dimensionality of the acquisition region makes greater demands on the NMR system and the data processing software, but with the advantage of faster image reconstruction.

10.2.2 Sensitive point and sensitive line methods

The methods described in this subsection are not used in modern medical MRI systems. However, as well as being of historical interest they have the advantage that the minimum of instrumentation is required; in particular, images can be reconstructed without the need for calculation-intensive data processing. So while slow, this might well be the chosen method for a preliminary study in a physics laboratory.

The methods make use of *oscillating* field gradients. An oscillating gradient will define a null plane where the magnetic field does not vary, as indicated in Figure 10.3. Away from the null plane the field will alternate so that the precession signal from this region will be destroyed, with any remaining artifacts modulated at the oscillation frequency. This can be removed with a filter, following which one sees only spins in the neighbourhood of the null plane.

The sensitive point method must be the simplest imaging technique to implement. Application of three perpendicular oscillating gradients defines a null point at the intersection of the three null planes so that after filtering only spins in the neighbourhood of this point will contribute to the precession signal. One thus maps the spin density as the sensitive point is moved around the specimen.

In the sensitive point method each captured FID contains information about a single point; this is a point method. The sensitive line method is a modification which permits the capture of an entire line in each free induction decay; it is thus a line method.

The use of only two orthogonal oscillating field gradients defines a null line. A static field gradient is applied in the third direction which encodes distance along the line to

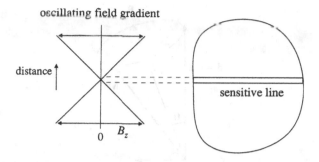

Figure 10.3 Sensitive line produced by oscillating field gradient.

frequency. Each FID contains information about the spins along the null line. Image reconstruction thus requires a one-dimensional Fourier transform of the FIDs, giving the spin density along the line. The spin density is mapped as the sensitive line is moved (remaining parallel to its original orientation) in the specimen.

In this method, assuming that the static field gradient is present when the 90° pulse is applied, a severe demand is placed on the pulse since it must tip *all* spins, resonating over a range of frequencies. In order to achieve this there must be adequate spectral power in the pulse over the frequency range; thus the pulse must be short and therefore of large amplitude. If that were not the case then only those spins in a given slice would be rotated. However, this very effect can be exploited as an aid to imaging.

10.2.3 Shaped pulses – selecting a slice

When a 90° pulse (or other tipping pulse) is applied in the presence of a magnetic field gradient, only those spins on resonance will be rotated by the required amount. In the frame rotating at the frequency of the RF pulse this will be solely the resonant spins: those for which the B_0 field has entirely vanished. These spins see just the B_1 field in the transverse plane, about which they precess. Thus in a field gradient only spins in a given plane will experience the full effect of the pulse. Other spins will also experience some field in the z direction and they precess around the resultant effective field.

Consider a spin a distance z along the gradient direction from the resonant reference position; this will precess around an effective field B_{eff} as indicated in Figure 10.4. The magnitude of B_{eff} is given by

$$B_{eff} = (B_1^2 + G^2 z^2)^{\frac{1}{2}}$$

and it points along a direction making an angle θ with the z axis, where

$$\sin\theta = \frac{B_1}{(B_1^2 + G^2 z^2)^{\frac{1}{2}}}$$

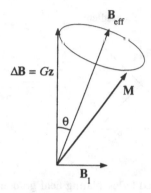

Figure 10.4 Nutation of an off-resonance spin.

$$= \frac{1}{(1 + G^2 z^2/B_1^2)^{\frac{1}{2}}}. \tag{10.8}$$

Off-resonance spins will rotate around the cone, finally making an angle between 0 and 2θ with the z axis. The exact calculation is a little involved, but we may assume that on average the tip angle is θ; in reality there will be oscillations superimposed on this. In that case Equation (10.8) gives the (approximate) angle the spins at position z are tipped.

The NMR signal is proportional to the sine of the tip angle, as given by Equation (10.8). This is indicated in Figure 10.5. The width of the sensitive region, which we may denote by Δz, is seen to scale with B_1. This supports the spectral power/uncertainty principle statement at the end of the previous section. There we argued that a large B_1 field allows the 90° pulse to act on a larger spatial region. Now we see that Δz is directly proportional to B_1.

Except in the limit of small tip angles, the spin system is not linear – for instance, doubling the magnitude of B_1 will not double the magnitude of the transverse magnetisation. It then follows that Fourier transform arguments about spectral power etc. are not rigorously applicable; they can only be used as a guide. It is the non-linear time domain analysis of the system which must be relied upon.

Inspection of Figure 10.5 indicates that selected region Δz is not all that sharply differentiated; the wings of the curve seem quite broad. They fall off as $1/z$, corresponding to the envelope of the sinc function which is the spectrum of a rectangular pulse. This behaviour may be improved substantially by 'shaping' the pulse whereby the B_1 field is turned on and off gradually rather than instantaneously.

When B_1 is turned on and off gradually the effective field B_{eff} gradually tilts away from the z axis and then it gradually returns. Since the spins are precessing about the *instantaneous* B_{eff} this means that the off-resonant spins – particularly those far from

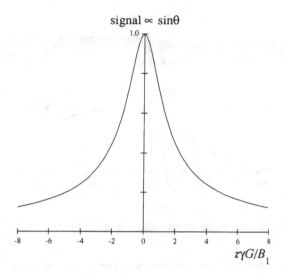

Figure 10.5 Approximate signal contributions from off-resonance spins.

resonance – will be returned very close to the z axis; thus they will contribute negligible transverse signal, thus a Gaussian envelope might be a good choice; the spatial sensitivity will then fall off as the Gaussian, $\exp[-(Gz/B_1)^2]$, which decays much more rapidly than Equation (10.8) for the rectangular pulse.

In the approximate framework of the Fourier description, which becomes exact in the limit of an excitation pulse of small tip angle, a sinc pulse (in the time domain) is attractive since its Fourier transform is a box function (in the frequency domain); all spins in a well-defined slice would be excited equally. In the non-linear case of a 90° pulse the non-linear spin equations must be solved to find the optimum pulse profile for localised slice selection. However, even in this case, the sinc profile is a good first attempt.

In Figure 10.6 we show the evolution of the transverse magnetisation during the application of a 90° sinc pulse $\sin(at)/t$ in the presence of gradient G. We see that at a distance $a/\gamma G$ from the resonant plane the transverse magnetisation has fallen to some 0.4 of its maximum value. However, the full effect of the sinc pulse is better appreciated from Figure 10.7. This shows the magnitude of the transverse magnetisation signal from planes at different distances along the direction of the field gradient. Observe that the sensitivity is quite uniform for most of the slice, falling off very rapidly at distance $z = a/\gamma G$.

We see from the above discussion how a suitably chosen RF pulse may excite only those spins in a narrow slice. This method of selecting a plane at the start of a procedure is used in many imaging methods.

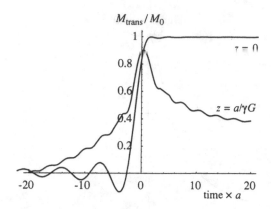

Figure 10.6 Effect of a 90° sinc pulse on two slices in a gradient.

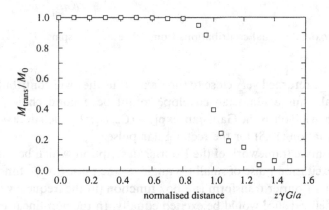

Figure 10.7 Effect of a 90° sinc pulse as a function of distance.

10.2.4 Back projection

The method of reconstructing an object by back projection has its origins in tomographic X-ray imaging. It is a highly intuitive method, requiring relatively little data processing. It can be implemented as a plane method if a tailored 90° pulse is used to select a slice, or it can be implemented as a volume method, although then the data storage requirements can be considerable.

In Figure 10.8 we show two objects, the black discs, together with two projection images which would be absorption spectra in the NMR case. In the back projection method these spectra are projected back, giving the shaded strips. The darker regions, where strips intersect, are likely to be where the original objects were placed. We see that in the case of just two projections there is an ambiguity in interpretation since as

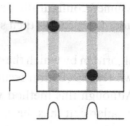

Figure 10.8 Reconstruction by back projection: from two projections.

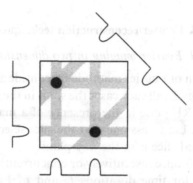

Figure 10.9 Reconstruction by back projection: from three projections.

well as the real objects there are two 'ghosts': intersections which do not correspond to real objects. The inclusion of a third projection improves matters, as we see in Figure 10.9. Now the two discs are located unambiguously as the only intersections of three strips. In general it requires at least n different projections to create an image of $n \times n$ pixels.

Unfortunately this very simple method of reconstructing objects from their projections is prone to distortion since regions of high density tend to produce star shaped areas. However, it is possible to correct for this by suitably filtering the spectra before back projecting them. Filtered back projection is particularly useful in the imaging of solids where T_2 is so short that there is no time for phase encoding, used in the Fourier reconstruction techniques treated in Section 10.3. However, for fluid and 'soggy' systems Fourier techniques are the most convenient.

10.2.5 Iterative reconstruction

This technique is frequently used in X-ray imaging systems. The idea is that one *assumes* a given object configuration. It is relatively simple to calculate the projections that would follow from the assumed configuration. The actual projection data are compared with those from the assumed configuration and according to the difference between them a correction is made. The process then starts again with the improved assumed configuration.

Of key importance here is the algorithm by which the corrections are calculated. This must be simpler than a full-blown reconstruction technique, but at the same time it should be reliably convergent. Although this method was used in the past, with the increase in computing power available it has been mainly superseded by more sophisticated techniques. An important advantage of this reconstruction method is that corrections, such as those for RF penetration, can be applied in a very simple way.

10.3 Fourier reconstruction techniques

10.3.1 Fourier imaging in two dimensions

We shall start the discussion of Fourier methods by considering the reconstruction of two-dimensional slice images. In all such cases the spins in the selected slice are initially excited by using a shaped RF pulse in the presence of a magnetic field gradient, as described in Section 10.2.3. Let us assume that the slice selection gradient is in the z direction so that the selected slice is in the x–y plane.

Now consider the application, consecutively or concurrently, of gradients G_x and G_y in the x and y directions for time durations t_x and t_y. Following this the phase accumulated, in the rotating frame, by a spin at position $\mathbf{r}$ in the x–y plane will be given by

$$\phi(\mathbf{r}) = \gamma(xG_x t_x + yG_y t_y),\tag{10.9}$$

which may be expressed as the vector dot product

$$\phi(\mathbf{r}) = \mathbf{k}\cdot\mathbf{r},\tag{10.10}$$

where the vector $\mathbf{k}$ is given by

$$\mathbf{k} = \gamma G_x t_x \mathbf{i} + \gamma G_y t_y \mathbf{j}.\tag{10.11}$$

The transverse magnetisation signal from this point will then be proportional to $\exp[i\phi(\mathbf{r})]$.

The density of spins as a function of position in the plane is given by $\rho(\mathbf{r})$ and the transverse magnetisation of the entire slice will be given by integrating the signal over the density function

$$F(t) = \int_{\text{slice}} \rho(\mathbf{r})\exp(i\mathbf{k}\cdot\mathbf{r})d^2r,\tag{10.12}$$

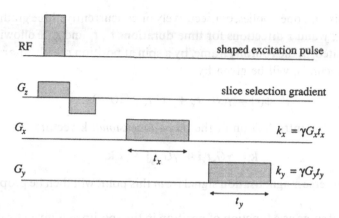

Figure 10.10 Two-dimensional Fourier imaging.

which is seen to be a Fourier transform relation. This indicates that the signal corresponding to a particular G_x, t_x, G_y, t_y gives the spatial Fourier transform of the density function $\rho(\mathbf{r})$ at a single point in k-space. The process of reconstructing the image then reduces to accumulating data over a grid in k-space, by varying combinations of G_x, t_x, G_y, t_y and then performing a discrete two-dimensional (inverse) Fourier transform:

$$\rho(\mathbf{r}) = \frac{1}{(2\pi)^2} \int F(\mathbf{k}) \exp(-i\mathbf{k}\cdot\mathbf{r}) d^2k. \tag{10.13}$$

The basic procedure for effecting this is shown in Figure 10.10. In principle, ignoring the effects of relaxation, the transverse signal may be captured at any time after the application of the gradients. One should note an important feature of the slice selection gradient G_z. Following the RF pulse the slice selection gradient has a negative part in order to reverse the dephasing caused by this gradient during its positive part.

We have seen that the precession signal for a given value of k gives the spatial Fourier transform of $\rho(\mathbf{r})$ at a single point in reciprocal space. In order to reconstruct an image it is therefore necessary to collect data from many points covering the reciprocal space before performing the Fourier transform back to real space. In practice one captures signals from a grid defined by the discrete increments of time and/or the discrete increments of the components of the field gradient. The different imaging methods we describe in the sections below differ only in the manner in which the k-space is covered.

10.3.2 Fourier imaging in three dimensions

For the acquisition of a three-dimensional image the procedure is initiated by a broad band excitation pulse so that transverse magnetisation is created for the entire specimen. In this case one applies, consecutively or concurrently, three gradients G_x, G_y and G_z in the x, y and z directions for time durations t_x, t_y and t_z. Following this the phase accumulated, in the rotating frame, by a spin at position $\mathbf{r}$, with co-ordinates x, y and z in the specimen, will be given by

$$\phi(\mathbf{r}) = \gamma(xG_xt_x + yG_yt_y + zG_zt_z). \tag{10.14}$$

This leads us to the introduction of the *three-dimensional* $\mathbf{k}$ vector

$$\mathbf{k} = \gamma G_xt_x\mathbf{i} + \gamma G_yt_y\mathbf{j} + {}_zt_z\mathbf{k}, \tag{10.15}$$

so that the transverse magnetisation signal from this point will then be proportional to $\exp(i\mathbf{k}\cdot\mathbf{r})$.

The density of spins as a function of position in the specimen is given by $\rho(\mathbf{r})$ and the transverse magnetisation of the entire volume will be given by integrating the signal over the density function

$$F(t) = \int_{\text{volume}} \rho(\mathbf{r})\exp(i\mathbf{k}\cdot\mathbf{r})d^3r, \tag{10.16}$$

a three-dimensional Fourier transform expression. As in the two-dimensional case, this indicates that the signal corresponding to a particular G_xt_x,G_yt_y,G_zt_z gives the spatial Fourier transform of the density function $\rho(\mathbf{r})$ at a single point in k-space. The process of reconstructing the image then reduces to accumulating data over a grid in the three-dimensional k-space, by varying combinations of G_xt_x,G_yt_y,G_zt_z and then performing a discrete three-dimensional (inverse) Fourier transform:

$$\rho(\mathbf{r}) = \frac{1}{(2\pi)^3} \int F(\mathbf{k})\exp(-i\mathbf{k}\cdot\mathbf{r})d^3k. \tag{10.17}$$

The basic procedure for effecting this is shown in Figure 10.11. In principle, ignoring the effects of relaxation, the transverse signal may be captured at any time after the application of the gradients.

As in the two-dimensional case, the precession signal for a given value of $\mathbf{k}$ gives the spatial Fourier transform of $\rho(\mathbf{r})$ at a single point in reciprocal space. In order to reconstruct an image it is therefore necessary to collect data from many points covering the reciprocal space before performing the Fourier transform back to real space. In practice one captures signals from a grid defined by the discrete increments of time and/or the discrete increments of the components of the field gradient. Since we are now considering a three-dimensional space this will involve acquisition of many more data points and a much greater computing requirement, rather than the (relatively) less demanding requirements of a slice-by-slice reconstruction. Very generally, a two-

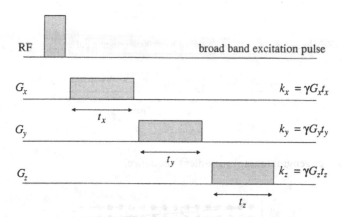

Figure 10.11 Three-dimensional Fourier imaging.

dimensional procedure is appropriate if only a single slice or a few slices are required, but once the number of slices becomes large it becomes more efficient, computer memory permitting, to use a three-dimensional method. In the following subsections we will consider various implementations of two- and three-dimensional Fourier imaging processes.

10.3.3 Fourier zeugmatography

One of the earliest Fourier techniques of three-dimensional image reconstruction used field gradients of a constant magnitude, applied successively in the x, y and z directions for varying times t_x, t_y and t_z to span the k-space. This is indicated in Figure 10.12. In this case the **k** vector is given by

$$\mathbf{k}(t_x,t_y,t_z) = \gamma G_x t_x \mathbf{i} + \gamma G_y t_y \mathbf{j} + \gamma G_z t_z \mathbf{k}$$

or

$$\mathbf{k}(t_x,t_y,t_z) = \gamma G(t_x \mathbf{i} + t_y \mathbf{j} + t_z \mathbf{k})$$

if the gradients are of equal magnitude. The points in the k_x–k_y plane are shown in Figure 10.13. Imagine that t_x is incremented by small amounts, and when it reaches its maximum value then t_y is incremented by one step and the t_x increments start again. In this way the k-space will be traversed as indicated by the lines in Figure 10.13 as time increases and in all one accumulates $n_x \times n_y \times n_z$ data points for processing by the fast Fourier transform calculator. In practice t_x and t_y would be varied, and for each value of this pair the FID in the presence of G_z would be captured at n_z time intervals t_z.

The Fourier zeugmatography procedure described here is the three-dimensional variant. Alternatively a single slice could be imaged using a selective 90° pulse in the

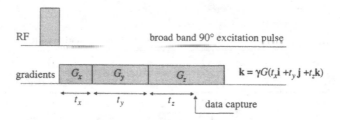

Figure 10.12 Fourier zeugmatography gradient sequence.

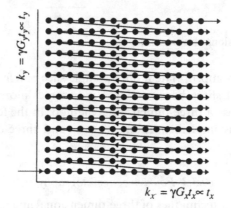

Figure 10.13 The k-space trajectory for Fourier zeugmatography.

presence of one gradient, and then the durations of the other *two* gradients would be varied. Processing of the data then only requires a two-dimensional Fourier transform.

10.3.4 Relaxation

During the evolution period $t_x + t_y + t_z$ there will be relaxation of the transverse magnetisation. Denoting the time constant by T_2 we see that the signal will have suffered a reduction by a factor $\exp[-(t_x + t_y + t_z)/T_2]$ at the moment of capture.

Since the spin density $\rho(\mathbf{r})$ is the Fourier transform of the k-space signal $F(\mathbf{k})$, it follows that since relaxation multiplies $F(\mathbf{k})$ by a decaying exponential, the effect is to convolute $\rho(\mathbf{r})$ or smear the image out. If the value of T_2 is known then the relaxation effect can be removed by correcting $F(t)$ by multiplying it by $\exp[+(t_x + t_y + t_z)/T_2]$, so that

$$\rho(\mathbf{r}) \propto \int F(\mathbf{t})\exp[+(t_x + t_y + t_z)/T_2]\exp(-i\gamma G\mathbf{r}\cdot\mathbf{t})\mathrm{d}^3t,$$

where $\mathbf{t}$ is the vector time $t_x\mathbf{i} + t_y\mathbf{j} + t_z\mathbf{k}$. Alternatively the total time $t_x + t_y + t_z$ can be

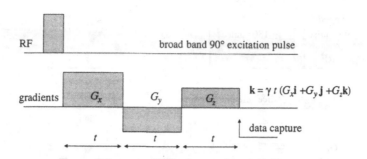

Figure 10.14 Spin warp imaging.

kept constant, in which case the relaxation effect factorizes out. However, that has the effect of distorting the Fourier transform as one is no longer integrating over the entire k-space, but only the plane corresponding to $t_x + t_y + t_z = $ const. The solution is to use a variant known as spin warp imaging.

10.3.5 Spin warp imaging

In this method the k-space is traversed by varying the magnitude of the gradients, keeping their duration constant. Since the durations of the three gradients are constant (and equal) the effect of relaxation is then simply to multiply the signal by the constant factor $\exp(-3t/T_2)$. The path in k-space is mapped out, in this case, as the magnitudes of the gradients are varied.

As in the Fourier zeugmatography procedure described above, this version of spin warp imaging is the three-dimensional variant. In the two-dimensional version a single slice is imaged using a selective 90° pulse in the presence of one gradient, and then the magnitudes of the other *two* gradients are varied.

10.3.6 Two-dimensional spin-echo imaging

A procedure for two-dimensional imaging using spin echoes is shown in Figure 10.15. The 180° pulse creates a spin echo which can then be digitised, without any problems of paralysis of the receiver due to the close proximity of the 90° excitation pulse. The initiation pulse is a selective, shaped pulse so that only the spins in the particular slice are tipped to the transverse plane. Another advantage of the spin-echo method is that relaxation effects from T_2 are eliminated if the time between the 90° pulse and the echo remains constant and relaxation effects from T_1 are eliminated if the repetition time of the sequence is kept constant.

The time duration of the x gradient is fixed; the value of k_x is changed through the variation of the magnitude of G_x. The y gradient, of fixed magnitude, is applied for the

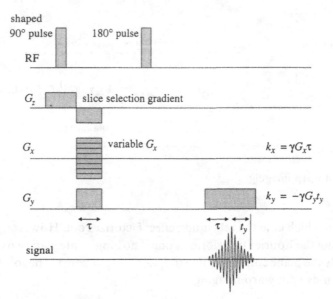

Figure 10.15 Two-dimensional spin-echo imaging.

same fixed duration. This sets a large value for k_y before the 180° pulse is applied. After application of the pulse the echo grows. It grows in the presence of an identical y gradient and so its effect gradually cancels until the centre of the echo, which corresponds to full cancellation. Thus as one proceeds through the echo k_y starts at its maximum value, decreases through zero at the echo centre, and proceeds to negative values at the latter side of the echo. Measuring t_y from the centre of the echo we see that $k_y = -\gamma G_y t_y$. The capture of each trace gives a sweep of k_y and successive traces are captured for different values of G_x, providing the variation of k_x. In this way the k_x–k_y plane is mapped out.

10.4 Gradient echoes

10.4.1 Creation of gradient echoes

We saw, in Section 3.4.6, that when signals are being acquired repetitively there can be a signal-to-noise advantage if excitation pulses of an angle smaller than 90° are used. In such cases the repetition time can be shortened without incurring a significant T_1 penalty. Now when an initiation pulse smaller than 90° is used not all the equilibrium magnetisation is rotated into the transverse plane; some is left pointing in the z direction. An undesirable effect of the 180° pulse used to create the spin echo is that it will also reverse the remnant M_z. This problem can, however, be avoided by creating the echo in a different way, by reversing the direction of the echo. The echo occurs when

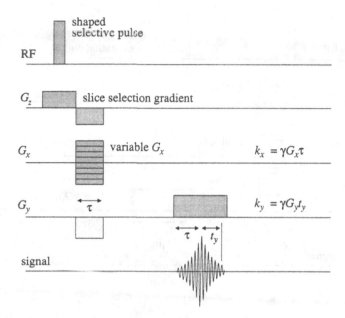

RF — shaped selective pulse

G_z — slice selection gradient

G_x — variable G_x $k_x = \gamma G_x \tau$

G_y — τ $k_y = \gamma G_y t_y$

τ t_y

signal

Figure 10.16 Two-dimensional gradient-echo imaging.

the positive and negative contributions from the gradient to the signal phase cancel.

Although the 'gradient echo', as it is called, does have the effect of recovering the magnetisation lost through dephasing in the field gradient without reversing the z component of the magnetisation, it cannot reduce the effect of the inhomogeneity of the static magnetic field, as does the conventional spin echo.

When small tip angle pulses are used, so that one is in the linear response régime, the profile for a slice-selective pulse does become precisely the sinc function; the conventional Fourier results then apply. Also, the tip angle provides a parameter which gives access to T_1 measurement/comparison in addition to the sequence repetition time.

10.4.2 Two-dimensional gradient-echo imaging

The procedure for two-dimensional gradient-echo imaging is very similar to that for two-dimensional conventional spin-echo imaging. The only difference is that the y gradient is reversed to produce the echo. This is shown in Figure 10.16.

The time duration of the x gradient is fixed; the value of k_x is changed through the variation of the magnitude of G_x. The y gradient, of fixed magnitude, is applied for the same fixed duration, but with a negative magnitude. This sets a large negative value for k_y. After the y gradient is reversed the echo grows. It grows in the presence of a positive y gradient and so its effect gradually cancels the negative gradient until the centre of the

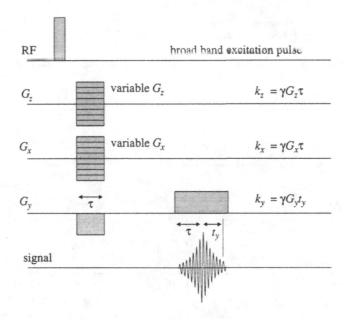

Figure 10.17 Three-dimensional gradient-echo imaging.

echo, which corresponds to full cancellation, is reached. Thus, as one proceeds through the echo, k_y starts at its maximum negative value, increases through zero at the echo centre and proceeds to positive values at the late side of the echo. Measuring t_y from the centre of the echo we see that $k_y = \gamma G_y t_y$. The capture of each trace gives a sweep of k_y and successive traces are captured for different values of G_x, providing the variation of k_x. In this way the k_x–k_y plane is mapped out.

10.4.3 Three-dimensional gradient-echo imaging

For the acquisition of a three-dimensional image the procedure is initiated by a broad band excitation pulse so that transverse magnetisation is created for the entire specimen. As shown in Figure 10.17, gradients are applied in the z and the x directions for a fixed time duration. The values of k_z and k_x are changed through the variation of the magnitude of G_z and G_x.

As in the two-dimensional case the y gradient, of fixed magnitude, is applied for the same fixed duration, but with a negative magnitude. This sets a large negative value for k_y and after the y gradient is reversed the echo grows. Measuring t_y from the centre of the echo we see that $k_y = \gamma G_y t_y$. The capture of each trace gives a sweep of k_y and successive traces are captured for different values of G_z and G_x, providing the variation of k_z and k_x. In this way the entire k-space is mapped out.

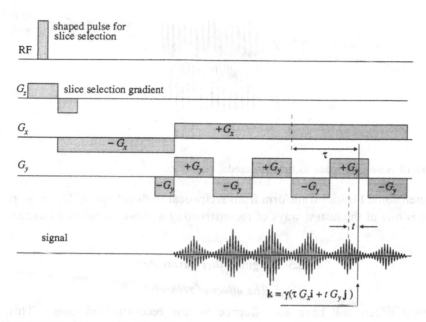

Figure 10.18 Echo planar imaging schematic.

10.4.4 Echo planar imaging

This technique permits the acquisition of information on an entire slice in a single FID (Mansfield, 1977). However, it is easily understood as the cascading of a sequence of projections in the plane; essentially it provides a raster scan of the slice in k-space.

Fundamental to this method is the use of gradient echoes, now produced by the reversal of both the x and the y gradients. The sequence is shown in Figure 10.18. The required slice in the x–y plane is selected by the shaped RF pulse and the slice selection gradient G_z. Then the negative lobes of the x and y gradients initiate k_x and k_y at their negative-most value. Full cancellation of the gradient dephasing effects occurs at the centre of the subsequent gradient pulses. Thus denoting by τ the time after the centre of the x gradient, with negative τ for times before the centre, and similarly denoting by t the time after the y gradient, we see that each point on the extended signal trace represents a point in the k_x–k_y plane:

$$k = \gamma(\tau G_x \mathbf{i} + t G_y \mathbf{j}).$$

Thus as time proceeds k scans through the reciprocal space plane forming a raster, as indicted in Figure 10.19. Having accumulated the precession signal as a function of k the image is resampled onto a rectangular grid and then reconstructed by a

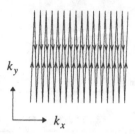

Figure 10.19 Scan of k-space as time proceeds.

two-dimensional Fourier transform from reciprocal to direct space. The echo planar method is one of the fastest ways of reconstructing a two-dimensional image.

10.5 Imaging other parameters

10.5.1 The effect of relaxation

Relaxation effects will have an influence on the reconstructed image. This was considered briefly in Section 10.3.4. Although such effects have the potential to degrade the image – in particular in limiting the spatial resolution – the effects can also be exploited to build an image of a relaxation time throughout the specimen instead of the conventional spin density.

We shall, for clarity, develop the arguments of the following sections for a one-dimensional model. However, before considering specifically the imaging of relaxation times we shall examine the simple degradation effect of a relaxation time which is uniform throughout the specimen. In the absence of relaxation a 'slice' of the specimen at height z contributes a distinct signal at its characteristic frequency $\omega = \gamma G_z$. However, when there is transverse relaxation the signal from this slice will be broadened in frequency by $\approx 1/T_2$. In other words it will appear to be spread over a spatial distance $\Delta z \approx 1/\gamma G T_2$. Thus any feature from the specimen will appear to be 'smeared' over this distance; the spatial resolution is thereby degraded.

This may be demonstrated qualitatively in the following way. In the absence of relaxation the Fourier relation between the free induction signal $F(t)$ and the spin density $\rho(z)$ is

$$F(t) = \int \rho(z)\exp(i\gamma Gzt)dz,$$

$$\rho(z) = \frac{\gamma G}{2\pi} \int F(t)\exp(-i\gamma Gzt)dt.$$

The effect of relaxation, say T_2, is to multiply $F(t)$ by a decaying exponential. The *observed* free induction signal $F'(t)$ and the naïvely inferred spin density $\rho'(z)$ are related by a similar Fourier pair

$$F'(t) = \int \rho'(z)\exp(i\gamma Gzt)dz,$$

$$\rho'(z) = \frac{\gamma G}{2\pi} \int F'(t)\exp(-i\gamma Gzt)dt,$$

where

$$F'(t) = F(t)R(t)$$

and

$$R(t) = \exp(-t/T_2).$$

Now the Fourier transform of a product of two functions is the *convolution* of the Fourier transforms of the individual functions (Appendix A). The Fourier transform of the relaxation function $R(t)$ is a Lorentzian of width T_2^{-1}. It then follows that the inferred spin density $\rho'(z)$ is the convolution of the real spin density $\rho(z)$ with a Lorentzian of width $\Delta z \approx 1/\gamma GT_2$. An example of the effect on a sharp edge detail is shown in Figure 10.20.

The correction of the degraded image for the effect of relaxation seems straightforward. We must *deconvolute* the image, which is conveniently performed in the time domain. All that must be done is to correct the free induction signal by dividing it by the relaxation function $\exp(-tT_2)$ before the Fourier transformation. This is easy – in principle. However there is a difficulty if we do not know the magnitude of the relaxation time since both over-correction and under-correction will give a distorted image.

10.5.2 Imaging T_2

We now turn to the situation in which the relaxation time T_2 varies throughout the specimen. Let us consider a conventional spin-echo imaging technique using a 180° RF pulse to create a spin echo. Then the signal from regions of shorter T_2 will be attenuated more when the period between the pulses is shorter. Thus by varying the pulse spacing an image of T_2 may be constructed.

We designate by T_E the time duration after which the echo occurs. Then an image reconstructed from such echoes will be distorted through the effect of relaxation, and we write this as $\rho(r,T_E)$. Assuming the relaxation is exponential we then have

$$\rho(r,T_E) = \rho(r,0)\exp[- T_E/T_2(r)] \tag{10.18}$$

where

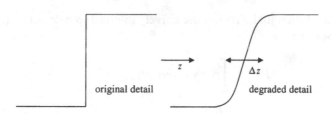

Figure 10.20 Resolution degradation of image.

$$\rho(\mathbf{r},0) = \rho(\mathbf{r}),$$

the true spin density in the specimen, and $T_2(\mathbf{r})$ is the position-dependent transverse relaxation time. This is based on the assumption that the repetition time between sequences is sufficiently long that there is no T_1 degradation of signals. It follows that by acquiring data at different values of T_E, a fit to Equation (10.18) for each point in the specimen will yield both the spin density $\rho(\mathbf{r})$ and the spatial variation of T_2.

In Figure 10.21 we show a one-dimensional profile of T_2 from solid and liquid ^{3}He in a cell 1 cm high. The solid is growing from a 'cold finger' at the top of the cell. The solid can be identified by its short T_2 and the liquid by its longer T_2. The signal from the interface region is blurred to a certain extent because the solid–liquid boundary is not quite perpendicular to the gradient direction.

In Figure 10.22 there are two images of a 5 mm slice of a human brain. They came from a two-echo sequence, image (a) being reconstructed from the first echo at 34 ms and image (b) being reconstructed from the second echo at 90 ms. The difference between the two is thus due to T_2 relaxation effects. In image (a) the bright areas correspond to both white matter and cerebrospinal fluid. However, since the cerebrospinal fluid has a long T_2, in image (b) it is only that which appears bright.

10.5.3 Imaging T_1

When T_1 varies throughout the specimen there will be distortion of the reconstructed image when the repetition time between sequences T_R is not much greater that T_1 since M_z will not have had time to grow to its equilibrium value. The analysis is very similar to that above for the T_2 case. We write the distorted spin density as $\rho(\mathbf{r},T_R)$ and assuming the relaxation is exponential, then

$$\rho(\mathbf{r},T_R) = \rho(\mathbf{r},\infty)\{1 - \exp[-T_R/T_1(\mathbf{r})]\} \qquad (10.19)$$

where

$$\rho(\mathbf{r},\infty) = \rho(\mathbf{r}),$$

the true spin density in the specimen, and $T_1(\mathbf{r})$ is the position dependent transverse relaxation time.

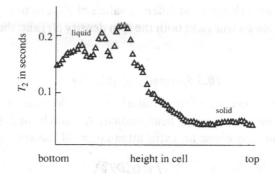

Figure 10.21 T_2 profile in 1 cm cell containing solid and liquid.

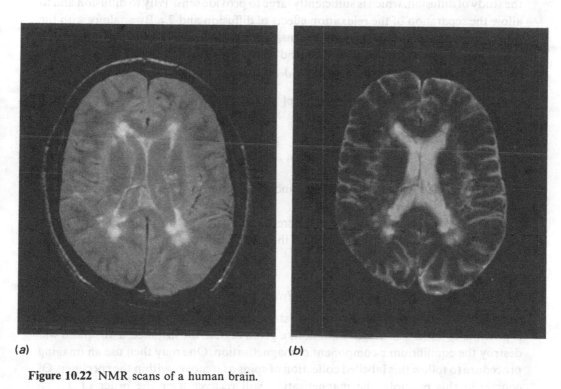

(a)   (b)

Figure 10.22 NMR scans of a human brain.

It follows that by acquiring data at different values of T_R, a fit to Equation (10.19) for each point in the specimen will yield both the spin density $\rho(\mathbf{r})$ and the spatial variation of T_2.

10.5.4 Imaging of diffusion

We saw, in section 4.5, that the effect of diffusion in a field gradient is made manifest through the application of a magnetic field gradient. A result from that section was that the effect of diffusion is to cause an extra attenuation of the spin echo of

$$\exp - \left(\frac{\gamma^2 G_d^2 D T_E^3}{12}\right).$$

Here we have denoted by G_d the gradient which has been applied specifically to study the diffusion. In the context of NMR imaging it is usual to apply an extra gradient for the study of diffusion, which is sufficiently large to provide sensitivity to diffusion and to allow the separation of the relaxation effects of diffusion and T_2. By analogy with the discussions in the previous two subsections, diffusive effects will distort the reconstructed image, and the extent of this will depend on the strength of the field gradient G_d. We write the distorted spin density as $\rho(\mathbf{r}, G_d)$, so then

$$\rho(\mathbf{r}, G_d) = \rho(\mathbf{r}, 0)\exp[-\gamma^2 G_d^2 D(\mathbf{r})T_E^3/12], \qquad (10.20)$$

where

$$\rho(\mathbf{r}, 0) = \rho(\mathbf{r}),$$

the true spin density in the specimen, and $D(\mathbf{r})$ is the positive dependent diffusion coefficient.

It follows that by acquiring data at different values of T_R, a fit to Equation (10.20) for each point in the specimen will yield both the spin density $\rho(\mathbf{r})$ and the spatial variation of the diffusion coefficient.

10.5.5 Studying fluid flow

There are two ways in which flow may be studied using NMR. Using selective pulses it is possible to 'label' the magnetisation in a given region; for instance, a 90° pulse will destroy the equilibrium z component of magnetisation. One may then use an imaging procedure to follow this labelled collection of spins as it moves within the specimen. Of course, in this example, the magnetisation will recover over the order of T_1, so ultimately this determines the utility of the method.

Alternatively a field gradient may be used for encoding the spatial information. The phase angle accumulated by a spin as it moves in this field gradient is

$$\phi(t) = \int \omega(\tau) d\tau,$$

where the instantaneous frequency is given by

$$\omega(\tau) = \gamma G z(\tau)$$

and the z co-ordinate increases with time as the spins move

$$z(\tau) = z_0 + v\tau,$$

where v is the velocity of the spins in the z direction.

It then follows that the accumulated phase angle is

$$\phi(t) = \gamma G z_0 t + \gamma G v t^2/2$$

and the transverse magnetisation is then proportional to

$$\exp(i\phi) = \exp(i\gamma G z_0 t)\exp(i\gamma G v t^2/2).$$

Thus it is possible to extract velocity information by varying the magnitude (and direction) of the extra field gradient, and extracting the component of t^2. The data processing procedures are thus similar to those used for relaxation time imaging and diffusion imaging. Comparing the discussions of the previous few subsections we see that relaxation effects are found from the term linear in time, velocity from the term quadratic in time, and diffusion from the term cubic in time. However, it must be emphasised that the treatment has been very elementary, touching only on the bare principles. In practice the instrumentation can be complex and the processing of the experimental data can require some very sophisticated algorithms and involved calculations.

Appendix A
Fourier analysis

A.1 The real Fourier transform

In essence Fourier theory tells us that we can make any waveform out of a suitable superposition of sinusoidal waves of the appropriate amplitudes and phases. A waveform which is finite in extent (or which is infinite but repetitive) may be constructed from a sum of discrete sinusoidal waves: a Fourier series. However, when the waveform is not repetitive and it is of infinite extent then a continuous distribution of sine waves is required: a Fourier integral. Thus we are saying that a function of time $F(t)$ may be expressed by the integral

$$F(t) = \int_0^\infty a(\omega)\sin[\omega t + \phi(\omega)]d\omega.$$

Here $a(\omega)$ is the amplitude of the component at (angular) frequency ω and $\phi(\omega)$ is the phase angle of this component. It is, however, more convenient to take account of the phases by using in-phase and quadrature components. Multiplying out the $\sin(\omega t + \phi)$ we obtain

$$F(t) = \int_0^\infty \{a(\omega)\sin[\phi(\omega)]\cos(\omega t) + a(\omega)\cos[\phi(\omega)]\sin(\omega t)\}d\omega,$$

which can be written as

$$F(t) = \frac{1}{\pi} \int_0^\infty [f'(\omega)\cos(\omega t) + f''(\omega)\sin(\omega t)]d\omega. \tag{A.1}$$

In this expression the spectral information is contained in the orthogonal amplitudes or

362

components $f'(\omega)$ and $f''(\omega)$. The factor $1/\pi$ appears by convention; it is a mathematical convenience. We note that in this description the time t can vary over the entire range $-\infty$ to $+\infty$, and that frequencies from zero to infinity may be required. If $F(t)$ is defined only for positive t then it is possible to express it as a sine-only or a cosine-only integral.

It remains, then, to find a procedure for evaluating the components $f'(\omega)$ and $f''(\omega)$. This is provided by the Fourier inversion integrals

$$\left.\begin{array}{l} f'(\omega) = \displaystyle\int_{-\infty}^{\infty} F(t)\cos(\omega t)dt, \\[4em] f''(\omega) = \displaystyle\int_{-\infty}^{\infty} F(t)\sin(\omega t)dt. \end{array}\right\} \tag{A.2}$$

For a proof of this the reader is referred to Bracewell (1979), Champeney (1973) or any good book on mathematical methods.

A.2 The complex Fourier transform

In many circumstances it proves convenient to represent oscillatory behaviour by the complex exponential $\exp(i\omega t)$ rather than by the real sines and cosines. The principal advantage is that such operations as time evolution, integrating, differentiation, phase changes etc. are effected simply by multiplication by (complex) numbers rather than cumbersome combinations of sines and cosines. With this motivation let us introduce a complex representation for the Fourier transform.

We define the complex quantity $f(\omega)$ by

$$f(\omega) = f'(\omega) + if''(\omega). \tag{A.3}$$

Then $f(\omega)$ may be expressed, using Equation (A.2), in the compact form

$$f(\omega) = \int_{-\infty}^{\infty} F(t)\exp(i\omega t)dt. \tag{A.4}$$

We shall now see how to express $F(t)$ directly in terms of this new, complex, $f(\omega)$. Note, however, that at this stage $F(t)$ is still a real quantity, as are the Fourier components $f'(\omega)$ and $f''(\omega)$. Thus as a preliminary we must find these components in terms of the complex $f(\omega)$. From Equation (A.3) we have

$$f'(\omega) = \frac{1}{2}[f(\omega) + f^*(\omega)],$$

$$f''(\omega) = \frac{1}{2i}[f(\omega) - f^*(\omega)],$$

where, of course, $f^*(\omega)$ is the complex conjugate of $f(\omega)$. From Equation (A.4) we see that (formally) complex conjugation is equivalent to a reversal of frequency ω:

$$f^*(\omega) = \int_{-\infty}^{\infty} F(t)\exp(-i\omega t)dt,$$

so that

$$f^*(\omega) = f(-\omega), \tag{A.5}$$

since $F(t)$ is real. The expressions for $f'(\omega)$ and $f''(\omega)$ then become

$$f'(\omega) = \frac{1}{2}[f(\omega) + f(-\omega)],$$

$$f''(\omega) = \frac{1}{2i}[f(\omega) - f(-\omega)],$$

and in this form they may be inserted in Equation (A.1) to give

$$F(t) = \frac{1}{2\pi}\int_0^{\infty}\left\{[f(\omega) + f(-\omega)]\cos(\omega t) + \frac{1}{i}[f(\omega) - f(-\omega)]\sin(\omega t)\right\}d\omega.$$

Rearranging, this expression may be written as

$$F(t) = \frac{1}{2\pi}\int_0^{\infty} f(\omega)[\cos(\omega t) - i\sin(\omega t)]d\omega + \frac{1}{2\pi}\int_0^{\infty} f(-\omega)[\cos(\omega t) + i\sin(\omega t)d\omega.$$

Now in the second integral we change variables from ω to $-\omega$ which then gives

$$F(t) = \frac{1}{2\pi}\int_0^{\infty} f(\omega)\exp(-i\omega t)d\omega + \frac{1}{2\pi}\int_{-\infty}^{0} f(\omega)\exp(-i\omega t)d\omega$$

and now the integrals may be combined, giving

$$F(t) = \frac{1}{2\pi} \int_{-\infty}^{\infty} f(\omega)\exp(-i\omega t)d\omega. \qquad (A.6)$$

This is the complex form for the Fourier integral, the inverse of which is given by Equation (A.4).

The derivation of the complex Fourier transform pair, Equations (A.4) and (A.6), has been made on the assumption that $F(t)$ is real. However, the linearity of the integral transform means that the relation will generalise to complex functions, although of course the complex conjugation relation, Equation (A.5), will not hold in that case. In summary then, we have the complex Fourier transform pair:

$$F(t) = \frac{1}{2\pi} \int_{-\infty}^{\infty} f(\omega)\exp(-i\omega t)dt,$$

$$f(\omega) = \int_{-\infty}^{\infty} F(t)\exp(i\omega t)dt. \qquad (A.7)$$

A.3 Fourier transform relations

Sine and cosine transforms

(1) $F(t) = \dfrac{2}{\pi} \displaystyle\int_{0}^{\infty} f_s(\omega)\sin(\omega t)d\omega$ ← If $F(0) \neq 0$ then work with $F(t) - F(0)$

(2) $F(t) = \dfrac{2}{\pi} \displaystyle\int_{0}^{\infty} f_c(\omega)\cos(\omega t)d\omega$

$F(t)$ defined for $0 \le t < \infty$
Required ω for $0 \le \omega < \infty$

(3) $f_s(\omega) = \displaystyle\int_{0}^{\infty} F(t)\sin(\omega t)dt$

(4) $f_c(\omega) = \displaystyle\int_{0}^{\infty} F(t)\cos(\omega t)dt$

(5) $F(t) = \dfrac{1}{\pi} \displaystyle\int_0^\infty \{f'(\omega)\cos(\omega t) + f''(\omega)\sin(\omega t)\}d\omega$ $\left.\vphantom{\begin{array}{c}a\\a\\a\\a\\a\\a\end{array}}\right|$ $F(t)$ defined for $-\infty < t < \infty$

(6) $f'(\omega) = \displaystyle\int_{-\infty}^\infty F(t)\cos(\omega t)dt$ $\qquad\qquad\qquad$ ω required for $0 \le \omega < \infty$

(7) $f''(\omega) = \displaystyle\int_{-\infty}^\infty F(t)\sin(\omega t)dt$

Complex form

(8) $F(t) = \dfrac{1}{2\pi} \displaystyle\int_{-\infty}^\infty f(\omega)\exp(-i\omega t)d\omega$

$\qquad\qquad\qquad\qquad$ $F(t)$ defined for $-\infty < t < \infty$
$\qquad\qquad\qquad\qquad$ Required ω for $-\infty < \omega < \infty$

(9) $f(\omega) = \displaystyle\int_{-\infty}^\infty F(t)\exp(i\omega t)dt$

Convolutions

(10) $\dfrac{1}{2\pi} \displaystyle\int_{-\infty}^\infty f(\omega)g(\omega)\exp(-i\omega t)d\omega = \displaystyle\int_{-\infty}^\infty F(\tau)G(t-\tau)d\tau$

$\qquad\qquad = F(t)\otimes G(t)$ $\qquad\qquad = \displaystyle\int_{-\infty}^\infty F(t-\tau)G(\tau)d\tau$

(11) $\displaystyle\int_{-\infty}^\infty F(t)G(t)\exp(i\omega t)dt$ $\qquad = \dfrac{1}{2\pi} \displaystyle\int_{-\infty}^\infty f(\omega')g(\omega-\omega')d\omega'$

$\qquad\qquad = f(\omega)\otimes g(\omega)$ $\qquad\qquad = \dfrac{1}{2\pi} \displaystyle\int_{-\infty}^\infty f(\omega-\omega')g(\omega')d\omega'$

Complex conjugation

(12) $h(\omega) = g^*(\omega) \Rightarrow H(t) = G^*(-t)$

(13) $H(t) = G^*(t) \Rightarrow h(\omega) = g^*(-\omega)$

Complex convolutions

$$(14)\quad \frac{1}{2\pi} \int_{-\infty}^{\infty} f(\omega)g^*(\omega)\exp(-i\omega t)d\omega = \int_{-\infty}^{\infty} F(\tau)G^*(t-\tau)d\tau$$

$$= F(t) \otimes G^*(t) \qquad\qquad = \int_{-\infty}^{\infty} F(t-\tau)G^*(-\tau)d\tau$$

$$(15)\quad \int_{-\infty}^{\infty} F(t)G^*(t)\exp(i\omega t)dt \qquad = \frac{1}{2\pi} \int_{-\infty}^{\infty} f(\omega')g^*(\omega-\omega')d\omega'$$

$$= f(\omega) \otimes g^*(\omega) \qquad\qquad = \frac{1}{2\pi} \int_{-\infty}^{\infty} f(\omega-\omega')g^*(-\omega')d\omega'$$

Parseval's Theorem

$$(16)\quad \int_{-\infty}^{\infty} F(t)F^*(t)dt = \frac{1}{2\pi} \int_{-\infty}^{\infty} f(\omega)f^*(\omega)d\omega$$

or

$$\int_{-\infty}^{\infty} |F(t)|^2 dt = \frac{1}{2\pi} \int_{-\infty}^{\infty} |f(\omega)|^2 d\omega$$

Uncertainty principle

$$(17)\quad \frac{\displaystyle\int_{-\infty}^{\infty} (t-\langle t\rangle)|F(t)|^2 dt}{\displaystyle\int_{-\infty}^{\infty} |F(t)|^2 dt} \times \frac{\displaystyle\int_{-\infty}^{\infty} (\omega-\langle\omega\rangle)|f(\omega)|^2 d\omega}{\displaystyle\int_{-\infty}^{\infty} |f(\omega)|^2 d\omega} \geq \frac{1}{4}$$

or

$$\Delta t \Delta \omega \geq \tfrac{1}{2}$$

where

$$\langle t \rangle = \frac{\displaystyle\int_{-\infty}^{\infty} t\,|F(t)|^2 dt}{\displaystyle\int_{-\infty}^{\infty} |F(t)|^2 dt} \quad \text{and} \quad \langle \omega \rangle = \frac{\displaystyle\int_{-\infty}^{\infty} \omega\,|f(\omega)|^2 d\omega}{\displaystyle\int_{-\infty}^{\infty} |f(\omega)|^2 d\omega}$$

A.4 Symmetry properties of Fourier transforms

We have seen in Equation (A.5) that the symmetry properties of a function are related to the complex conjugation properties of its Fourier transform. It is often helpful to know how some symmetry of one function is reflected in its transform. This may be derived from the Fourier transform integrals, Equation (A.7). In Table A.1 we summarise some of the possibilities.

A.5 Dirac's delta function

Dirac introduced the delta function for use in quantum mechanics, although soon after this physicists found it was of very general utility. However, at the same time mathematicians dismissed at as unrigorous; it did not satisfy the mathematical requirements to qualify as a function. It was not until 1950, when Laurent Schwartz established the theory of distributions or generalised functions, that Dirac's 'function' finally obtained the mathematical seal of approval. Our treatment follows that of Cohen-Tannoudji *et al.* (1977).

Very crudely, the delta function $\delta(x)$ is zero everywhere except at $x = 0$, where it is infinite. Furthermore, the function has unit area. We may understand this as the limit of a normalised box function $\delta^{(\varepsilon)}(x)$ of width 2ε as shown in Figure A.1,

$$\delta^{(\varepsilon)}(x) = 0 \qquad |x| > \varepsilon,$$

$$= \frac{1}{2\varepsilon} \qquad |x| < \varepsilon,$$

as ε goes to zero: i.e. we define

Table A.1. *Symmetry of Fourier transforms.*

$F(t)$	$f(\omega)$
Real and even	Real and even
Real and odd	Imaginary and odd
Imaginary and even	Imaginary and even
Complex and even	Complex and even
Complex and odd	Complex and odd
Real and asymmetrical	Real part even, imaginary part odd
Imaginary and asymmetrical	Real part odd, imaginary part even
Real even and imaginary odd	Real
Real odd and imaginary even	Imaginary
Even	Even
Odd	Odd

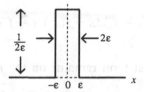

Figure A.1 Box function whose limit leads to a delta function.

$$\delta^{(\varepsilon)}(x) = \underset{\varepsilon \to 0}{\mathscr{L}}\ \delta^{(\varepsilon)}(x).$$

So although not strictly a function, it may be understood as the *limit* of a respectable mathematical function.

The most important property of the delta function is its 'sifting' property; by integrating it can extract the value of a function at a given point. Consider the integral

$$\int_{-\infty}^{\infty} f(x)\delta(x - x_0)dx,$$

where $f(x)$ is an arbitrary, but differentiable, function. We investigate this by examining the corresponding integral with $\delta^{(\varepsilon)}(x - x_0)$ and finally taking the $\varepsilon \to 0$ limit. The integral is then

$$\int_{-\infty}^{\infty} f(x)\delta^{(\varepsilon)}(x-x_0)dx = \frac{1}{2\varepsilon}\int_{x_0-\varepsilon}^{x_0+\varepsilon} f(x)dx.$$

Since $f(x)$ is differentiable, it may be expanded as a Taylor series about x_0:

$$f(x) = f(x_0) + f'(x_0)(x-x_0) + f''(x_0)\frac{(x-x_0)^2}{2} + \dots.$$

We integrate this with respect to x, term by term. And since

$$\int_{x_0-\varepsilon}^{x_0+\varepsilon} dx = 2\varepsilon, \quad \int_{x_0-\varepsilon}^{x_0+\varepsilon} (x-x_0)dx = 0, \quad \int_{x_0-\varepsilon}^{x_0+\varepsilon} (x-x_0)^2 dx = \frac{2\varepsilon^3}{3},$$

we then find

$$\int_{-\infty}^{\infty} f(x)\delta^{(\varepsilon)}(x-x_0)dx = f(x_0) + \frac{\varepsilon^2}{6}f''(x_0) + \dots.$$

In the limit $\varepsilon \to 0$ only the first term remains on the right hand side, and thus we conclude that

$$\int_{-\infty}^{\infty} f(x)\delta(x-x_0)dx = f(x_0). \tag{A.8}$$

In this way we see how the delta function can be used to extract the value of a function at a given point.

The box function is only one of many which can be used to approximate the delta function. Other functions suitable for $\delta^{(\varepsilon)}(x)$ include

$$\delta^{(\varepsilon)} = \frac{1}{2\varepsilon}\exp(-|x|/\varepsilon) \qquad\qquad \text{exponential}$$

$$\delta^{(\varepsilon)} = \frac{1}{\pi}\frac{\varepsilon}{x^2+\varepsilon^2} \qquad\qquad \text{Lorentzian}$$

$$\delta^{(\varepsilon)} = \frac{1}{\varepsilon\pi^{1/2}}\exp(-x^2/\varepsilon^2) \qquad\qquad \text{Gaussian}$$

$$\delta^{(\varepsilon)} = \frac{1}{\pi}\frac{\sin(x/\varepsilon)}{x} \qquad\qquad \text{sinc}$$

$$\delta^{(\varepsilon)} = \frac{\varepsilon}{\pi} \frac{\sin^2(x/\varepsilon)}{x^2}. \qquad \text{sinc}^2$$

The sifting property (and other properties of the delta function) can be derived using any of these approximating functions.

Of particular importance are the Fourier transform relations involving the delta function. From the Fourier expression

$$F(t) = \frac{1}{2\pi} \int_{-\infty}^{\infty} f(\omega) \exp(-i\omega t) d\omega,$$

if we take $f(\omega) = \delta(\omega - \omega_0)$, then using the sifting relation we find

$$F(t) = \frac{1}{2\pi} \exp(-i\omega_0 t), \qquad (A.9)$$

indicating that the Fourier transform of a delta function centred at a given frequency is a complex sinusoid at that frequency. In particular, if $\omega_0 = 0$, then the Fourier transform of a delta function at the origin is simply a constant.

Using the inverse Fourier transform

$$f(\omega) = \int_{-\infty}^{\infty} F(t) \exp(i\omega t) dt$$

we find the Fourier representation for the frequency delta function:

$$\delta(\omega - \omega_0) = \frac{1}{2\pi} \int_{-\infty}^{\infty} \exp[i(\omega - \omega_0)t] dt. \qquad (A.10)$$

Conversely, if we start with a delta function in the time domain, $F(t) = \delta(t - t_0)$ then we obtain the Fourier transform

$$f(\omega) = \exp(i\omega t_0). \qquad (A.11)$$

Thus the Fourier transform of a delta function centred at a given time is a complex sinusoid at that time. Transforming back to the time domain we find the Fourier representation for the time delta function:

$$\delta(t - t_0) = \frac{1}{2\pi} \int_{-\infty}^{\infty} \exp[-i\omega(t - t_0)] d\omega. \qquad (A.12)$$

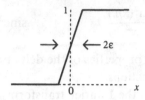

Figure A.2 Integral of box function whose limit leads to a step function.

A.6 Heaviside's step function

The integral of the delta function $\delta(x)$ is the Heaviside step function $\theta(x)$, which is zero for negative x and unity for positive x:

$$\theta(x) = \begin{cases} 0 & |x| < 0, \\ 1 & |x| > 0. \end{cases} \tag{A.13}$$

This may be seen most clearly through integrating the box approximating function $\delta^{(\varepsilon)}(x)$ and then taking the limit $\varepsilon \to 0$. This is indicated in Figure A.2. As ε becomes smaller the rising line becomes steeper and steeper and in the limit $\varepsilon \to 0$ it becomes a vertical step.

The Fourier transform of the step function $\theta(t)$, which we denote by $\Theta(\omega)$, is indeed in the next section when we discuss questions relating to causality. We must be very careful in evaluating the Fourier transform of $\theta(t)$ to avoid divergence of the integral at the upper limit. To ensure a finite result we shall use an approximating function $\theta^{(\varepsilon)}(t)$, given by

$$\begin{aligned} \theta^{(\varepsilon)}(x) &= 0 & |x| < 0, \\ &= \exp(-\varepsilon t) & |x| > 0. \end{aligned}$$

The Fourier transform of this, $\Theta^{(\varepsilon)}(\omega)$, is then

$$\Theta^{(\varepsilon)}(\omega) = \frac{i}{\omega + i\varepsilon}.$$

It is the small positive quantity ε which ensures convergence of the Fourier integral. But we cannot set it to zero yet. If we did so, then the expression i/ω would blow up in any integral covering $\omega = 0$, as we need in the following section. So we must treat this limit with care.

Firstly we separate the real and imaginary parts:

$$\Theta^{(\varepsilon)}(\omega) = \frac{\omega}{\omega^2 + \varepsilon^2} - \frac{i\varepsilon}{\omega^2 + \varepsilon^2}.$$

Now the imaginary part here is observed to be proportional to the Lorentzian approximation to the delta function. We see it is equal to $-i\pi\delta(\omega)$. The real part of the expression is essentially $1/\omega$, but the small ε part in the denominator ensures that it never actually blows up. So as ε goes to zero the real part goes to $1/\omega$ but always excluding *just* the divergent point $\omega = 0$. We refer to this as the *principal part* of the function $1/\omega$ and we use the notation $\mathscr{P}(1/\omega)$ to denote this. Thus we have, in the $\varepsilon \to 0$ limit

$$\Theta(\omega) = i\mathscr{P}\frac{1}{\omega} + \pi\delta(\omega). \tag{A.14}$$

A.7 Kramers–Kronig relations

In Chapter 3 we encountered the susceptibility function in the time domain, $X(t)$, and in the frequency domain, $\chi(\omega)$, which is a complex function. The pair $X(t)$ and $\chi(\omega)$ are related through Fourier transformation. It turns out that the real and the imaginary parts of $\chi(\omega)$ are not independent. This follows as a consequence of causality – in a physical system the effect must come *after* the cause. This means that the time susceptibility function must be zero for negative times. We may express this (in what will be a useful way) by saying that $X(t)$ must be equal to itself multiplied by the step function

$$X(t) = X(t)\theta(t).$$

The point about the causality condition expressed in this way is that we can immediately transform to the frequency domain. The Fourier transform of the product of two functions is the *convolution* of the Fourier transforms of the two functions. Thus in the frequency domain the causality condition becomes

$$\chi(\omega) = \chi(\omega) \otimes \Theta(\omega)$$

or

$$\chi(\omega) = \frac{1}{2\pi} \int_{-\infty}^{\infty} \chi(\omega')\Theta(\omega - \omega')d\omega'.$$

Using the expression for $\Theta(\omega)$ from Equation (A.14), this becomes

$$\chi(\omega) = \frac{-i}{2\pi} \mathscr{P} \int_{-\infty}^{\infty} \frac{\chi(\omega')}{\omega - \omega'}d\omega' + \frac{1}{2}\chi(\omega),$$

or

$$\chi(\omega) = \frac{-i}{\pi} \mathscr{P} \int\limits_{-\infty}^{\infty} \frac{\chi(\omega')}{\omega - \omega'} d\omega'. \tag{A.15}$$

Here the principal part of the integral excludes the singular point at $\omega' = \omega$, which may be ensured by the limiting procedure

$$\mathscr{P} \int\limits_{-\infty}^{\infty} = \mathscr{L}_{\varepsilon \to 0} \left\{ \int\limits_{-\infty}^{\omega - \varepsilon} + \int\limits_{\omega + \varepsilon}^{\infty} \right\}.$$

Now Equation (A.15) is simply the mathematical embodiment of the causality condition. The value of the equation, however, stems from the i factor. Because of this, when we express the susceptibility in terms of its real and imaginary parts.

$$\chi(\omega) = \chi'(\omega) + i\chi''(\omega).$$

Equation (A.15) gives the real part as an integral over the imaginary part and the imaginary part as an integral over the real part

$$\left. \begin{aligned} \chi'(\omega) &= \frac{1}{\pi} \mathscr{P} \int\limits_{-\infty}^{\infty} \frac{\chi''(\omega')}{\omega - \omega'} d\omega', \\[2em] \chi''(\omega) &= -\frac{1}{\pi} \mathscr{P} \int\limits_{-\infty}^{\infty} \frac{\chi'(\omega')}{\omega - \omega'} d\omega'. \end{aligned} \right\} \tag{A.16}$$

These relations, connecting the real and imaginary parts of the susceptibility, are known as the Karmers–Kronig relations. In the more general context, these relations are valid for the real and imaginary parts of the Fourier transform of *any* causal function. In the general case this is called a Hilbert transform.

Appendix B
Random functions

B.1 Mean behaviour of fluctuations

The mean value of a thermodynamic quantity X is constant for a system in equilibrium; this is essentially the definition of equilibrium. However, on the microscopic scale X will fluctuate with time, as dictated by the equations of motion of the system. The deviations of X from the mean will average to zero: equally likely to be positive or negative, but the *square* of the deviations of X from the mean will have a non-zero average. For convenience let us redefine X by subtracting off its mean value; our newly defined X has zero mean: $\langle X \rangle = 0$. But as mentioned above, $\langle X^2 \rangle \neq 0$, and in general, certainly for even n, we will have $\langle X^n \rangle \neq 0$. These 'moments' can often be calculated without too much difficulty. However, they give no indication about the time dependence of the fluctuations.

Box B.1 Ensembles and averages

It is not immediately obvious what is meant by taking an average in this case. If many copies of the system are imagined, each having the same values for the macroscopic observables, then one can consider the average evaluated over this collection of copies. This imaginary collection of copies is referred to as an *ensemble* and the average is called an *ensemble average*.

What can we say about the time variation of the fluctuations without completely solving the equations of motion for the system? Can we define a quantity which describes the mean time evolution of the variations? At a particular time t_0 we may observe that the variable X has the value a^i:

$$X(t_0) = a^i. \tag{B.1}$$

This will subsequently develop in time so that the observed mean becomes zero. We can

375

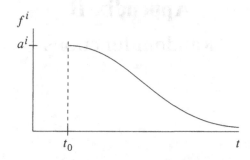

Figure B.1 Mean regression of a random fluctuation from an initial value.

talk of the mean time dependence of X from this value a^i by taking a subensemble from a complete ensemble – the subensemble consisting of all those elements which at time t_0 have the value a^i, i.e. the elements of the subensemble obey

$$X^i_j(t_0) = a^i.$$

The upper index of X labels the subensemble and the lower index indicates which element within the subensemble. We can then define the mean regression from the initial value a^i at time t_0 by the average

$$f^i(t_0 + t) = \frac{1}{n_i} \sum_{j=1}^{n_i} X^i_j(t_0 + t), \tag{B.2}$$

where n_i is the number of elements in the ith subensemble.

For sufficiently long times t, the value of $f^i(t_0 + t)$ will go to zero since there will then be no means of distinguishing that this subensemble is not a representative one. Thus we expect the function $f^i(t_0 + t)$ to behave as in Figure B.1. This function describes 'on the average' how X varies if at time t_0 it had the value a_i. However, since we are considering a system in equilibrium – i.e. it has time translation invariance – it follows that the time evolution of $f^i(t_0 + t)$ is not peculiar to the time t_0. Whenever X is observed to take on the value a^i the subsequent mean time dependence will be given by f^i. So we may ignore t_0 and we shall put it to zero in our future discussions.

The behaviour does, however, refer specifically to the particular value a^i. To find the mean regression of any fluctuation we may think of averaging over all the initial values a^i with appropriate weights, but this is simply an average over the complete ensemble and this, we know, must give zero. Physically we can see that this is so since there will be many positive and negative values of f^i which will average to zero as in Figure B.2. Mathematically this follows since

$$\langle f(t) \rangle = \sum_i w_i f^i(t) = \sum_i \frac{w_i}{n_i} \sum_j X^i_j(t), \tag{B.3}$$

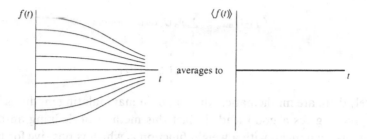

Figure B.2 Average behaviour of many different initial states.

where w_i is the weight factor for the ith subensemble. Now this weight factor is the proportion of the whole ensemble which the ith subensemble represents, i.e. n_i/N, where N is the total number of elements in the ensemble: $N = \Sigma_i n_i$. Thus

$$\langle f(t) \rangle = \sum_i \frac{n_i}{Nn_i} \sum_j X_j^i(t)$$

$$= \frac{1}{N} \sum_{i,j} X_j^i(t),$$

the common mean over the full ensemble which is zero.

Instead of the direct mean we could evaluate the mean of the squares of the regressions of the subensembles. In this way positive and negative variations will all contribute without cancelling each other. We then have

$$\langle f^2(t) \rangle = \sum_i w_i \{f^i(t)\}^2$$

$$= \sum_i \frac{w_i}{n_i} \left\{ \sum_j X_j^i(t) \right\}^2$$

$$= \sum_i \frac{w_i}{n_i} \sum_{j,k} X_j^i(t) X_k^i(t)$$

$$= \frac{1}{N} \sum_{i,j,k} X_j^i(t) X_k^i(t),$$

but this involves cross term between members of the ensemble. The idea of correlations between different elements of an ensemble is completely unphysical, the mathematics is becoming complicated, so we shall reject this approach.

Another possibility is to take the average of the *magnitude* of the $f^i(t)$, defining

$$\langle f(t)\rangle_{\text{mag}} = \sum_i w_i |f^i(t)|$$

$$= \frac{1}{N}\sum_i \left|\sum_j X_i^{\,j}(t)\right|. \tag{B.4}$$

Unfortunately there are mathematical difficulties in manipulating modulus functions, but this approach gives a good lead. In fact this method of defining an average is equivalent to taking a mean with a weight function ε_i which is positive for positive a^i and negative for negative a^i:

$$\langle f(t)\rangle_{\text{mag}} = \frac{1}{N}\sum_i \varepsilon_i \sum_j X_j^i(t)$$

$$= \frac{1}{N}\sum_{i,j} \varepsilon_i X_j^i(t). \tag{B.5}$$

Written in this form the expression for the average behaviour of the regression seems promising. The only difficulty is with the strange weight function

$$\varepsilon_i = \begin{cases} 1 & \text{if } a^i > 0, \\ -1 & \text{if } a^i < 0. \end{cases}$$

Instead of this discontinuous function ε multiplying the components, there is a much more straightforward weight function which satisfies the fundamental requirement of respecting the sign of the a^i: we could use the a^i themselves. In other words, in evaluating the average behaviour of the natural fluctuations we weight each element of the ensemble by its initial value. We have then

$$\frac{1}{N}\sum_{i,j} a^i X_j^i(t)$$

or, since $X_j^i(0) = X^i(0) = a^i$, we can write it as

$$\frac{1}{N}\sum_{i,j} X_j^i(0)X_j^i(t)$$

and in this form we have achieved an important advance in that the average is over the complete ensemble (with no cross terms) and we can write it as

$$\langle X(0)X(t)\rangle.$$

B.2 The autocorrelation function

The above expression is the mean over the ensemble where each element is weighted in

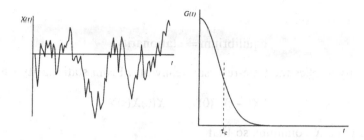

Figure B.3 Random function and its autocorrelation function.

proportion to its initial value. This function of the randomly varying quantity $X(t)$ is known as the *autocorrelation function*. We shall denote it by the symbol $G(t)$:

$$G(t) = \langle X(0)X(t) \rangle. \tag{B.6}$$

Figure B.3 shows the behaviour of a typical random function together with its autocorrelation function. In the sense described above, the function $G(t)$ describes the mean time variation of the fluctuations in $X(t)$. Observe the smooth behaviour of the autocorrelation function. In a sense this function has distilled the fundamental essence of the random function $X(t)$ without its wealth of unimportant fine detail.

The zero time value of $G(t)$ has an immediate interpretation. From Equation (B.6) we have

$$G(0) = \langle X^2 \rangle,$$

the mean square value of the random function, one of the 'moments' discussed at the start of this section. For long times, as we have argued above, $G(t)$ must go to zero.

If X is not defined so that its average $\langle X \rangle$ is zero then we must replace the definition in Equation (B.6) by

$$
\begin{aligned}
G(t) &= \langle (X(0) - \langle X \rangle)(X(t) - \langle X \rangle) \rangle \\
&= \langle X(0)X(t) \rangle - \langle X(0)\langle X \rangle \rangle - \langle \langle X \rangle X(t) \rangle + \langle \langle X \rangle \langle X \rangle \rangle \\
&= \langle X(0)X(t) \rangle - \langle X \rangle^2,
\end{aligned}
\tag{B.7}
$$

since

$$\langle \langle X \rangle X(t) \rangle = \langle X \rangle \langle X(t) \rangle = \langle X \rangle^2.$$

The time-translation invariance property, which we stated was a property of equilibrium systems (to be derived rigorously in the quantum case in Section B.6), now becomes the stationarity principle:

$$\langle X(\tau)X(t + \tau) \rangle = \langle X(0)X(t) \rangle: \tag{B.8}$$

i.e.

$$\text{equilibrium} \rightarrow \text{stationarity},$$

and stationarity implies the time-reversal behaviour. From stationarity we have:

$$\langle X(-\tau)X(0)\rangle = \langle X(0)X(\tau)\rangle$$

but classically the X commute, so that

$$\langle X(-\tau)X(0)\rangle = \langle X(\tau)X(0)\rangle$$

or

$$G(t) = G(-t). \tag{B.9}$$

B.3 A paradox

The correlation function $G(t)$ has the physical significance that a 'large' fluctuation will be expected on average to die out according to $G(t)$. A large fluctuation will occur infrequently and on the 'large' scale X may be assumed to decay to its equilibrium value.

We say that $G(t)$ traces the mean decay of a fluctuation. Now according to this view one would expect $G(-t)$ to tell us of the past history of the fluctuation, but we have the time-reversal rule $G(t) = G(-t)$. Our zero of time always seems to coincide with the peak value of the fluctuation – but how can this be consistent with time-translation invariance? There seems to be a paradox; the time origin is important. The solution is that the zero of time *is* unique because according to the values of X at this time are the subensembles selected: i.e. the weighting factors of the averaging procedure are selected at the zero of time.

B.4 The Wiener–Kintchine Theorem

We shall now examine the Fourier transform of the correlation function $G(t)$. The definition of the autocorrelation function is taken as

$$G(t) = \langle X^*(0)X(t)\rangle, \tag{B.10}$$

where we have allowed for the possibility of a complex X by taking the complex conjugate of the zero-time term. This ensures that $G(0)$ will be real. Now the variable $X(t)$ is assumed stationary and this means that strictly speaking the Fourier transform of $X(t)$ does not exist as the integral

$$x(\omega) = \int_{-\infty}^{\infty} X(t)\exp(i\omega t)dt \qquad (B.11)$$

must diverge. This equation must thus be regarded with caution. The road to rigour is to redefine $X(t)$ to be zero outside some time interval T and to take the limit $T \to \infty$ at the end of the calculation. Assuming that this or something equivalent is done so that Equation (B.11) has some validity, the inverse relation will be

$$X(t) = \frac{1}{2\pi} \int_{-\infty}^{\infty} x(\omega)\exp(-i\omega t)d\omega. \qquad (B.12)$$

We now substitute this into Equation (B.10), but take account of stationarity by writing Equation (B.10) as

$$G(t) = \langle X^*(\tau)X(t + \tau)\rangle \qquad (B.13)$$

and where the end result must be independent of τ. From Equation (B.12)

$$\left.\begin{array}{l} X^*(t) = \dfrac{1}{2\pi} \displaystyle\int_{-\infty}^{\infty} x^*(\omega)\exp(+i\omega t)d\omega, \\[3mm] X(t + \tau) = \dfrac{1}{2\pi} \displaystyle\int_{-\infty}^{\infty} x(\omega')\exp[-i\omega'(t + \tau)]d\omega', \end{array}\right\} \qquad (B.14)$$

so that Equation (B.13) becomes

$$G(t) = \frac{1}{4\pi^2} \int_{-\infty}^{\infty} d\omega \int_{-\infty}^{\infty} d\omega' \langle x^*(\omega)x(\omega')\rangle\exp[-i(\omega' - \omega)\tau]\exp(-i\omega t). \qquad (B.15)$$

In this form we identify $G(t)$ as the Fourier transform of a function $g(\omega)$ which is given by

$$g(\omega) = \frac{1}{2\pi} \int_{-\infty}^{\infty} \langle x^*(\omega)x(\omega')\rangle\exp[-i(\omega' - \omega)\tau]d\omega'. \qquad (B.16)$$

Now the stationarity principle means that this expression must be independent of the time variable τ. This can only be so if the correlation function $\langle x^*(\omega)x(\omega')\rangle$ is proportional to the delta function $\delta(\omega - \omega')$, for in this case the argument of the exponential becomes zero and there is no resultant τ dependence. Thus the stationarity of $X(t)$ requires that

$$\langle x^*(\omega)x(\omega')\rangle = a(\omega)\delta(\omega - \omega'). \tag{B.17}$$

To determine the coefficient $a(\omega)$ we substitute into Equation (B.16), to give

$$g(\omega) = \frac{1}{2\pi} \int_{-\infty}^{\infty} a(\omega)\delta(\omega - \omega')\exp[-i(\omega' - \omega)]d\omega'$$

$$= \frac{a(\omega)}{2\pi} \int_{-\infty}^{\infty} \delta(\omega')\exp(i\omega'\tau)d\omega'$$

$$= \frac{a(\omega)}{2\pi}.$$

Thus we have the relation between the coefficient $a(\omega)$ and the $g(\omega)$, giving

$$\langle x^*(\omega)x(\omega')\rangle = 2\pi g(\omega)\delta(\omega - \omega') \tag{B.18}$$

or

$$g(\omega)\delta(\omega - \omega') = \frac{1}{2\pi}\langle |x(\omega)|^2\rangle. \tag{B.19}$$

So we see that the Fourier components of $G(t)$ are essentially the square of the Fourier components of $X(t)$. Thus $g(\omega)$ is often referred to as the *power* spectrum of the random function $X(t)$. There are clearly no mathematical difficulties with the definition of the power spectrum $g(\omega)$ as there were with the (amplitude) spectrum $x(\omega)$, Equation (B.11).

The above result is known as the Wiener–Kintchine Theorem (Wiener, 1933), which states that the power spectrum of a random variable is the Fourier transform of its autocorrelation function.

B.5 Quantum variables

Let us now consider the quantum mechanical formulation of the ideas discussed above. Now the observable X is represented by a quantum operator, and the averaging process, indicated by the angle brackets $\langle \rangle$, may be performed according to the rules of quantum statistical mechanics, discussed in Section 5.2.2. There we saw that the average (expectation value) of an operator $\mathscr{A}$ was given by the expression

$$A = \mathrm{Tr}\{\mathscr{A}\exp(-\beta\mathscr{H})\}/\mathrm{Tr}\{\exp(-\beta\mathscr{H})\}, \tag{B.20}$$

where $\mathscr{H}$ is the Hamiltonian operator for the system. In finding a quantum analogy for Equation (B.10), which defines the autocorrelation function, the operator $\mathscr{A}$ in Equation (B.20) must be replaced by $X^+(0)X(t)$. Here X^+ is the Hermitian conjugate of

X. Unfortunately there is a further slight complication in the quantum case because in general the operators $X^+(0)$ and $X(t)$ do not commute with each other. It is therefore customary, when required, to define the quantum correlation function in terms of the anti-commutator or symmetrised product:

$$\{A,B\} = (AB + BA)/2.$$

Thus the quantum autocorrelation function is

$$G(t) = \langle\langle\{X^+(0),X(t)\}\rangle\rangle, \tag{B.21}$$

where the thermal average is given by

$$\langle...\rangle = \mathrm{Tr}\{...\exp(-\beta\mathcal{H})\}/\mathrm{Tr}\{\exp(-\beta\mathcal{H})\}. \tag{B.22}$$

B.6 Thermal equilibrium and stationarity

An important property of the correlation functions we have used is that of *stationarity*. This applied both in the classical case where randomly fluctuating magnetic fields were being considered, and in the quantum case where time dependent interactions obeyed a Heisenberg equation of motion. In this section we show that quantum correlation functions are stationary for a system in equilibrium; it is thermal equilibrium which leads to stationarity. Since the classical case may be regarded as the $\hbar \to 0$ limit of the quantum case we may regard the proof as applying to both cases.

To be most general we consider the cross-correlation function of two variables A and B. If these have the quantum operators $\mathcal{A}$ and $\mathcal{B}$, then their Heisenberg time evolution is

$$\mathcal{A}(t) = \exp\left(\frac{i}{\hbar}\mathcal{H}t\right)\mathcal{A}\exp\left(-\frac{i}{\hbar}\mathcal{H}t\right),$$

$$\mathcal{B}(t) = \exp\left(\frac{i}{\hbar}\mathcal{H}t\right)\mathcal{B}\exp\left(-\frac{i}{\hbar}\mathcal{H}t\right),$$

where $\mathcal{H}$ is the system Hamiltonian. The correlation function is

$$G_{AB}(t,\tau) = \langle\mathcal{A}(t)\mathcal{B}(\tau)\rangle$$

$$= \left\langle \exp\left(\frac{i}{\hbar}\mathcal{H}t\right)\mathcal{A}\exp\left(-\frac{i}{\hbar}\mathcal{H}t\right)\exp\left(\frac{i}{\hbar}\mathcal{H}\tau\right)\mathcal{B}\exp\left(-\frac{i}{\hbar}\mathcal{H}\tau\right)\right\rangle \tag{B.23}$$

and, for the moment, we shall not worry about symmetrising the product.

When the system is in thermal equilibrium the average is performed over a Boltzmann distribution of states, as in Equation (B.22). The correlation function will thus be

$$G_{AB}(t,\tau) = \frac{\mathrm{Tr}\left\{\exp\left(-\frac{\mathscr{H}}{kT}\right)\exp\left(\frac{i}{h}\mathscr{H}t\right)\mathscr{A}\exp\left(-\frac{i}{h}\mathscr{H}t\right)\exp\left(\frac{i}{h}\mathscr{H}\tau\right)\mathscr{B}\exp\left(-\frac{i}{h}\mathscr{H}\tau\right)\right\}}{\mathrm{Tr}\left\{\exp\left(-\frac{\mathscr{H}}{kT}\right)\right\}}.$$

Now in the numerator trace all exponential factors commute since they contain precisely the same Hamiltonian operator. We can then shuffle these around; in particular we can move the time-evolution propagators pat the Boltzmann operator. In this way the correlation function can be transformed into

$$G_{AB}(t,\tau) = \frac{\mathrm{Tr}\left\{\exp\left(-\frac{\mathscr{H}}{kT}\right)\exp\left(\frac{i}{h}\mathscr{H}(t-\tau)\right)\mathscr{A}\exp\left(-\frac{i}{h}\mathscr{H}(t-\tau)\right)\mathscr{B}\right\}}{\mathrm{Tr}\left\{\exp\left(-\frac{\mathscr{H}}{kT}\right)\right\}}.$$

So it *does* depend on time only through the difference $t - \tau$. This expression can be interpreted as having all the time evolution contained in the $\mathscr{A}$ operator, so that an alternative expression for $G_{AB}(t,\tau)$ is

$$G_{AB}(t,\tau) = \langle \mathscr{A}(t-\tau)\mathscr{B}(0)\rangle \tag{B.24}$$

and stationarity is established.

Note that the shuffling of the time-evolution propagators would not have been possible if they had not commuted with the Boltzmann distribution operator and that commutation was only possible because the Boltzmann operator was the exponential of the *same* Hamiltonian operator, which was a consequence of the system being in thermal equilibrium. Thus we conclude that it is thermal equilibrium which leads to stationarity.

Clearly the conclusions of this section would continue to be true if it was the symmetrised product of the operators $\mathscr{A}$ and $\mathscr{B}$ which had been used. In that case, however, the correlation function is forced to be real and it is then invariant under time reversal.

A good general discussion of random functions is Wang and Uhlenbeck (1945) included in the excellent reprint book: Wax (1954). See also Yaglom (1962).

Appendix C

Interaction picture

C.1 Heisenberg and Schrödinger pictures

If a system is in a quantum state described by the ket $|0\rangle$ then the expectation value A of an arbitrary $\mathscr{A}$ may be expressed as

$$A = \langle 0| \mathscr{A} |0\rangle \tag{C.1}$$

and the time evolution of A is given by

$$A(t) = \langle 0| \exp(i\mathscr{H}t/\hbar)\mathscr{A}\exp(-i\mathscr{H}t/\hbar)|0\rangle, \tag{C.2}$$

where $\mathscr{H}$ is the Hamiltonian for this system.

Equation (C.2) may be viewed in two different ways. One can refer to the expectation value of the time dependent operator $\mathscr{A}(t)$ in the state $|0\rangle$, i.e.

$$A(t) = \langle 0| \mathscr{A}(t)|0\rangle,$$

where

$$\mathscr{A}(t) = \exp(i\mathscr{H}t/\hbar)\mathscr{A}\exp(-i\mathscr{H}t/\hbar). \tag{C.3}$$

Here the time evolution is associated with the operator $\mathscr{A}$, and this occurs according to Heisenberg's equation of motion

$$\frac{d\mathscr{A}}{dt} = -\frac{i}{\hbar}[\mathscr{A},\mathscr{H}]. \tag{C.4}$$

Alternatively the time evolution may be invested in the quantum state, where the appropriate description is in terms of the expectation value of the operator $\mathscr{A}$ in the time dependent state $|t\rangle$, i.e.

$$A(t) = \langle t| \mathscr{A} |t\rangle,$$

385

where

$$|t\rangle = \exp(-i\mathcal{H}t/\hbar)|0\rangle. \tag{C.5}$$

Here the operator $\mathcal{A}$ remains constant while the quantum state (ket) evolves according to Schrödinger's equation:

$$\frac{d|t\rangle}{dt} = -\frac{i}{\hbar}\mathcal{H}|t\rangle. \tag{C.6}$$

Mathematically these two views of time evolution in quantum mechanics are equivalent. The former is called the Heisenberg picture and this is the fundamental approach used in this book. In terms of the Heisenberg description many of the analogies between quantum and classical mechanics become apparent, as we have seen in Chapter 2. Very generally, Heisenberg operators go over to classical variables and the commutators of quantum mechanics go over to the Poisson brackets of classical mechanics.

The Schrödinger description of evolution is used in most elementary approaches to quantum mechanics and it emphasises the wave-like aspects of quantum states, in terms of such ideas as interference and superposition.

C.2 Interaction picture

The two pictures, those of Schrödinger and Heisenberg, correspond to the placing of the evolution generator $\exp(i\mathcal{H}t/\hbar)$ either with the quantum state $|0\rangle$ or with the operator $\mathcal{A}$. There is, however, a whole range of intermediate cases where the evolution is separated into two parts so that it is shared between the states and the operators. One term is associated with the quantum state: giving it a Schrödinger-like evolution. Such a view of the time behaviour of a system is called an interaction picture of Dirac picture. It can be of particular use when the Hamiltonian separates naturally into two parts, one of which may have only a trivial effect, such as the Zeeman Hamiltonian which simply causes precession. The interaction picture also lends itself naturally to the perturbation calculations where one part of a Hamiltonian is very much smaller than the other.

Let us write the system Hamiltonian as the sum of two terms:

$$\mathcal{H} = \mathcal{H}_1 + \mathcal{H}_2. \tag{C.7}$$

The evolution operator then takes the form

$$\exp(i\mathcal{H}t/\hbar) = \exp[i(\mathcal{H}_1 + \mathcal{H}_2)t/\hbar]. \tag{C.8}$$

In general the Hamiltonians $\mathcal{H}_1$ and $\mathcal{H}_2$ will not commute with each other so that factorisation of the exponential into

$$\exp(i\mathcal{H}_1 t/\hbar)\exp(i\mathcal{H}_2 t/\hbar) \tag{C.9}$$

is not permissible.

Assume, however, for the moment that the operators $\mathcal{H}_1$ and $\mathcal{H}_2$ *do* commute so that the factorisation above *is* allowed. Then the expectation value for the operator $\mathcal{A}$ may be written as

$$A(t) = \langle 0 | \exp(i\mathcal{H}_1 t/\hbar)\exp(i\mathcal{H}_2 t/\hbar)\mathcal{A}\exp(-i\mathcal{H}_2 t/\hbar)\exp(-i\mathcal{H}_1 t/\hbar)|0\rangle,$$

which symbolically may be expressed

$$A(t) = \langle t | \mathcal{A}(t) | t\rangle, \tag{C.10}$$

where $\mathcal{H}_1$ here generates the evolution of the state:

$$|t\rangle = \exp(-i\mathcal{H}_1 t/\hbar)|0\rangle \qquad \qquad \frac{\partial |t\rangle}{\partial t} = -\frac{i}{\hbar}\mathcal{H}_1|t\rangle,$$

while $\mathcal{H}_2$ is involved with the evolution of the operator:

$$\mathcal{A}(t) = \exp(i\mathcal{H}_2 t/\hbar)\mathcal{A}\exp(-i\mathcal{H}_2 t/\hbar) \qquad \qquad \frac{\partial \mathcal{A}}{\partial t} = -\frac{i}{\hbar}[\mathcal{A},\mathcal{H}].$$

Clearly when we are using the notation of Equation (C.10) we must keep our wits about us to be clear which part of the Hamiltonian is involved in the time evolution of the state and which with the operator.

C.3 General case

The above discussion, for the special case where $\mathcal{H}_1$ and $\mathcal{H}_2$ commute, is instructive in introducing the basic ideas of separating the time evolution of a system into two parts. In particular we saw that the crucial point in going to the interaction picture was the factorisation of the time-evolution operator. We now turn to the general case where the two Hamiltonians no longer commute. The factorisation in (C.9) is then certainly not allowed. While that simple factorisation is not correct, we shall still attempt a factorisation, accepting that it will be somewhat more complicated.

We shall demand an equality of the form:

$$\exp[i(\mathcal{H}_1 + \mathcal{H}_2)t/\hbar] = F(t)\exp(i\mathcal{H}_2 t/\hbar), \tag{C.11}$$

which can be regarded as a definition of $F(t)$. Our task, then, is to find a convenient expression for the interaction evolution operator $F(t)$. Differentiating Equation (C.11) with respect to time we obtain

$$\exp[i(\mathcal{H}_1 + \mathcal{H}_2)t/\hbar][i(\mathcal{H}_1 + \mathcal{H}_2)/\hbar] = F'(t)\exp(i\mathcal{H}_2 t/\hbar) + F(t)\exp(i\mathcal{H}_2 t/\hbar)[i\mathcal{H}_2/\hbar]$$
$$= F'(t)\exp(i\mathcal{H}_2 t/\hbar) + \exp[i\mathcal{H}_1 + \mathcal{H}_2)t/\hbar][i\mathcal{H}_2/\hbar].$$

Subtracting the last term of the right hand side from both sides and reexpressing the left hand side in terms of $F(t)$ using Equation (C.11) we have

$$F(t)\exp(i\mathcal{H}_2 t/\hbar)(i\mathcal{H}_1/\hbar) = F'(t)\exp(i\mathcal{H}_2 t/\hbar)$$

or

$$F'(t) = F(t)\exp(i\mathcal{H}_2 t/\hbar)(i\mathcal{H}_1/\hbar)\exp(-i\mathcal{H}_2 t/\hbar),$$

which is a differential equation $F(t)$ of familiar form. Before we integrate this let us take a look at the multiplier of $F(t)$ on the right hand side of the equation. In particular, observe that $\exp(i\mathcal{H}_2 t/\hbar)\mathcal{H}_1\exp(-i\mathcal{H}_2 t/\hbar)$ has an interesting form; it is the Hamiltonian $\mathcal{H}_1$ with a 'Heisenberg' time evolution generated by $\mathcal{H}_2$. We can therefore write this as $\mathcal{H}_1(t)$ whereupon the equation for $F(t)$ becomes

$$F'(t) = (i/\hbar)F(t)\mathcal{H}_1(t),$$

which can be integrated, using time-ordered exponentials, to give

$$F(t) = \exp\left[\frac{i}{\hbar}\int_0^t \mathcal{H}_1(\tau)d\tau\right].$$

The requirement that $F(0)$ be the unit operator determines the constant of integration.

Appendix D
Magnetic fields and canonical momentum

D.1 The problem

Magnetic fields are a little difficult to incorporate into the formalism of mechanics. While a force may be considered directly, in terms of its effect on Newton's Second Law of Motion, it is often convenient to work in terms of a potential energy function, the gradient of which gives the value of the force. Electric and gravitational forces are treated in this way, and the time evolution of such systems follows from considerations such as energy conservation. At the simplest level working with an energy function is preferable since it is a scalar function rather than a vector and one is therefore free to work in any co-ordinate system or frame of reference. These ideas are conventionally treated within the formalism of the Hamiltonian description of classical mechanics. In this discussion we wish to avoid this complication, restricting ourselves to the basic physical ideas involved. However, it should be realised that when going over to quantum mechanics it is essentially the energy function which becomes the Hamiltonian operator; force is not a well-defined quantum concept.

Physically we see that there is a difficulty with the energy related to magnetic fields because the force on a charged particle in such a field is perpendicular to its motion. Thus motion of the particle does not involve doing work against the field. Mathematically the problem is that the curl of the magnetic force is not zero, so it cannot be expressed as the gradient of a scalar potential. We must find a way of modifying the 'energy' formulation of Newton's Second Law:

$$m\frac{d^2\mathbf{r}}{dt^2} = -\operatorname{grad} V \qquad (D.1)$$

to incorporate magnetic effects in a natural and straightforward way.

D.2 Change of reference frame

One way of doing this is to transform to a new frame of reference where the charge

389

under consideration is stationary. After all, the *behaviour* of an object is independent of the reference frame used even though its mathematical description may be different. We shall start from the expression for the Lorentz force on the particle of charge q,

$$\mathbf{F} = q(\mathbf{E} + \mathbf{v} \times \mathbf{B}) \tag{D2}$$

and write the electric and magnetic fields in terms of the usual scalar and vector potentials:

$$\mathbf{E} = -\operatorname{grad} V - \frac{\partial \mathbf{A}}{\partial t}, \qquad \mathbf{B} = \operatorname{curl} \mathbf{A}.$$

Now we shall change to a frame of reference moving with the charge, i.e. to a frame with velocity $\mathbf{v}$. We first established that the (total) force experienced by a body is invariant under such a transformation, since

$$\mathbf{F}' = m\frac{d}{dt}\left(\frac{d\mathbf{r}}{dt} + \mathbf{v}\right) = m\frac{d^2\mathbf{r}}{dt^2} = \mathbf{F},$$

where $\mathbf{v}$ is a uniform velocity.

In the new frame S' moving with velocity $\mathbf{v}$ away from the old frame S the force on the charge is given by

$$\mathbf{F} = \mathbf{F}' = q\{-\operatorname{grad} V' - \partial \mathbf{A}'/\partial t\}.$$

Observe that the curl $\mathbf{A}$ term has vanished here; we must be on the right track. Now the grad V term which is related to the distribution of other charges around it is clearly unchanged by the transformation. Let us examine the $\partial \mathbf{A}'/\partial t$ term in some detail. This time derivative must now be evaluated in the frame moving with the charge. So as well as the old $\partial \mathbf{A}'/\partial t$ term there is a new contribution due to the motion:

$$\frac{\partial \mathbf{A}'}{\partial t} = \frac{\partial \mathbf{A}}{\partial t} + \frac{\partial \mathbf{A}}{\partial x}\frac{dx}{dt} + \frac{\partial \mathbf{A}}{\partial y}\frac{dy}{dt} + \frac{\partial \mathbf{A}}{\partial z}\frac{dz}{dt}.$$

But this is, by the chain rule of differentiation, the expression for the *total* differential of $\mathbf{A}$:

$$\partial \mathbf{A}'/\partial t = d\mathbf{A}/dt.$$

Thus ignoring the contribution from the electrostatic potential, the Lorentz force takes the particularly simple form

$$\mathbf{F} = -q\left(\frac{d\mathbf{A}}{dt}\right). \tag{D.3}$$

Newton's equation of motion then appears as

$$m(d^2r/dt^2) = -q(dA/dt) \tag{D.4}$$

or

$$\frac{d}{dt}(mv + qA) = 0. \tag{D.5}$$

D.3 Canonical momentum

We see that in the presence of a magnetic field the quantity mv is no longer conserved. Usually this would imply the presence of an external potential. However, since we know that the magnetic effect cannot be represented by the addition of a potential, we can resolve the difficulty by generalising the definition of momentum. We define the canonical momentum by

$$p = mv + qA. \tag{D.6}$$

We see that this momentum is conserved in the absence of a potential field and, of course, in the absence of a magnetic field it reduces to the familiar form for mechanical momentum.

The conclusion from these considerations is that if the equations of motion are written in potential energy–momentum form:

$$\frac{dp}{dt} = -\text{grad}V \tag{D.7}$$

then the effect of magnetic fields can be incorporated through the introduction of the canonical momentum. The only thing to bear in mind is that the kinetic energy is still $mv^2/2$, so that in terms of the canonical momentum it now becomes

$$T = (p - qA)^2/2m. \tag{D.8}$$

This was the way that the magnetic field was introduced in Section 5.5.7 in the discussion of the origin of chemical shifts.

In the transition from classical to quantum mechanics, where observables are replaced by operators, it is the canonical momentum which is represented by the momentum operator $-i\hbar\text{grad}$. So the generalised rule for converting to quantum mechanics is

$$-i\hbar\text{grad} = mv + qA. \tag{D.9}$$

Appendix E
Alternative classical treatment of relaxation

E.1 Introduction

In Chapter 4 we treated the question of magnetic relaxation in a classical way. In particular we considered the case of mobile spins, subject to fluctuating magnetic fields. We started from the classical 'gyroscope' equations for the precession of a magnetic moment in a magnetic field and by a reasoned sequence of approximations we established that the transverse magnetisation relaxed in an exponential manner and that the relaxation time T_2 was given by

$$1/T_2 = M_2 \tau_c.$$

The merit of that treatment was that it provided a powerful and intuitive model of wide applicability. We were able to use it to treat diffusion in Chapter 4 and in Chapter 7 its quantum mechanical generalisation led to the Kubo–Tomita Theory of relaxation.

It was not possible, however, using that procedure, to treat either longitudinal relaxation or the non-adiabatic contribution to the transverse relaxation within the classical framework. For that reason we present in this appendix an alternative classical description of relaxation. The same approximations as in Chapter 4 will be used, but in a different order. This will allow the treatment of longitudinal and non-adiabatic transverse relaxation as well as the adiabatic transverse case, in a unified way. The inspiration for this treatment was a paper by Rorschach (1986).

E.2 Equations of motion

We start from the basic equation of motion for a magnetic moment in a magnetic field

$$\dot{\boldsymbol{\mu}} = \gamma \boldsymbol{\mu} \times \mathbf{B},$$

where the magnetic field $\mathbf{B}$ is made up of the applied static field B_0 in the z direction, together with the fluctuating field components $b_x(t)$, $b_y(t)$ and $b_z(t)$:

$$\mathbf{B} = b_x(t)\mathbf{i} + b_y(t)\mathbf{j} + [b_z(t) + B_0]\mathbf{k}.$$

In component form the equations of motion are

$$\dot{\mu}_x = \gamma\{\mu_y[b_z(t) + B_0] - \mu_z b_y(t)\},$$
$$\dot{\mu}_y = -\gamma\{\mu_x[b_z(t) + B_0] - \mu_z b_x(t)\},$$
$$\dot{\mu}_z = \gamma\{\mu_x b_y(t) - \mu_y b_x(t)\}.$$

In the absence of the fluctuating fields $b_x(t)$, $b_y(t)$ and $b_z(t)$ the equations of motion

$$\dot{\mu}_x = \gamma\mu_y B_0,$$
$$\dot{\mu}_y = -\gamma\mu_x B_0,$$
$$\dot{\mu}_z = 0$$

have solution

$$\begin{rcases} \mu_x(t) = \mu_x(0)\cos(\omega_0 t) + \mu_y(0)\sin(\omega_0 t), \\ \mu_y(t) = -\mu_x(0)\sin(\omega_0 t) + \mu_y(0)\cos(\omega_0 t), \\ \mu_z(t) = \mu_z(0), \end{rcases} \quad (\text{E.1})$$

where $\omega_0 = \gamma B_0$; this is the usual Lamor precession.

This motivates our transformation to the rotating frame. Within the spirit of the interaction picture we incorporate the effects of the fluctuating fields by allowing the 'constants' $\mu_x(0)$, $\mu_y(0)$ and $\mu_z(0)$ to vary with time. In other words, we write the solution of the full equations of motion as

$$\mu_x(t) = m_x(t)\cos(\omega_0 t) + m_y(t)\sin(\omega_0 t),$$
$$\mu_y(t) = -m_x(t)\sin(\omega_0 t) + m_y(t)\cos(\omega_0 t),$$
$$\mu_z(t) = m_z(t).$$

Here $m_x(t)$, $m_y(t)$ and $m_z(t)$ are the components of the magnetic moment in the rotating frame. They obey the equations of motion

$$\begin{rcases} \dot{m}_x(t) = \gamma\{-m_z[b_x\sin(\omega_0 t) + b_y\cos(\omega_0 t)] + b_z m_y\}, \\ \dot{m}_y(t) = \gamma\{-b_z m_x + m_z[b_y\sin(\omega_0 t) + b_x\cos(\omega_0 t)], \\ \dot{m}_z(t) = \gamma\{m_y[b_y\sin(\omega_0 t) - b_x\cos(\omega_0 t)] + m_x[b_x\sin(\omega_0 t) + b_y\cos(\omega_0 t)]\}. \end{rcases} \quad (\text{E.2})$$

It is the solutions of these equations which will be investigated in this appendix.

E.3 Adiabatic T_2

It is the z component of the fluctuating magnetic field which has the dominant influence on the transverse relaxation. Considering the adiabatic case, where the y and z components of the fluctuating fields may be ignored, the equations of motion, Equation (E.2), reduce to

$$\left.\begin{array}{ll} \dot{m}_x(t) = & \gamma b_z m_y, \\ \dot{m}_y(t) = & -\gamma b_z m_x, \\ \dot{m}_z(t) = & 0. \end{array}\right\} \quad \text{(F 3)}$$

It is a straightforward matter to solve these, giving

$$m_x(t) = m_x(0)\cos\left[\gamma \int_0^t b_z(\tau)d\tau\right] + m_y(0)\sin\left[\gamma \int_0^t b_z(\tau)d\tau\right],$$

$$m_y(t) = -m_x(0)\sin\left[\gamma \int_0^t b_z(\tau)d\tau\right] + m_y(0)\cos\left[\gamma \int_0^t b_z(\tau)d\tau\right],$$

$$m_z(t) = m_z(0).$$

Using the procedures of Chapter 4, one would perform an ensemble average over the spins in the system using a Gaussian function for the distribution of phase accumulated by the spins. Essentially one is performing the averaging procedures on the *solution* to the spin equations of motion. The interest in this appendix, however, is in performing the various procedures directly on the equations of motion so that we obtain equations of motion for the averaged, macroscopic magnetisation.

We start by integrating the $\dot{m}_y$ equation of Equation (E.3)

$$m_y(t) = m_y(0) - \gamma \int_0^t b_z(\tau)m_x(\tau),$$

which is then substituted into the $\dot{m}_x$ equation of Equation (E.3):

$$\dot{m}_x(t) = \gamma b_z(t)m_y(0) - \gamma^2 \int_0^t b_z(t)b_z(\tau)m_x(\tau)d\tau.$$

This equation is exact, and it remains exact when we perform an average over all the particles in the system,

$$\frac{d}{dt}\langle m_x(t)\rangle = \gamma\langle b_z(t)m_y(0)\rangle - \gamma^2 \int_0^t \langle b_z(t)b_z(\tau)m_x(\tau)\rangle d\tau.$$

We now proceed with our sequence of approximations. First is the separation of time scales. We know, particularly from the discussions of Chapter 4, that when the variation of $b_z(t)$ is rapid, the variation of $m_x(t)$ is slow. In that case the b and the m can be regarded as statistically independent and they may be averaged separately:

$$\frac{d}{dt}\langle m_x(t)\rangle = \gamma\langle b_z(t)\rangle\langle m_y(0)\rangle - \gamma^2 \int_0^t \langle b_z(t)b_z(\tau)\rangle\langle m_x(\tau)\rangle d\tau. \qquad \text{(E.4)}$$

Now the average of $b_z(t)$ in the first term on the right hand side is zero. It has to be, since if it were not then its mean value would be subtracted off and incorporated into the static field $\mathbf{B}_0$. Thus the first term on the right hand side vanishes. Turning to the second term, we observe that $\langle b_z(t)b_z(\tau)\rangle$ is the autocorrelation function of the fluctuating field $b_z(t)$. This is a stationary random function, depending only on the time difference $t - \tau$. Using the notation of Chapter 4 and to be consistent with the definition of Equation (4.31) we write this correlation function as

$$G_{zz}(t - \tau) = \gamma^2\langle b_z(t)b_z(\tau)\rangle.$$

Then by a change of variables, Equation (E.4) can be written as

$$\frac{d}{dt}\langle m_x(t)\rangle = -\int_0^t G_{zz}(\tau)\langle m_x(t - \tau)\rangle d\tau.$$

Now $b_z(t)$ varies randomly, as indicated in Figure 4.5, and $G(t)$ will decay to zero, as shown in Figure 4.6, over a time scale τ_c, the correlation time. During the time that $G(t)$ falls to zero, $\langle m_x(t - \tau)\rangle$ will have changed negligibly (separation of time scales). Therefore we may take the magnetisation out of the integral, as $\langle m_x(t)\rangle$. Then if we consider times greater than τ_c, over which $\langle m_x(t)\rangle$ will vary, the upper limit of the integral can be extended to infinity since its argument is already zero. Thus we obtain

$$\frac{d}{dt}\langle m_x(t)\rangle = -\langle m_x(t)\rangle \int_0^\infty G_{zz}(\tau)d\tau,$$

which has an exponential solution

$$\langle m_x(t)\rangle = \langle m_x(0)\rangle\exp(-t/T_2),$$

where the relaxation time T_2 is given by

$$1/T_2 = \int_0^\infty G_{zz}(\tau)d\tau. \qquad \text{(E.5)}$$

We can identify the zero-time value of the autocorrelation function as the second moment M_2 and *define* τ_c the correlation time as the area under $G(t)$ divided by its zero-time value so that

$$1/T_2 = M_2\tau_c,$$

the conventional expression.

   This discussion has been carried out in some detail, using the familiar case of the adiabatic T_2, so that the application and the implications of the various approximations are clear. In the following sections we will use similar procedures in treating the less familiar examples of longitudinal relaxation and the non-adiabatic T_2 from the classical standpoint.

E.4 Treatment of T_1

It is a fundamental assumption that the components of the fluctuating field $b_x(t)$, $b_y(t)$ and $b_z(t)$ are statistically independent. This means that averages of products such as $\langle b_x(t)b_y(\tau)\rangle$ are zero. Since the fundamental equations of motion are linear in the fields, this means that the effects of the three components may be treated separately and the results added. This provides a considerable simplification. As a byproduct of the previous section we saw that the z component of the fluctuating field had no effect on the longitudinal relaxation; m_z remained constant under its influence. In this section we will then consider the effect of the transverse components $b_x(t)$ and $b_y(t)$, and we know we can treat them separately. Thus we start by considering just the effect of $b_x(t)$; we set $b_y(t)$ and $b_z(t)$ to zero. Then Equation (E.2) reduces to the set

$$\left.\begin{aligned}
\dot{m}_x(t) &= -\gamma m_z b_x \sin(\omega_0 t),\\
\dot{m}_y(t) &= \gamma m_z b_x \cos(\omega_0 t),\\
\dot{m}_z(t) &= \gamma[-m_y b_x \cos(\omega_0 t) + m_x b_x \sin(\omega_0 t)].
\end{aligned}\right\} \tag{E.6}$$

In this case we integrate the first two equations:

$$m_x(t) = m_x(0) - \gamma \int_0^t m_z(\tau)b_x(\tau)\sin(\omega_0\tau)d\tau$$

$$m_y(t) = m_y(0) + \gamma \int_0^t m_z(\tau)b_x(\tau)\cos(\omega_0\tau)d\tau$$

and substitute them into the third, giving

$$\dot{m}_z(t) = -\gamma m_y(0)b_x(t)\cos(\omega_0 t) - \gamma^2 \int_0^t b_x(t)b_x(\tau)\cos(\omega_0 t)\cos(\omega_0 \tau)m_z(\tau)d\tau$$

$$+ \gamma m_x(0)b_x(t)\sin(\omega_0 t) - \gamma^2 \int_0^t b_x(t)b_x(\tau)\sin(\omega_0 t)\sin(\omega_0 \tau)m_z(\tau)d\tau.$$

At this stage the equation is exact if there is only b_x present. Next we take the average over all spins in the specimen. As in the previous section, firstly we invoke the separation of time scales to permit averaging over the magnetic moment and field independently. Then since b_x averages to zero and we identify $\gamma^2\langle b_x(t)b_x(\tau)\rangle$ as the autocorrelation function of the x components of the field, $G_{xx}(t - \tau)$, the equation of motion for m_z becomes

$$\frac{d}{dt}\langle m_z(t)\rangle = -\int_0^t G_{xx}(t - \tau)\cos[\omega_0(t - \tau)]\langle m_z(\tau)\rangle d\tau,$$

or, by a change of integration variable:

$$\frac{d}{dt}\langle m_z(t)\rangle = -\int_0^t G_{xx}(\tau)\cos(\omega_0 \tau)\langle m_z(t - \tau)\rangle d\tau.$$

As in the transverse case, during the time that the $G(t)$ falls to zero, $\langle m_z(t - \tau)\rangle$ will have changed negligibly (separation of time scales). Therefore we may take the magnetisation out of the integral, as $\langle m_z(t)\rangle$, and if we consider times greater than τ_c, over which $\langle m_z(t)\rangle$ will vary, then the upper limit of the integral can be extended to infinity since its argument is already zero. Thus we obtain

$$\frac{d}{dt}\langle m_z(t)\rangle = -\langle m_z(t)\rangle \int_0^\infty G_{xx}(\tau)\cos(\omega_0 \tau)d\tau.$$

This is the effect of the x component of the fluctuating magnetic fields. There will be a similar effect from the y component, which could be established by performing the above calculations for b_y instead of b_x. As we argued at the start of this section, these effects are additive, but there is no effect from b_z, so the total effect of the fluctuating fields on the longitudinal magnetisation is contained in the equation

$$\frac{d}{dt}\langle m_z(t)\rangle = -\langle m_\perp(t)\rangle \int_0^\infty [G_{xx}(\tau) + G_{yy}(\tau)]\cos(\omega_0\tau)d\tau,$$

which has an exponential solution

$$\langle m_z(t)\rangle = \langle m_z(0)\rangle\exp(-t/T_1),$$

where the relaxation time T_1 is given by

$$1/T_1 = \int_0^\infty [G_{xx}(\tau) + G_{yy}(\tau)]\cos(\omega_0\tau)d\tau. \tag{E.7}$$

E.5 Non-adiabatic T_2

We return now to consider transverse relaxation; this time we are interested in the effects of the transverse magnetic fields. As in the previous section, since the effects of b_x and b_y are independent, we will treat them separately and we start with a fluctuating field with x component only. Thus again we are working with Equation (E.6), and here it proves convenient to express the set in complex form through the introduction of a complex $m(t)$:

$$m(t) = m_x(t) + im_y(t).$$

Then Equation (E.6) becomes

$$\dot{m}(t) = i\gamma m_z(t)b_x(t)\exp(i\omega_0 t),$$

$$\dot{m}_z(t) = \frac{i\gamma b_x(t)}{2}[m(t)\exp(-i\omega_0 t) - m^*(t)\exp(i\omega_0 t)].$$

We integrate the second equation

$$m_z(t) = m_z(0) + \frac{i\gamma}{2}\int_0^t b_x(\tau)[m(\tau)\exp(-i\omega_0\tau) - m^*(\tau)\exp(i\omega_0\omega)]d\tau$$

and substitute it into the first, giving

$$\dot{m}(t) = i\gamma m_z(0)b_x(t)\exp(i\omega_0 t) - \frac{\gamma^2}{2}\int_0^t b_x(t)b_x(\tau)[m(\tau)\exp(-i\omega_0\tau) - m^*(\tau)\exp(i\omega_0 t)]\exp(i\omega_0 t)d\tau.$$

Again we note that at this stage the equation is exact if there is only b_x present. Next we take the average over all spins in the specimen. As before, firstly we invoke the separation of time scales to permit averaging over magnetic moment and field independently. Then since b_x averages to zero and we identify $\gamma^2\langle b_x(t)b_x(\tau)\rangle$ as the autocorrelation function of the x components of the field, $G_{xx}(t-\tau)$, the equation of motion for the complex m becomes

$$\frac{d}{dt}\langle m(t)\rangle = -\frac{1}{2}\int_0^t G_{xx}(t-\tau)\exp[-i\omega_0(t-\tau)]\langle m(\tau)\rangle d\tau$$

$$+\frac{1}{2}\int_0^t G_{xx}(t-\tau)\exp[-i\omega_0(t+\tau)]\langle m^*(\tau)\rangle d\tau.$$

This gives the relaxation of the complex transverse magnetisation caused by fluctuating fields in the z direction. We may calculate the similar effect from fields in the y direction, but in that case the sign of the second term is reversed. The total effect from fields in the x and the y directions is then

$$\frac{d}{dt}\langle m(t)\rangle = -\frac{1}{2}\int_0^t [G_{xx}(t-\tau)+G_{yy}(t-\tau)]\exp[-i\omega_0(t-\tau)]\langle m(\tau)\rangle d\tau +$$

$$\frac{1}{2}\int_0^t G_{xx}(t-\tau)-G_{yy}(t-\tau)]\exp[-i\omega_0(t+\tau)]\langle m^*(\tau)\rangle d\tau.$$

Assuming the system has axial symmetry about the z direction then the xx and the yy correlation functions will be equal, and the second term will vanish. Making a change of integration variable, invoking the separation of time scales to remove the magnetisation from the integral and extending the upper limit of the integral to infinity we obtain the equation of motion for $\langle m(t)\rangle$:

$$\frac{d}{dt}\langle m(t)\rangle = -\frac{1}{2}\langle m(t)\rangle\int_0^\infty [G_{xx}(\tau)+G_{yy}(\tau)]\exp(-i\omega_0\tau)d\tau.$$

Although this equation has an exponential solution we observe that the 'time constant' will, in general, be complex. The behaviour of the complex magnetisation can thus be expressed as

$$\langle m(t)\rangle = \langle m(0)\rangle\exp(-t/T_2^{na})\exp(i\Delta t),$$

where

$$\frac{1}{T_2^{na}} - i\Delta = \frac{1}{2}\int_0^\infty [G_{xx}(\tau) + G_{yy}(\tau)]\exp(-i\omega_0\tau)d\tau. \tag{E.8}$$

Here T_2^{na} is the non-adiabatic contribution to the transverse relaxation, while Δ represents a small frequency shift:

$$\frac{1}{T_2^{na}} = \frac{1}{2}\int_0^\infty [G_{xx}(\tau) + G_{yy}(\tau)]\cos(\omega_0\tau)d\tau,$$

$$\Delta = \frac{1}{2}\int_0^\infty [G_{xx}(\tau) + G_{yy}(\tau)]\sin(\omega_0\tau)d\tau. \tag{E.9}$$

The shift is the classical analogue of the effect treated quantum mechanically in Section 7.3.5. It is sometimes believed that the shift is fundamentally quantum mechanical in origin; this is not the case.

Because of the statistical independence of the fluctuating magnetic fields $b_x(t)$, $b_y(t)$ and $b_z(t)$, we know that we can add their effects. Thus to the non-adiabatic $1/T_2$ we can add the adiabatic part obtained in Section E.2, giving the total transverse relaxation rate

$$\frac{1}{T_2} = \int_0^\infty G_{zz}(\tau)d\tau + \frac{1}{2}\int_0^\infty [G_{xx}(\tau) + G_{yy}(\tau)]\cos(\omega_0\tau)d\tau. \tag{E.10}$$

E.6 Spectral densities

In Chapter 7 we introduced the spectral density functions as the Fourier transform of the autocorrelation functions. By analogy with Equation (7.44) we here define

$$J_{\alpha\alpha}(\omega) = \int_{-\infty}^\infty G_{\alpha\alpha}(t)\exp(i\omega t)dt. \tag{E.11}$$

Then the expressions for T_1, the adiabatic T_2 and the full T_2 may be written as

$$\left.\begin{aligned}
1/T_1 &= \tfrac{1}{2}[J_{xx}(\omega_0) + J_{yy}(\omega_0)]; \\
1/T_2^{ad} &= \tfrac{1}{2}J_{zz}(0), \\
1/T_2 &= \tfrac{1}{2}J_{zz}(0) + \tfrac{1}{4}[J_{xx}(\omega_0) + J_{yy}(\omega_0)],
\end{aligned}\right\} \tag{E.12}$$

which are the classical/local field model equivalents to the quantum expressions, Equations (7.45) and (7.48), but note that we have, in the derivation of the non-adiabatic T_2, made an assumption of axial symmetry of the fluctuating fields – i.e., $J_{xx} = J_{yy}$. For a fully isotropic system, where we can write $J_{xx}(\omega)$, $J_{yy}(\omega)$ and $J_{zz}(\omega)$ all equal to a single $J(\omega)$ the relaxation times become

$$
\left.
\begin{aligned}
1/T_1 &= J(\omega_0), \\
1/T_2^{ad} &= \tfrac{1}{2}J(0), \\
1/T_2 &= \tfrac{1}{2}J(0) + \tfrac{1}{2}J(\omega_0).
\end{aligned}
\right\}
\tag{E.13}
$$

As expected, we only have the single-frequency terms here. The double-frequency terms which appear from the quantum treatment of the dipolar interaction result from the simultaneous flip of two spins; this model does not cope with such transitions.

In the isotropic case the expression for the shift simplifies to

$$
\Delta = \int_0^\infty G(\tau)\sin(\omega_0\tau)d\tau,
\tag{E.14}
$$

but this cannot be expressed (simply) in terms of the spectral density $J(\omega)$. A relation is possible, however, using a Hilbert transform (Appendix A).

Appendix F
$G_m(t)$ for rotationally invariant systems

F.1 Introduction

In Chapter 7 we introduced the dipolar autocorrelation functions $G_m(t)$, where m is the spin-flip index taking values 0, 1 and 2. Various properties of the $G_m(t)$ were treated in that chapter. However, we simply stated there that for a rotationally invariant system, as a normal fluid would be, the functions become independent of the index m and the relaxation properties of the system are all contained in a single function $G(t)$.

The dipolar autocorrelation functions are defined in Equation (7.35):

$$G_m(t) = \left(\frac{\mu_0}{4\pi}\right)^2 \frac{12\pi\hbar^2\gamma^4}{5N} \sum_{i \neq j} \frac{Y_2^{m*}[\Omega_{ij}(t)]\, Y_2^m[\Omega_{ij}(0)]}{r_{ij}^3(t) r_{ij}^3(0)},$$

where Ω_{ij} is the direction the line joining the ith and the jth spins makes with the z axis. If the system is rotationally invariant then all directions will be equally likely. Thus, for such systems, we are led to making an average over all orientations of the $\mathbf{r}_{ij}$ vector.

F.2 Rotation of spherical harmonics

If the vector $\mathbf{r}$ is rotated through an angle ω to a new vector $\mathbf{r}'$, in other words if the old orientation Ω is transformed into the new orientation Ω', then the spherical harmonic Y_2^m transforms as

$$Y_2^m(\Omega') = \sum_{m' = -2}^{2} Y_2^{m'}(\Omega)\mathscr{D}_{m'm}^2(\omega), \tag{F.1}$$

into a linear combination of the five second order spherical harmonics. The coefficients are given by the elements of the matrix $\mathscr{D}$; details may be found in the book by Edmonds (1957), although he uses Euler angles to provide a specific representation for the (three-dimensional) rotation which we symbolically denote by ω.

In performing the averaging of the autocorrelation function over all orientations it is only the product of the spherical harmonics which must be considered; only they contain orientation information, since the separation r_{ij} remains fixed. The product transforms as

$$Y_2^{m*}[\Omega'(t)]\,Y_2^m[\Omega'(0)] = \sum_{m',m''=-2}^{2} Y_2^{m'*}[\Omega(t)]\,Y_2^{m''}[\Omega(0)]\mathscr{D}_{m'm}^2{}^*(\omega)\mathscr{D}_{m''m}^2(\omega),$$

which must be averaged over orientations ω. From the structure of the above expression we see that it is the product of the rotation matrices which must be averaged. By reference to Edmonds (p. 62) we find

$$\langle \mathscr{D}_{m'm}^2{}^*(\omega)\mathscr{D}_{m''m}^2(\omega)\rangle_\omega = \frac{\delta_{m'm''}}{5}, \tag{F.2}$$

so that the average of the spherical harmonics is

$$\langle Y_2^{m*}[\Omega'(t)]\,Y_2^m[\Omega'(0)]\rangle_\omega = \frac{1}{5}\sum_{m'=-2}^{2} Y_2^{m'*}[\Omega(t)]\,Y_2^{m'}[\Omega(0)]. \tag{F.3}$$

This is independent of the index m, so that the $G_m(t)$ will be so, as required.

F.3 Simplification

A further simplification is possible by use of the addition theorem for spherical harmonics (Edmonds, p. 63). That states

$$P_2\{\cos[\Omega(t)]\} = \frac{4\pi}{5}\sum_{m=-2}^{2} Y_2^{m*}[\Omega(t)]\,Y_2^m[\Omega(0)],$$

where $\Omega(t)$ is the (scalar) angle between the directions $\Omega(t)$ and $\Omega(0)$ and P_2 is the second order Legendre polynomial: $P_2(z) = (3z^2 - 1)/2$. Thus we find the orientational average of the product of spherical harmonics to be

$$\langle Y_2^{m*}[\Omega'(t)]\,Y_2^m[\Omega'(0)]\rangle_\omega = \frac{1}{4\pi}P_2\{\cos[\Omega(t)]\}$$

or

$$\langle Y_2^{m*}[\Omega'(t)]\,Y_2^m[\Omega'(0)]\rangle_\omega = \{3\cos^2[\Omega(t)] - 1\}/8\pi \tag{F.4}$$

and the expression for $G(t)$ is then

$$G(t) = \left(\frac{\mu_0}{4\pi}\right)^2 \frac{3\hbar^2\gamma^4}{10N} \sum_{i\neq j} \frac{\{3\cos^2[\Omega_{ij}(t)] - 1\}}{r_{ij}^3(t)r_{ij}^3(0)}. \tag{F.5}$$

Here $\Omega_{ij}(t)$ is the angle through which the vector r_{ij} has turned in time t.

Appendix G

$P(\Omega, \Omega_0, t)$ for rotational diffusion

G.1 Rotational diffusion

In this appendix we derive the form for the probability function for rotational diffusion. This is the quantity $P(\Omega, \Omega_0, t)$ giving the probability that an internuclear vector in a tumbling molecule, starting from an initial orientation Ω_0, will have rotated to a new orientation Ω in a time t. The tumbling motion is assumed to be described by the rotational diffusion equation

$$\frac{\partial P}{\partial t} = \frac{D}{a^2} \nabla^2 P, \tag{G.1}$$

where D is the rotational diffusion coefficient and ∇^2 here is the rotational Laplacian – the 'radius' a is fixed. Thus in spherical polar co-ordinates ∇^2 is given by

$$\nabla^2 P = \frac{1}{a^2 \sin\theta} \frac{\partial}{\partial\theta}\left(\sin\theta \frac{\partial P}{\partial\theta}\right) + \frac{1}{a^2 \sin^2\theta} \frac{\partial^2 P}{\partial\phi^2}.$$

The rotational diffusion equation may be regarded as describing the diffusion of a point over the surface of a sphere of radius a.

G.2 General solution

The general solution to Equation (G.1) can be written as a linear combination of spherical harmonics

$$P(\Omega, t) = \sum_{l=1}^{\infty} \sum_{m=-l}^{l} a_m^l(t) Y_l^m(\Omega), \tag{G.2}$$

the time dependence being in the coefficients a_m^l. These are found by substituting this solution into the diffusion equation and making use of the orthogonality relations for the spherical harmonics. Since the effect of the rotational Laplacian is

$$\nabla^2 Y_l^m(\Omega) = -l(l+1)Y_l^m(\Omega),$$

we obtain the equation for the a_m^l coefficients:

$$\frac{\partial}{\partial t}a_m^l(t) = -\frac{D}{a^2}l(l+1)a_m^l(t).$$

This has the solution

$$a_m^l(t) = a_m^l(0)\exp(-t/\tau_l), \tag{G.3}$$

where

$$\frac{1}{\tau_l} = \frac{D}{a^2}l(l+1). \tag{G.4}$$

Equation (G.3) gives us the time dependence of the expansion coefficients. We substitute this into the general expression, equation (G.2), giving

$$P(\Omega,t) = \sum_{l,m} a_m^l(0)Y_l^m(\Omega)\exp(-t/\tau_l). \tag{G.5}$$

This is the general solution to the rotational diffusion equation. It is given in terms of the set of initial coefficients $a_m^l(0)$ which must be determined from the boundary conditions.

G.3 Particular solution

The function which we require, $P(\Omega,\Omega_0,t)$, is the solution which tells us that at time $t = 0$, $\Omega = \Omega_0$. In other words, we require the particular solution for which

$$P(\Omega,0) = \sum_{l,m} a_m^l(0)Y_l^m(\Omega) = \delta(\Omega - \Omega_0).$$

Now the delta function may be expanded in spherical harmonics as

$$\delta(\Omega - \Omega_0) = \sum_{l,m} Y_l^{m*}(\Omega_0)Y_l^m(\Omega),$$

so that we immediately identify the coefficients as

$$a_m^l(0) = Y_l^{m*}(\Omega_0)$$

and the solution for our probability function is then

$$P(\Omega,\Omega_0,t) = \sum_{l,m} Y_l^{m*}(\Omega_0) Y_l^m(\Omega) \exp(-t/\tau_l).$$ \hfill (G.6)

This is the required result, as used in Chapter 7, in the discussion of relaxation through rotational motion.

Problems

Chapter 1

1.1 Using the expression $d\mathbf{F} = I d\mathbf{l} \times \mathbf{B}$ for the force exerted on a current element by a magnetic field $\mathbf{B}$, show that a circular current loop of area A carrying current I behaves like a dipole (i.e. it experiences a torque) μ given by $\mu = IA$ where the direction of $\mathbf{A}$ is perpendicular to the plane of the loop.

1.2 Consider a particle of mass m executing damped simple harmonic motion, and subject to a driving force $mF_0\cos(\omega t)$.
  Show that P, the power absorbed from the force, is given by

$$P = mF_0{}^2\omega\chi''/2,$$

where χ'' is the quadrature response per unit force (see Equations (1.9) and (1.10)). Note that the in-phase response does not involve energy transfer and for this reason the $\chi''(\omega)$ curve is loosely called 'the absorption'.

1.3 A damped simple harmonic oscillator (Equation (1.7)) is subject to a very narrow pulse of amplitude F_0 (force per unit mass) lasting for a time τ. Show that:

(a) the initial effect of the pulse is to displace the oscillator from equilibrium a distance $F_0\tau/2\beta$,
(b) subsequently the oscillator amplitude decays exponentially.

Discuss how short τ must be to obtain the result in (a). Explain how you could treat the problem by considering the pulse as a Fourier synthesis of the form $\int_{-\infty}^{\infty} f(\omega)\cos(\omega t)d\omega$, and therefore the displacement of the oscillator as a synthesis of its responses to sinusoids of frequency ω and amplitude $f(\omega)$.

1.4 Show that Equations (1.9) and (1.10) represent the results of regarding the force per unit mass $F(t)$ to be the real part of $F_0\exp(i\omega t)$, and the response to be the real part of

$\chi F_0 \exp(i\omega t)$, where $\chi = \chi' + i\chi''$. Show that χ could also be expressed as $|\chi|\exp(i\theta)$ and relate $|\chi|$ and θ to χ' and χ''.

1.5 Calculate the NMR signal obtainable from a volume of water placed in the earth's magnetic field ($B_0 \approx 0.5 \times 10^{-4}$ T). Assume the equilibrium magnetisation has been rotated into the transverse plane.

Whether the signal can be observed will depend on the competing noise from the detection system. The irreducible minimum noise (discussed in Chapter 3) is the thermal noise voltage from the pick-up coil, given by

$$[1.7 \times 10^{-20}(\omega_0 L/Q)\Delta f]^{1/2} \text{ V}.$$

In this formula, Q is the quality factor of the coil (say 100), L its self-inductance, given approximately by

$$\mu_0 (NA)^2/\text{coil volume},$$

and Δf the bandwidth of the detection system (assumed to be 1 Hz since this is approximately the natural linewidth of water).

Find the sample volume required to achieve a voltage signal to noise ratio of 100.

Chapter 2

2.1 The characteristic time for a relaxation function is a measure of its 'width'. Thus we may define a relaxation time which applies equally to exponential and non-exponential decays as the 'area' of the function divided by its 'height':

$$T = \frac{1}{F(0)} \int_0^\infty F(t)dt.$$

Show that for the case of exponential relaxation, this definition corresponds to the conventional time constant. What is it in the case of a Gaussian relaxation?

2.2 Starting from the Bloch equations, Equations (2.9), show that if a transverse field of magnitude B_1 is rotating at an angular frequency of ω then in the frame rotating with this field the equations of motion can be written

$$M'_x = M_y \Delta - M_x/T_2,$$
$$M'_y = -M_y \Delta + \gamma M_z B_1 - M_y/T_2,$$
$$M'_z = -\gamma M_y B_1 + (M_0 - M_z)/T_1,$$

where $\Delta = \omega_0 - \omega$, the 'distance' from resonance.

The steady state solution of these equations corresponds to the situation in which the rotating frame derivatives M'_x, M'_y and M'_z are zero. Show that in this case

$$M_x = \frac{M_0 T_2^2 (\omega - \omega_0)\gamma B_1}{1 + (\omega - \omega_0)^2 T_2^2 + \gamma^2 B_1^2 T_1 T_2},$$

$$M_y = \frac{M_0 T_2 \gamma B_1}{1 + (\omega - \omega_0)^2 T_2^2 + \gamma^2 B_1^2 T_1 T_2},$$

$$M_z = \frac{M_0 [1 + (\omega - \omega_0)^2 T_2^2]}{1 + (\omega - \omega_0)^2 T_2^2 + \gamma^2 B_1^2 T_1 T_2}.$$

These are the general solutions of the Bloch equations.

2.3 Using the result of the previous question, show that the imaginary part of the transverse susceptibility is given by

$$\chi''(\omega) = \frac{\chi_0 \omega_0 T_2}{1 + (\omega - \omega_0)^2 T_2^2 + \gamma^2 B_1^2 T_1 T_2}.$$

In the *linear response* limit this reduces to

$$\chi''(\omega) = \frac{\chi_0 \omega_0 T_2}{1 + (\omega - \omega_0)^2 T_2^2}.$$

Why has the term vanished from the denominator?

2.4 As was discussed in Section 3.5.1, the Kramers–Kronig relations lead to the result that the static susceptibility is related to the imaginary part of the dynamic susceptibility by

$$\chi_0 = \frac{1}{\pi} \int_{-\infty}^{\infty} \frac{\chi''(\omega)}{\omega} d\omega.$$

Verify this for the Bloch expression for the linear response $\chi''(\omega)$ above. Why will it not hold for the general (non-linear) response?

Recall that since the resonance line will usually be very narrow in the vicinity of ω_0, the susceptibility expression is approximately

$$\chi_0 = \frac{1}{\pi \omega_0} \int_{-\infty}^{\infty} \chi''(\omega) d\omega,$$

which provides a way of finding the static susceptibility. Unfortunately this will not hold in the presence of saturation. Show that in this case, for the Bloch susceptibility, a pair of Kramers–Kronig relations exists between $\chi'(\omega)$ and $\chi''(\omega)(1 + \gamma^2 T_1 T_2)^{\frac{1}{2}}$. What does this imply about the effect of saturation on susceptibility measurements?

Chapter 3

3.1 By comparing the admittances of the following two circuits at a frequency ω:

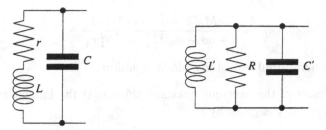

Figure P.1 .

show that they are identical (at a given frequency) provided

$$L' = L, \ C' = C, \ R = \omega^2 L^2/r, \ \text{if} \ Q(\omega L/r) \gg 1.$$

3.2 An oscillator is made by connecting the tuned circuit in the above problem to an amplifier and providing feedback.

Assuming that there is no phase shift in the amplifier or feedback resistor Z, at what frequency will the circuit oscillate when the feedback reaches the critical value?

If an NMR sample is placed on the coil and the magnetic field $\mathbf{B}_0$ is swept through magnetic resonance, show that the real part of the susceptibility χ' will cause a frequency shift, and obtain an expression for its magnitude.

3.3 In the 'linear régime' the solution to the Bloch equations indicates that the absorption signal size is proportional to the magnitude of the applied transverse field

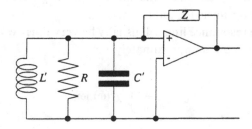

Figure P.2 .

B_1. However, as this field gets larger the saturation factor $\gamma^2 B_1^2 T_1 T_2$ in the denominator gets larger, which reduces the signal. What is the largest absorption signal which may be observed, and what is the required magnitude of B_1?

3.4 Taking the Bloch expression for the complex transverse magnetic susceptibility

$$\chi'(\omega) = \frac{\chi_0 \omega_0 (\omega - \omega_0) T_2^2}{1 + (\omega - \omega_0)^2 T_2^2},$$

$$\chi''(\omega) = \frac{\chi_0 \omega_0 T_2}{1 + (\omega - \omega_0)^2 T_2^2},$$

demonstrate the locus of χ in the complex plane traces out a circle, centred on the imaginary axis and passing through the origin. Show that the radius of the circle is proportional to $\chi_0 \omega_0 T_2$.

Chapter 4

4.1 The steps involved in showing that $G(t)$ in Equation (4.63),

$$G(t) = \frac{\gamma^2 G^2}{q^2} \langle \cos[q \Delta z(t)] \rangle,$$

relaxes exponentially as Equation (4.66)

$$G(t) = \frac{\gamma^2 G^2}{q^2} \exp(-q^2 D t)$$

may be recast as a conventional NMR problem studying the 'relaxation function' $G(t)$. Thus derive the result using the 'Gaussian phase' procedure. What is the corresponding 'local field autocorrelation function'? What is the 'motional narrowing' criterion which ensures that the relaxation is exponential?

4.2 Consider a sphere of radius r containing a uniform density of magnetic moments. If this sphere is placed in a magnetic field gradient G then the spins at co-ordinate z will experience a magnetic field Gz and so they will resonate at an angular frequency of γGz. Spins at different heights will resonate at different frequencies and the frequency spectrum of the resonance will be related to the shape of the specimen.

What is the area of a slice of the sphere at height z? Express this in terms of the frequency $\omega = \gamma Gz$ and hence show that the observed absorption spectrum is given by

$$I(\omega) \propto \begin{cases} 1 - (\omega/\gamma Gr)^2 & \omega < \gamma Gr \\ 0 & \omega > \gamma Gr. \end{cases}$$

Sketch the shape of this. How will the absorption profile appear if the specimen has a spherical hole in the middle?

Show that the FID for the original sphere is

$$F(t) = \frac{3}{(\gamma Grt)^3}[\sin(\gamma Grt) - \gamma Grt\cos(\gamma Grt)].$$

4.3 Section 4.2.3 gave two views of the Gaussian averaging procedure. This problem provides a third. Write the expression for the FID

$$F(t) = \langle \exp[i\phi(t)] \rangle$$

as the exponential for another function $\Psi(t)$. In other words, let us define $\Psi(t)$ by

$$F(t) = \exp[\Psi(t)].$$

Now show that Equation (4.27) may be obtained by expanding $\Psi(t)$ to leading order in ϕ.

Observe that writing $F(t) = \exp[\Psi(t)]$ ensures that $F(t)$ will converge at long times; it 'feeds in' the physical boundary condition to the mathematical procedure.

Chapter 5

5.1 Use the method of Section 5.5.7 to study a possible mechanism for an electron-mediated interaction between nuclear moments. Consider the magnetic vector potential at an electron position to be made up of contributions from the dipolar fields of two nuclear moments and show that the contribution to the energy which is bilinear in the two magnetic moments is of the form

$$E' = \frac{e^2}{m}\left(\frac{\omega_0}{4\pi}\right)^2 \frac{(\boldsymbol{\mu}_i \times \mathbf{r}_i)\cdot(\boldsymbol{\mu}_j \times \mathbf{r}_j)}{r_i^3 r_j^3},$$

where i and j label the nuclei. Show that this results in a spin-spin interaction Hamiltonian given by

$$\mathscr{H}_{ij}' = \sum_{\alpha,\beta=x,y,z} I_i^\alpha J_{\alpha\beta} I_j^\beta,$$

where

$$J_{\alpha\beta} = \frac{e^2}{m}\left(\frac{\mu_0}{4\pi}\right)^2 \gamma_i\gamma_j \sum_{\text{states } s}\left\langle s\left|\left(\frac{\mathbf{r}_i\cdot\mathbf{r}_j\delta_{\alpha\beta} - r_{i,\alpha}r_{j,\beta}}{r_i^3 r_j^3}\right)\right|s\right\rangle$$

and the sum is over the electron states.

Although this is a possible spin–spin interaction mechanism, in practice it is very small and it is usually masked by other effects.

5.2 In Box 5.1 Curie's Law was calculated explicitly for spin $\frac{1}{2}$ moments from the expression

$$\langle M_z \rangle = N_v \hbar^2 \gamma^2 \beta B_z \frac{\text{Tr}\{(I_z^i)^2\}}{\text{Tr}\{1\}}.$$

The result for general spin I was quoted as

$$\langle M_z \rangle = \frac{N_v \hbar^2 \gamma^2 I(I+1)}{3kT} B_z,$$

which followed from evaluating the traces $\text{Tr}\{I_z^2\}$ and $\text{Tr}\{1\}$.

In the I_z representation the matrices representing I_z and 1 are diagonal:

$$I_z = \begin{pmatrix} I & & & & \\ & I-1 & & & \\ & & \ddots & & \\ & & & -I+1 & \\ & & & & -I \end{pmatrix} \quad \text{and} \quad 1 = \begin{pmatrix} 1 & & & \\ & 1 & & \\ & & \ddots & \\ & & & 1 \\ & & & & 1 \end{pmatrix}$$

and I_z^2 also will be diagonal. Thus by summing the appropriate series obtain the required results $\text{Tr}\{I_z^2\} = I(I+1)(2I+1)/3$ and $\text{Tr}\{1\} = I(I+1)$.

5.3 Using the Heisenberg equation of motion in the interaction picture/rotating frame, set up the appropriate Hamiltonian for a 90° pulse and a 180° pulse. Show that the time evolution of the magnetisation following such pulses is as expected from the classical description of Section 2.5.4.

Chapter 6

6.1 By repeated application of the trigonometric identity

$$\sin(x) = 2\sin(x/2)\cos(x/2)$$

show that

$$\sin(x) = 2^N \sin(x/2^N) \prod_{k=1}^{N} \cos(x/2^k).$$

Rearrange this to give:

$$\prod_{k=1}^{N} \cos(x/2^k) = \frac{\sin(x)/x}{\sin(x/2^N)/(x/2^N)},$$

as required in Section 6.2.3.

6.2 Equation (6.44) gives the volume of sphere in n dimensions:

$$V_n = \frac{2\pi^{n/2}r^n}{n\Gamma(n/2)},$$

where Γ represents the gamma function. Show that when $n = 1$ this give the circumference of a circle, when $n = 2$ this gives the area of a circle and when $n = 3$ this gives the volume of a sphere.

6.3 Extend the derivation of Equation (6.48) to include the orientation dependence of the dipolar interaction.

6.4 Consider the fourth approximant to the transverse relaxation function:

$$F_4(t) = 1 - M_2 t^2/2 + M_4 t^4/4!.$$

In Section 6.3.7 we saw that the area *above* this curve is forbidden. So if this function becomes negative then the relaxation function must be negative at some stage of its evolution. Thus by examining the existence of roots of the polynomial equation $F_4(t) = 0$ show that at least one zero must occur in $F(t)$ if $M_4/M_2^2 < 3/2$.

6.5 Exponential relaxation is unphysical at very short times since it is in conflict with the microscopic equations of motion. In the frequency domain this means that a Lorentzian spectrum is unphysical (at least in the wings). In section 6.3.4 we considered an apparently exponential relaxation by truncating the wings of the spectral function. There the truncation function was a 'box': unity for $|\Delta| < \Delta_c$ amd zero for $|\Delta| > \Delta_c$.

A more general treatment effects the truncation by multiplying the Lorentzian by an arbitrary function $g(\Delta)$. In this case the spectral function is given by

$$I(\Delta) = \frac{T}{\pi(1 + \Delta^2 T^2)} g(\Delta),$$

where T is the time constant of the relaxation. Assuming that both $I(\Delta)$ and $g(\Delta)$ are even functions, and taking M_n and m_n as the moments of $I(\Delta)$ and $g(\Delta)$ defined in the usual manner, show that

$$m_{2n} = \frac{M_{2n} + M_{2n+2}T^2}{1 + M_2 T^2}.$$

In the realistic case $g(\Delta)$ will decay appreciably only at frequencies much greater than T^{-1}, so that the Lorentzian function is *visibly* changed. Then the above relation may be approximated by

$$m_{2n} = \frac{M_{2n+2}}{M_2},$$

for $n > 0$.

By analogy with the treatment in Section 6.3.4 show that the relaxation time may be related to M_2 and M_4 rather like Equation (6.74), but now

$$T = \frac{C \, M_4^{1/2}}{\pi \, M_2^{3/2}},$$

where the constant C depends on the *shape* of the truncation function. In conformity with Equation (6.74) you should see that $C = 2\sqrt{3}$ for the box truncation. Show also, that $C = (2\pi)^{\frac{1}{2}}$ for a Gaussian and $C = 2$ for an exponential truncation function.

Chapter 7

7.1 On a log–log plot of T_1 against τ_c show that the Lorentzian $J(\omega)$ leads to slopes of -1 and $+1$ to the left and right of the T_1 minimum.

7.2 Show that the translational diffusion model treated in Section 7.4.3 leads to a slope of $+\frac{1}{2}$ for $\tau_c \gg \omega_0^{-1}$ on a log–log plot of T_1 against τ_c.

7.3 Show that for a Gaussian correlation function the parameter k defined in Section 7.5.2 satisfies $3K = 1.97$ while for an exponential it satisfies $3K = 3$.

7.4 Derive the expression for the moments of a motionally averaged transverse relaxation, Equation (7.104),

$$M_{2(n+1)} = \frac{M_2}{\tau_c^{2n}} \left(\frac{\pi}{4}\right)^n \frac{(2n)!}{n!}.$$

The trick is to use Equation (7.106), which had not been established when Equation (7.104) was stated.

7.5 Demonstrate Equation (7.33):

$$\Gamma_{ijkl} = N \frac{\mathrm{Tr}\{T_{ij}{}^m T_{kl}^m\}}{\mathrm{Tr}\{I_z^2\}} = \frac{3}{2}(\delta_{ik}\delta_{jl} + \delta_{il}\delta_{jk}).$$

You have the choice of index m to make this easier. Having split the expression into a sum of different spin traces, each trace can be evaluated in the most appropriate representation.

7.6 The derivation of the T_1 sum rules in Section 7.5.3, Equation (7.109), was complicated by the inclusion of the double frequency term $J(2\omega)$ in the expression for T_1. Obtain the corresponding sum rule expressions when there is only the single frequency term

$$1/T_1 = J(\omega),$$

as would apply, for example, when the relaxation occurred through motion in an inhomogeneous magnetic field.

7.7 This problem considers the relaxation of tumbling molecules in a slightly different way from that treated in Section 7.4.1. Imagine one spin of a molecule fixed at the origin, and consider the motion of another spin; it will diffuse on the surface of a sphere. The probability that the spin will move through an angle α (subtended at the centre) is given by

$$\rho(\alpha,t) = \frac{1}{4\pi^2} \sum_{l=0}^{\infty} (2l + 1)\{\cos(l\alpha) - \cos[(l + 1)\alpha]\}\exp(-t/\tau_l),$$

where

$$\frac{1}{\tau_l} = \frac{D}{a^2}l(l + 1).$$

This formula was derived by W. H. Furry, *Phys. Rev.* (1957) **107**, 7 and quoted and used by P. S. Hubbard, *Phys. Rev* (1958) **109**, 1153.

Use this probability together with the result of Appendix E to obtain the exponential expression for $G(t)$, as given in Equations (7.64) and (7.65).

Chapter 8

8.1 Demonstrate the identity of Equation (8.3):

$$\prod_{i=n}^{\infty} \left[1 - \frac{b^2t^2}{(n\pi)^2} \right] = \frac{\sin(bt)}{bt}.$$

Firstly one must take the logarithm of both sides. The logarithm on the left hand side is then expanded as series in $(bt/n\pi)^2$. Each term may then be equated to terms of the expansion of the right hand side. Note that one needs to use the Riemann zeta function to remove the various powers of π.

8.2 In Problem 6.4 a condition for the existence of zeros in the FID function was examined. How can this be generalised to higher order approximants? Apply this to the calcium fluoride relaxation. What are the implications for the zeros of this function?

Chapter 9

9.1 Prove the relation $\langle p|q \rangle = \text{Tr}\{|q\rangle\langle p|\}$. This may be done by expanding both $\langle p|$ and $|q\rangle$ in terms of a complete orthonormal set of bras and kets.

9.2 Consider an ensemble of spin 1 particles, described by a 3×3 spin density operator. In this case the dipole moment components proportional to the expectation values of I_x, I_y and I_z are no longer sufficient to specify the system. Show that the quadrupole moment could be used in addition, to provide a complete specification.

9.3 The procedures of Section 9.3.8 can be used to show how Zeeman order can be transformed into dipolar order through the application of two pulses. Let us start with a spin system in equilibrium containing just Zeeman order. Its density operator will be

$$\sigma_{eq} = (1 + \beta_{eq}\hbar\omega_0 I_z)/\text{Tr}1,$$

where β_{eq} is the equilibrium Zeeman inverse temperature. Firstly we apply a 90° pulse about the x axis, creating transverse magnetisation, and transform the density operator to

$$\sigma = (1 - \beta_{eq}\hbar\omega_0 I_y)/\text{Tr}1.$$

Next we allow this state to evolve under the influence of the (truncated) dipolar Hamiltonian D_0 for a time τ. At this stage the spin density operator will be

$$\sigma(\tau) = [1 - \beta_{eq}\hbar\omega_0\exp(-iD_0\tau)I_y\exp(iD_0\tau)]/\text{Tr}1.$$

Next we apply a pulse of angle θ about the y axis. As in Section 9.3.8 the effect of the rotation is to transform the dipolar Hamiltonian D_0, leaving I_y unchanged:

$$\sigma' = [1 - \beta_{eq}\hbar\omega_0\exp(-i\tilde{D}_0\tau)I_y\exp(i\tilde{D}_0\tau)]/\text{Tr}1.$$

We assert that the resultant state contains dipolar order. So by waiting a further time of the order of T_2 the system will have relaxed to a quasi-equilibrium state

$$\sigma_{qe} = (1 + \beta_Z\hbar\omega_0 I_z - \beta D_0)/\text{Tr}1,$$

where β_Z is the new Zeeman inverse temperature and β is the dipole inverse temperature. During this further time the Zeeman temperature and the dipole temperature will remain unchanged. This means that the spin temperatures β_Z and β can be found from the state σ' using Equation (9.25)

$$\beta_z = -\text{Tr}\{\sigma' \mathcal{H}_z\}\text{Tr}1/\text{Tr}\{\mathcal{H}_z^2\},$$
$$\beta = -\text{Tr}\{\sigma' D_0\}\text{Tr}1/\text{Tr}\{D_0^2\}.$$

Show, using reflection symmetry, that the state has no Zeeman order: in other words show that $\beta_z = 0$.

Using the methods of Section 9.3.8, applying the rotation to D_0, show that the resultant dipolar temperature is

$$\beta = -\beta_{eq}\sin\theta\cos\theta\frac{\text{Tr}\{\mathcal{H}_z^2\}}{\text{Tr}\{D_0^2\}}\frac{d}{dt}F(t).$$

9.4 Using Equation (9.14) write down an expression for the entropy of a spin $\frac{1}{2}$ system in terms of the expectation values of the three components of I. Following a 90° pulse about the y axis show that the entropy is given, in terms of the transverse relaxation function, by

$$S = -k\left\{\left[\frac{1 + F(t)}{2}\right]\ln\left[\frac{1 + F(t)}{2}\right] + \left[\frac{1 - F(t)}{2}\right]\ln\left[\frac{1 - F(t)}{2}\right]\right\}.$$

Plot the time variation of the entropy: (a) for exponential relaxation, and (b) for the relaxation given by the exactly soluble model interaction of Section 6.2.3.

Is this behaviour consistent with the Second Law of Thermodynamics?

Chapter 10

10.1 In Problem 4.2 we considered a sphere of radius r containing a uniform density of magnetic moments, placed in a magnetic field gradient G. We saw that the FID relaxed according to

$$F(t) = \frac{3}{(\gamma Grt)^3}[\sin(\gamma Grt) - \gamma Grt\cos(\gamma Grt)].$$

Plot this function and consider how observation of the zeros of $F(t)$ may be utilised to calibrate the field gradient.

Calculate the relaxation function for a cylindrical specimen oriented (a) along the z axis, and (b) perpendicular to the z axis.

10.2 The magnetic field a distance z along the axis of a circular current loop of radius a, carrying a current of I amperes, is given by

$$B = \frac{\mu_0 I}{2}\frac{a^2}{z^2 + a^2}.$$

A reasonably homogeneous magnetic field may be produced in the region between two such coils, when arranged co-axially, if the current flows in the same direction in both coils.

(a) Show that the odd derivatives with respect to z of the magnetic field vanish at the mid-point between the coils.
(b) Show that when the coils are separated a distance equal to their radius, the second derivative of the magnetic field vanishes at the centre.
(c) Show that the magnetic field at the mid point is then given by

$$B_0 = \frac{8}{5\sqrt{5}} \frac{\mu_0 I}{a}.$$

(d) For this separation the first non-vanishing derivative of the magnetic field is then the fourth. Show that in the close vicinity of the centre the magnetic field varies as

$$B = B_0\left[1 - \frac{144}{125}\left(\frac{z}{a}\right)^4\right].$$

It can be seen that when z/a is small the deviation of B from its constant value is small so that the field is fairly homogeneous.

(e) What is the length of the region over which the magnetic field varies by less than 1 in 10^5?

This configuration of coils is called a Helmholtz pair and it is often used where relatively low magnetic fields are required. Some of the first MRI magnets were constructed in this way.

10.3 If the current is reversed in one of the coils of a Helmholtz pair then the fields will cancel at the centre. All even derivatives will vanish. A magnetic field gradient can be produced in this way. However, a greater linearity in the field variation can be achieved by changing the separation to cause the third derivative of the field to vanish.

(a) Show that the third derivative of the field vanishes at the centre when the coils are separated by $\sqrt{3}$ times their radius.
(b) Show that for this separation the field gradient at the centre is given by

$$G_0 = \frac{\partial B}{\partial z} = -\frac{48\sqrt{3}}{49\sqrt{7}}\frac{\mu_0 I}{a^2}.$$

(d) For this separation the first non-vanishing derivative (above the first) of the magnetic field is then the fifth. In other words the first non-vanishing derivative of the gradient is the fourth. Show that in the close vicinity of the centre the field gradient varies as

$$G = G_0\left[1 - \frac{880}{343}\left(\frac{z}{a}\right)^4\right].$$

It can be seen that when z/a is small the deviation of G from its constant value is small so that the field gradient is fairly uniform. This configuration of coils is called a Maxwell pair. The terms anti-Helmholtz and Holtz-helm are also used, but sometimes these refer to these coils having the inferior Helmholtz separation.

10.4 The Maxwell equations give relations between the various derivatives of a magnetic field. As a consequence field variations in one direction are necessarily associated with variations in other directions. Investigate the variation of the components of the magnetic field in the vicinity of both the Helmholtz and the Maxwell pair. Remember to exploit the cylindrical symmetry of the fields; it is appropriate to work in cylindrical polar co-ordinates.

References

Chapter 1

A. Abragam (1961) *The Principles of Nuclear Magnetism*, Oxford University Press, Oxford.

E. R. Andrew (1955) *Nuclear Magnetic Resonance*, Cambridge University Press, Cambridge.

F. Bloch, W. W. Hansen and M. Packard (1946) *Phys. Rev.*, **70**, 474.

P. A. M. Dirac (1930) *The Principles of Quantum Mechanics*, Oxford University Press, Oxford.

W. Gerlach & O. Stern (1924) *Ann. Phys., Lpz.* **74**, 673.

E. L. Hahn (1950) *Phys. Rev.*, **77**, 279.

E. M. Purcell, H. C. Torrey and R. V. Pound (1946) *Phys. Rev*, **79**, 37.

I. I. Rabi, S. Millman, P. Kusch and J. R. Zacharias (1939) *Phys. Rev.*, **55**, 526.

F. N. H. Robinson (1973) *Macroscopic Electromagnetism*, Pergamon Press, Oxford.

O. Stern (1921) *Z. Phys.*, **7**, 249.

Chapter 2

F. Bloch (1946) *Phys. Rev.*, **70**, 460.

C. Cohen-Tannoudji, B. Diu and F. Laloë (1977) *Quantum Mechanics*, John Wiley, New York.

P. A. M. Dirac (1930) *The Principles of Quantum Mechanics*, Oxford University Press, Oxford.

I. I. Rabi, N. F. Ramsey and J. Schwinger (1954) *Rev. Mod. Phys.*, **26**, 167

Chapter 3

E. R. Andrew (1955) *Nuclear Magnetic Resonance*, Cambridge University Press, Cambridge.

B. I. Bleaney and B. Bleaney (1976) *Electricity and Magnetism*, Oxford University Press, Oxford.

H. Y. Carr and E. M. Purcell (1954) *Phys. Rev.*, **94**, 630.

J. W. Cooley and J. W. Tukey (1965) *Math. Comput.*, **19**, 297.
R. R. Ernst and W. A. Anderson (1966) *Rev. Sci. Instrum.*, **37**, 93.
I. S. Gradshteyn and I. M. Ryzhik (1965) *Tables of Integrals, Series and Products*, Academic Press, New York.
E. L. Hahn (1950) *Phys. Rev.*, **77**, 279.
D. I. Hoult and R. E. Richards (1976) *J. Mag. Res.*, **24**, 71.
F. Reif (1965) *Fundamentals of Statistical and Thermal Physics*, McGraw-Hill, New York.
F. N. H. Robinson (1959) *J. Sci. Inst.*, **36**, 481.
F. N. H. Robinson (1974) *Noise and Fluctuations in Electronic Devices and Circuits*, Oxford University Press, Oxford.
H. C. Torrey (1952) *Phys. Rev.*, **85**, 365.

Chapter 4

P. W. Anderson (1954) *J. Phys. Soc. Jap.*, **9**, 316.
P. W. Anderson and P. R. Weiss (1953) *Rev. Mod. Phys.*, **25**, 269.
H. Y. Carr and E. M. Purcell (1954) *Phys. Rev.*, **94**, 630.
B. Cowan (1977) *J. Phys. C.*, **10**, 3383.
B. Cowan (1996) *Instrument Science and Technology*, **7**, 690.
J. Kärger, H. Pfeifer and W. Heink (1988) *Adv. Mag. Res.*, **12**, 1.
R. Kubo (1961) in *Fluctuation, Relaxation and Resonance in Magnetic Systems*, ed. D. ter Haar, Oliver and Boyd, Edinburgh P. 23.
S. Meiboom and D. Gill (1958) *Rev. Sci. Inst.*, **29**, 688.
W. H. Press, S. A. Teukolsky, W. T. Vetterling and B. P. Flannery (1992) *Numerical Recipes*, Cambridge University Press, Cambridge.
F. Reif (1965) *Fundamentals of Statistical and Thermal Physics*, McGraw-Hill, New York.
H. C. Torrey (1952) *Phys. Rev.*, **85**, 365.

Chapter 6

A. Abragam (1961) *The Principles of Nuclear Magnetism*, Oxford University Press, Oxford.
M. Abramowitz and I. A. Stegun (1965) *Handbook of Mathematical Functions*, Dover, New York.
P. W. Anderson (1951) *Phys. Rev.*, **82**, 342.
E. R. Andrew and R. Bersohn (1950) *J. Chem. Phys.*, **18**, 159. And (1950) **20**, 924.
P. Borckmans and D. Walgraef (1968) *Phys. Rev.*, **167**, 282.
C. Cohen-Tannoudji, B. Diu and F. Laloë (1977) *Quantum Mechanics*, John Wiley, New York.
S. P. Heims (1965) *Am. J. Phys.*, **33**, 722.
K. Huang (1987) *Statistical Mechanics*, John Wiley, New York.
C. Kittel and E. Abrahams (1953) *Phys. Rev.*, **90**, 238.
E. Kreyszig (1993) *Advanced Engineering Mathematics*, John Wiley, New York.

I. J. Lowe and R. E. Norberg (1957) *Phys. Rev.*, **107**, 46.
M. Mehring (1976) *High Resolution NMR Spectroscopy in Solids*, Springer, Berlin.
D. S. Metzger and J. R. Gaines (1966) *Phys. Rev.*, **147**, 644.
H. Mori (1965) *Prog. Theor. Phys.*, **34**, 399.
G. Pake (1948) *J. Chem. Phys.*, **16**, 327.
J. H. Van Vleck (1948) *Phys. Rev.*, **74**, 1168.
I. Waller (1932) *Z. Phys.*, **79**, 370.
R. W. Zwanzig (1961) *Lectures in Theoretical Physics*, Vol 3, Wiley-Interscience, New York

Chapter 7

M. Abramowitz and I. A. Stegun (1965) *Handbook of Mathematical Functions*, Dover, New York.
Y. Ayant, E. Belorizky, J. Alzion and J. Gallice (1975) *J. de Phys.*, **36**, 991.
B. Cowan and M. Fardis (1991) *Phys. Rev. B*, **44**, 4304.
H. Cramer (1946) *Mathematical Methods of Statistics*, Princeton University Press, Uppsala and Princeton.
P. S. Hubbard (1958) *Phys. Rev.*, **109**, 1153.
R. Kubo and K. Tomita (1953) *Proc. Int. Conf. Theor. Phys., Kyoto and Tokyo* Science Council of Japan, Ueno Park, Tokyo. 779.
R. Kubo and K. Tomita (1954) *J. Phys. Soc. Jap.*, **9**, 888.
R. Lenk (1977) *Brownian Motion and Spin Relaxation*, Elsevier, Amsterdam.
H. C. Torrey (1953) *Phys. Rev.*, **92**, 962.
H. C. Torrey (1954) *Phys. Rev.*, **96**, 690.

Chapter 8

A. Abragam (1961) *The Principles of Nuclear Magnetism*, Oxford University Press, Oxford.
D. E. Barnaal and I. J. Lowe (1966) *Phys. Rev.*, **148**, 328.
M. Bassou (1981) Ph.D. Thesis, University of Paris, Orsay.
B. Beal, R. P. Giffard, J. Hatton, M. G. Richards and P. M. Richards (1964) *Phys. Rev. Lett.*, **12**, 394.
M. E. R. Bernier and G. Guerrier (1983) *Physica*, **121B**, 202.
N. Bloembergen (1948) *Nuclear Magnetic Relaxation*, Benjamin, New York.
N. Bloembergen, E. M. Purcell and R. V. Pound (1948) *Phys. Rev.*, **73**, 679.
P. Borckmans and D. Walgraef (1968) *Phys. Rev. Lett.*, **21**, 1516.
P. Borckmans and D. Walgraef (1968) *Phys. Rev.*, **167**, 282.
C. R. Bruce (1957) *Phys. Rev.*, **107**, 43.
G. W. Canters and C. S. Johnson (1972) *J. Mag. Res.*, **6**, 1.
D. Ceperley and G. Jacucci (1987) *Phys. Rev. Let.* **58**, 1648.
D. M. Ceperley and G. Jacucci (1988) *Phys. Rev. Lett.*, **58**, 1648.
M. Chapellier, M. Bassou, M. Deroret, J. M. Delreill and N. S. Sullivan (1985) *J. Low Temp. Phys.*, **59**, 45.

B. Cowan, W. J. Mullin and E. Nelson (1989) *J. Low Temp. Phys.*, **77**, 181.

M. Engelsberg and I. J. Lowe (1974) *Phys. Rev. B*, **10**, 822.

W. A. B. Evans and J. G. Powles (1967) *Phys. Lett.*, **24A**, 218.

G. Ia. Glebashev (1957) *Zh. Eksp. Teor. Fiz.*, **32**, 82 [*Sov. Phys.-JETP*, **5**, (1957)].

R. Gordon (1968) *J. Math. Phys.*, **9.**, 1087.

J. F. Harmon (1970) *Chem. Phys. Lett.*, **71**, 207.

S. J. Knak Jensen and E. Kjaersgaard Hansen (1973) *Phys. Rev. B*, **7**, 2910.

W. P. Kirk and E. D. Adams (1972), Proc. 13th. Conf. Low Temp. Phys., **149**, Plenum, New York.

A. Landesman (1973) *Ann. de Phys.*, **8**, 53.

R. Lenk (1971) *J. Phys. C*, **4**, L21.

I. J. Lowe and M. Englesberg (1974) *Rev. Sci. Inst.*, **45**, 631.

I. J. Lowe and R. E. Norberg (1957) *Phys. Rev.*, **107**, 46.

I. J. Lowe, K. W. Vollmers and M. Punkkinen (1973), *Proc. First Specialised Colloque Ampère, Krakow*, North Holland, Amsterdam.

K. Luszczynski, J. A. E. Kail and J. G. Powles (1960) *Proc. Phys. Soc.*, **75**, 243.

P. Mansfield (1966) *Phys. Rev.*, **151**, 199.

K. Matsumoto, T. Abe and T. Izuyama (1989) *J. Phys. Soc. Jap.*, **58**, 1149.

F. Noack and G. Preissing (1967) *Proc. XIV Colloque Ampère*, North Holland, Amsterdam.

G. Pake and E. M. Purcell (1948) *Phys. Rev.*, **74**, 1184. And **75**, 534.

G. W. Parker (1970) *Phys. Rev. B*, **2**, 2543.

G. W. Parker and F. Lado (1973) *Phys. Rev. B*, **8**, 3081.

J. G. Powles and B. Carazza (1970) *Proc. Int. Conf. Mag. Resonance, Melbourne 1969*, Plenum, New York.

G. Preissing, F. Noack, R. Kosfeld and B. Gross (1971) *Z. Phys.*, **246**, 84.

E. M. Purcell, N. Bloembergen and R. V. Pound (1946) *Phys. Rev.*, **70**, 988.

A. G. Redfield (1959) *Phys. Rev.*, **116**, 315.

H. A. Reich (1963) *Phys. Rev.*, **129**, 630.

M. G. Richards, J. Hatton and R. P. Giffard (1965) *Phys. Rev.*, **139**, A91.

R. C. Richardson, E. Hunt and H. Meyer (1965) *Phys. Rev.*, **138**, A1326.

W. C. Thomlinson, J. F. Kelly and R. C. Richardson (1972) *Phys. Lett.*, **38A**, 531.

J. H. Van Vleck (1948) *Phys. Rev.*, **74**, 1168.

Chapter 9

A. Abragam (1961) *The Principles of Nuclear Magnetism*, Oxford University Press, Oxford.

A. Abragam and M. Goldman (1982) *Nuclear Magnetism: Order and Disorder*, Oxford University Press, Oxford.

F. Bloch (1957) *Phys. Rev.*, **105**, 1206.

D. M. Brink and G. R. Satchler (1962) *Angular Momentum*, Oxford University Press, Oxford.

L. J. Burnett and J. F. Harmon (1972) *J. Chem. Phys.*, **57**, 1293.

U. Fano (1957) *Rev. Mod. Phys.*, **29**, 74.

M. Goldman (1970) *Spin Temperature and NMR in Solids*, Oxford University Press, Oxford.

M. Goldman (1977) *Phys. Rep.*, **32C**, 1.

D. ter Haar (1961) *Rep. Prog. Phys.*, **24**, 304.

P. S. Hubbard (1961) *Rev. Mod. Phys.*, **33**, 249.

L. D. Landau and E. M. Lifshitz (1969) *Statistical Physics*, Pergamon Press, Oxford.

I. J. Lowe, K. W. Vollmers and M. Punkkinen (1973) *Proc. First Specialised Colloque Ampère, Krakow*, North Holland, Amsterdam.

J. von Neumann (1955) *Mathematical Foundations of Quantum Mechanics*, Princeton University Press, New Jersey (Princeton).

B. Provotorov (1961) *Zh. Eksp. Teor. Fiz.*, **41**, 1582. [*Sov. Phys.-JETP*, **14**, 1126 (1962)].

A. G. Redfield (1955) *Phys. Rev.*, **98**, 1787.

A. G. Redfield (1957) *IBM J. Res. Develop.*, **1**, 19.

A. G. Redfield (1965) *Adv. Mag. Res.*, **1**, 1.

R. K. Wangsness and F. Bloch (1953) *Phys. Rev.*, **89**, 728.

Chapter 10

G. N. Hounsfield (1973) *Br. J. Radiol.*, **46**, 1016.

P. C. Lauterbur (1973) *Nature*, **242**, 190.

P. Mansfield (1977) *J. Phys.*, **C10**, L55.

P. Mansfield and P. K. Grannell (1973) *J. Phys.*, **C6**, L422.

P. Mansfield and P. G. Morris (1982) *Adv. Mag. Res. Supplement 2*.

P. G. Morris (1986) *Nuclear Magnetic Resonance Imaging in Medicine and Biology*, Oxford University Press, Oxford.

D. D. Stark and W. G. Bradley (1992) *Magnetic Resonance Imaging*, Mosby Year Book Inc., New York.

F. W. Wehrli (1991), *Fast-Scan Magnetic Resonance Principles and Applications*, Raven Press, New York.

Appendices

R. N. Bracewell (1979) *The Fourier Transform and its Applications*, McGraw-Hill, New York.

D. C. Champeney (1973), *Fourier Transforms and their Physical Applications*, Academic Press, London.

C. Cohen-Tannoudji, B. Diu and F. Laloë (1977) *Quantum Mechanics*, John Wiley, New York.

A. R. Edmonds (1957) *Angular Momentum in Quantum Mechanics*, Princeton University Press, New Jersey (Princeton).

H. E. Rorschach (1986) *J. Mag. Res.* **67**, 519.

L. Schwartz (1950) *Théorie des Distributions*, Hermann et Cie, Paris.

M. C. Wang and G. E. Uhlenbeck (1945) *Rev. Mod. Phys.*, **17**, 323.

N. Wax (1954) *Selected Papers on Noise and Stochastic Processes*, Dover, New
    York.
N. Wiener (1933) *The Fourier Integral and Certain of its Applications*, Cambridge
    University Press, Cambridge.
A. M. Yaglom (1962) *An Introduction to the Theory of Stationary Random
    Functions*, Dover, New York.

Index